SYMBOLIC LOGIC

SYMBOLIC LOGIC

FIFTH EDITION

IRVING M. COPI

University of Hawaii

Macmillan Publishing Co., Inc.

NEW YORK

Collier Macmillan Publishers

LONDON

Printed in the United States of America

Macmillan Publishing Co., Inc.
866 Third Avenue, New York, New York 10022

Collier Macmillan Canada, Ltd.

Library of Congress Cataloging in Publication Data

Copi, Irving M.
Symbolic logic.

Includes index.
1. Logic, Symbolic and Mathematical. I. Title.
BC135.C58 1979 511'.3 78-15671
ISBN 0-02-324980-3

Printing: 1 2 3 4 5 6 7 8 Year: 9 0 1 2 3 4 5

For Amelia

PREFACE

The general approach of this book to logic remains the same as in earlier editions. Following Aristotle, we regard logic from two different points of view: On the one hand, logic is an instrument or organon for appraising the correctness of reasoning; on the other hand, the principles and methods of logic used as organon are interesting and important topics to be themselves systematically investigated. This dual approach to logic is especially appropriate for modern symbolic logic. Through the development of its special symbols, logic has become immeasurably more powerful an instrument for analysis and deduction. And the principles and methods of symbolic logic are fruitfully investigated through the study of logistic systems.

Chapters 1 through 5 present the standard notations, methods, and principles of symbolic logic for *use* in determining the validity or invalidity of arguments. Successively more complex modes of argumentation are examined: first those whose validity turns on truth-functional compounds of simple statements, next those involving the simplest kinds of quantification, then more complex kinds of multiple quantification, and finally relational arguments. The standard methods of truth tables, rules of inference, conditional and indirect modes of proof, and quantification theory by way of 'natural deduction' techniques are introduced. The logic of relations is developed in a separate chapter that includes identity theory, definite descriptions, predicates of higher type, and quantification of predicate variables. Well over a hundred new exercises have been added in this edition to help the student acquire a practical mastery of the material.

A transitional Chapter 6 explains the theory of deductive (or axiomatic) systems, which are the standard instruments for systematically investigating both mathematical and logical structures. To provide immediate contact with a very simple deductive system, what was Appendix B of the previous edition is brought forward to occupy the first two sections of a new Chapter 7. Also, to acquaint the student with a more complex, powerful, and important axiom system, the bulk of the new Chapter 7 presents the standard Zermelo-Fraenkel axiom system for set theory. Then the last three chapters offer a systematic treatment of the logical principles *used* in the first five chapters of the book. A propositional calculus is developed according to the highest modern standards of rigor, and proved to be consistent and complete. Alternative notations and axiomatic foundations for propositional calculi are presented, and then a first-order function (or predicate) calculus is developed. The latter is shown to be equivalent to the 'natural deduction' methods of the first part of the book, and is also proved to be consistent and complete.

There are three appendices: The first proves the deductive incompleteness of the 19 Rules of Inference presented in the first two sections of Chapter 3. This material occupied Section 3.4 of the previous edition, but it seemed to me (and to some others who used the book) that its presence there tended to interrupt the natural continuity desirable in a textbook. The second appendix develops Normal Forms and Boolean Expansions, which provide an algebraic treatment of truth-functional logic. The third appendix expounds the ramified theory of logical types as a method of resolving semantical paradoxes.

The biggest change in this new edition is the presence of a new chapter on set theory. Of course, only an introduction to that vast subject matter can be given in a single chapter. But it has seemed worthwhile to include that material for the following reasons: (1) A number of people who have used earlier editions suggested that some example of an axiomatic system should be included prior to the metalogically developed propositional and predicate calculi of the last three chapters. I think it an excellent suggestion and have responded by including both the very simple axioms for Boolean Algebra in Section 7.2, and the more elaborate and powerful **ZF** axioms in subsequent sections of the new Chapter 7. (2) Some of the reasoning used to prove the deductive completeness of first-order logic in the final chapter makes use of set theoretic results, so it is appropriate to present and explain the axiomatic basis for the methods utilized in the book's last chapter. (3) At least part of the importance of symbolic logic derives from its role in investigations into the foundations of mathematics. Set theory is another basic tool in this area: not an alternative but another equally essential instrument to be used in conjunction with symbolic logic in foundational studies. (4) Set theory is a discipline, clearly significant in its own right, in which the notations and methods of symbolic logic are immediately, fruitfully, and importantly applicable. So the presentation of set theory, in however limited a fashion, cannot but enhance the reader's appreciation of those notations and methods. (5) Set theory is an area that the student of symbolic logic is uniquely prepared to enter immediately after mastering quantification theory, and more and more symbolic logicians—especially those in mathematics departments—are pursuing their researches in specialized and advanced parts of set theory. (6) Finally, set theory is intrinsically interesting and important in its own right, and I hope the brief introduction to it in this new chapter will open the door for many students to enter into what Hilbert called 'the paradise which Cantor has created for us.'[1]

In addition to the major change involving the addition of a new chapter,

[1]On the Infinite' by David Hilbert, in *Philosophy of Mathematics*, P. Benacerraf and H. Putnam, eds., Prentice-Hall, Inc., Englewood Cliffs, New Jersey, 1964, p. 141.

and those involved in relocating one appendix into that chapter and relegating the former Section 3.4 to the status of an appendix, there are other changes in this new edition that I hope will make the book more useful. These include some historical information about the development of symbolic logic, a more precise characterization of components and truth-functional components of compound statements, a clearer distinction between substitution (for variables) and replacement (of equivalents), a simpler discussion of the rule of Indirect Proof and a clearer statement of its relationship with the strengthened principle of Conditional Proof, the addition of a needed restriction on the preliminary version of Universal Generalization, a new section on proving the invalidity of arguments whose propositions contain more than one quantifier, and later (in Chapter 5) an explanation of why that method cannot be used for relational arguments. Finally, with some help from friendly critics, a number of stylistic improvements have been made.

As before, several teachers and students of logic have been kind enough to suggest ways to improve this book. I have given earnest consideration to all of the advice offered, even though I was not able to incorporate all of the changes proposed. For their helpful communications I am especially grateful to Professor Robert L. Armstrong, of The University of West Florida; Professor Eugene Atkin, of Central College, Iowa; Professor Walter Bass, of Indiana State University; Professor Robert W. Beard, of Florida State University; Professor Marc Beaulieu, of Collège de Maisonneuve, Montréal, Quebec; Professor Richard Beaulieu, of Paris; Anthony Belfiglio, of Chester State College, Pennsylvania; Professor R. T. Blackwood, of Hamilton College, New York; Professor Alex Blum, of Bar-Ilan University, Israel; Professor Herbert G. Bohnert, of Michigan State University; Professor Lorin Browning, of The College of Charleston, South Carolina; Professor Suchitra Dhar, of the University of Northern Colorado; Professor James Edwards, of the University of Glasgow, Scotland; Professor Earl Eugene Eminhizer, of Youngstown State University, Ohio; Professor P. T. Geach, of The University of Leeds, England; Professor James A. Gould, of The University of South Florida; Gary Leach, of Mississauga, Ontario; Professor Brendan P. Minogue, of Youngstown State University, Ohio; Terry L. Rankin, of Eastern Kentucky University; Professor Elizabeth Muir Ring, of Hamilton College, New York; Jennifer M. Shotwell, of Swarthmore College, Pennsylvania; Ronald C. Smith, of Asheville, North Carolina; Professor Bangs L. Tapscott, of The University of Utah; Professor Norman L. Thomas, of Simpson College, Iowa; Professor Burke Townsend, of the University of Montana; and Professor Frank Williams, of Eastern Kentucky University.

Professor Paul Halmos, of Indiana University, and Professor Robert R. Stoll, of Oberlin College, read an early draft of the new Chapter 7 on set theory, and made a number of criticisms and very helpful suggestions. I have followed their advice where I could, and any errors that remain are of course my responsibility. The many omissions in the chapter are a matter of my own deliberate choice because I could not hope to cover in a single chapter all of the topics that fill hundreds of pages in the typical book on the subject.

This new edition has benefited also from comments made by Dr. Linda Alexander, Michael Bybee, and Dr. Raymond Corsini. Warm thanks are due to Kenneth J. Scott and Elaine W. Wetterau of the Macmillan Publishing Company for their expert editorial labor and unfailing helpfulness.

Finally, most of all I thank my wife for help and encouragement in preparing this new edition.

I. M. C.

CONTENTS

1

Introduction: Logic and Language

1.1 What Is Logic?

It is easy to find answers to the question 'What is Logic?' According to Charles Peirce, 'Nearly a hundred definitions of it have been given'.[1] But Peirce goes on to write: 'It will, however, generally be conceded that its central problem is the classification of arguments, so that all those that are bad are thrown into one division, and those which are good into another . . .'.

The study of logic, then, is the study of the methods and principles used in distinguishing correct (good) from incorrect (bad) arguments. This definition is not intended to imply, of course, that only the student of logic can make the distinction. But the study of logic will *help* one to distinguish between correct and incorrect arguments, and it will do so in several ways. First of all, the proper study of logic will approach it as an art as well as a science, and the student will do exercises in all parts of the theory being learned. Here, as anywhere else, practice will help to make perfect. In the second place, the study of logic, especially symbolic logic, like the study of any other exact science, will tend to increase proficiency in reasoning. And finally, the study of logic will give students certain techniques for testing the validity of all arguments, including their own. This knowledge is of value because when mistakes are easily detected, they are less likely to be made.

Logic has frequently been defined as the science of reasoning. That definition, although it gives a clue to the nature of logic, is not quite accurate. Reasoning is that special kind of thinking called inferring, in which conclusions are drawn from premisses. As thinking, however, reasoning is not the special province of logic, but part of the psychologist's subject matter as well. Psychologists who examine the reasoning process find it to be extremely complex and highly emotional, consisting of awkward trial and error procedures that are illuminated by sudden—and sometimes apparently irrelevant—flashes of insight. These are all of importance to psychology. Logicians, however, are not interested in the actual process of reasoning, but rather with the correctness of

[1]Logic, in *Dictionary of Philosophy and Psychology*, James Mark Baldwin, ed., Macmillan Publishing Co., Inc., New York, 1925.

the completed reasoning process. Their question is always: Does the conclusion that is reached *follow* from the premisses used or assumed? If the premisses provide adequate grounds for accepting the conclusion, if asserting the premisses to be true warrants asserting the conclusion to be true also, then the reasoning is correct. Otherwise, the reasoning is incorrect. Logicians' methods and techniques have been developed primarily for the purpose of making this distinction clear. Logicians are interested in all reasoning, regardless of its subject matter, but only from this special point of view.

1.2 The Nature of Argument

Inferring is an activity in which one proposition is affirmed on the basis of one or more other propositions that are accepted as the starting point of the process. The logician is not concerned with the *process* of inference, but with the propositions that are the initial and end points of that process, and the relationships between them.

Propositions are either true or false, and in this they differ from questions, commands, and exclamations. Grammarians classify the linguistic formulations of propositions, questions, commands, and exclamations as declarative, interrogative, imperative, and exclamatory sentences, respectively. These are familiar notions. It is important to distinguish between declarative sentences and the propositions they may be uttered to assert. A declarative sentence is always part of a language, the language in which it is spoken or written, whereas propositions are not peculiar to any of the languages in which they may be expressed. Another difference between declarative sentences and propositions is that the same sentence may be uttered in different contexts to assert different propositions. (For example, the sentence 'I am hungry' may be uttered by different persons to make different assertions.)

The same sort of distinction can be drawn between sentences and *statements*. The same statement can be made using different words, and the same sentence can be uttered in different contexts to make different statements. The terms 'proposition' and 'statement' are not exact synonyms, but in the writings of logicians they are used in much the same sense. In this book, both terms will be used. In the following chapters, we shall also use the term 'statement' (especially in Chapters 2 and 3) and the term 'proposition' (especially in Chapters 4 and 5) to refer to the sentences in which statements and propositions are expressed. In each case the context should make clear what is meant.

Corresponding to every possible inference is an *argument*, and it is with these arguments that logic is chiefly concerned. An argument may be defined as any group of propositions or statements, of which one is claimed to follow from the others, which are alleged to provide grounds for the truth of that one. In ordinary usage, the word 'argument' also has other meanings, but in logic it has the technical sense explained. In the following chapters we shall use the word 'argument' also in a derivative sense to refer to any sentence or collection of sentences in which an argument is formulated or expressed. When we do, we shall be presupposing that the context is sufficiently clear to ensure that unique statements are made or unique propositions are asserted by the utterance of those sentences.

Every argument has a structure, in the analysis of which the terms 'premiss' and 'conclusion' are usually employed. The *conclusion* of an argument is that proposition which is affirmed on the basis of the other propositions of the argument. These other propositions, which are affirmed as providing grounds or reasons for accepting the conclusion, are the *premisses* of that argument.

We note that 'premiss' and 'conclusion' are relative terms, in the sense that the same proposition can be a premiss in one argument and a conclusion in another. Thus the proposition *All humans are mortal* is premiss in the argument

All humans are mortal.
Socrates is human.
Therefore, Socrates is mortal.

and conclusion in the argument

All animals are mortal.
All humans are animals.
Therefore, all humans are mortal.

Any proposition can be either a premiss or a conclusion, depending upon its context. It is a premiss when it occurs in an argument in which it is assumed for the sake of proving some other proposition. And it is a conclusion when it occurs in an argument that is claimed to prove it on the basis of other propositions that are assumed.

It is customary to distinguish between *deductive* and *inductive* arguments. All arguments involve the claim that their premisses provide some grounds for the truth of their conclusions, but only a *deductive* argument involves the claim that its premisses provide *absolutely conclusive* grounds. The technical terms 'valid' and 'invalid' are used in place of 'correct' and 'incorrect' in characterizing deductive arguments. A deductive argument is *valid* when its premisses and conclusion are so related that it is absolutely impossible for the premisses to be true unless the conclusion is true also. The task of deductive logic is to clarify the nature of the relationship that holds between premisses and conclusion in a valid argument, and to provide techniques for discriminating valid from invalid arguments.

Inductive arguments involve the claim only that their premisses provide *some* grounds for their conclusions. Neither the term 'valid' nor its opposite 'invalid' is properly applied to inductive arguments. Inductive arguments differ among themselves in the degree of likelihood or probability that their premisses confer upon their conclusions. Inductive arguments are studied in inductive logic. In this book, however, we shall be concerned only with deductive arguments, and shall use the word 'argument' to refer to deductive arguments exclusively.

1.3 Truth and Validity

Truth and falsehood characterize propositions or statements and may also be said to characterize the declarative sentences in which they are formulated. Arguments, however, are not properly characterized as being either true or false but as valid or invalid.[2] There is a connection between the validity or invalidity of an argument and the truth or falsehood of its premisses and conclusion, but the connection is by no means a simple one.

Some valid arguments contain true propositions only, as, for example,

> All bats are mammals.
> All mammals have lungs.
> Therefore, all bats have lungs.

An argument may contain false propositions exclusively and still be valid, as, for example,

> All trout are mammals.
> All mammals have wings.
> Therefore, all trout have wings.

This argument is valid because *if* its premisses were true, its conclusion would have to be true also, even though, in fact, they are all false. These two examples show that although some valid arguments have true conclusions, not all of them do. The validity of an argument does not, therefore, guarantee the truth of its conclusion.

When we consider the argument

> If I am President, then I am famous.
> I am not President.
> Therefore, I am not famous.

we can see that although both premisses and conclusion are true, the argument is invalid. Its invalidity becomes obvious when it is compared with another argument of the same form:

[2] Some logicians use the term 'valid' to characterize statements that are *logically true*. We shall adopt that usage in Chapter 10, Section 10.6. Until then, however, we apply the terms 'valid' and 'invalid' to arguments exclusively.

If Rockefeller is President, then Rockefeller is famous.
Rockefeller is not President.
Therefore, Rockefeller is not famous.

This argument is clearly invalid because its premisses are true but its conclusion is false. The two latter examples show that although some invalid arguments have false conclusions, not all of them do. The falsehood of its conclusion does not guarantee the invalidity of an argument. But the falsehood of its conclusion does guarantee that *either* the argument is invalid *or* at least one of its premisses is false.

An argument must satisfy two conditions to establish the truth of its conclusion. It must be valid, and all of its premisses must be true. Such an argument is termed 'sound'. To determine the truth or falsehood of premisses is the task of scientific inquiry in general, since premisses may deal with any subject matter at all. But determining the validity or invalidity of arguments is the special province of deductive logic. The logician is interested in the question of validity even for arguments that might be unsound because their premisses might happen to be false.

A question might be raised about the legitimacy of that interest. It might be suggested that logicians should confine their attention to arguments that have true premisses only. It is often necessary, however, to depend upon the validity of arguments whose premisses are not known to be true. Scientists test their theories by deducing from them conclusions that predict the behavior of observable phenomena in the laboratory or observatory. The conclusion is then tested directly by observation of experimental data and if it is true, the results confirm the theory from which the conclusion was deduced. If the conclusion is false, the results disconfirm or refute the theory. In either case, the scientist is vitally interested in the validity of the argument by which the testable conclusion is deduced from the theory being investigated, for if that argument is invalid, his whole procedure is without point. Although an oversimplification of scientific method, our example shows that questions of validity are important even for arguments whose premisses are not true.

1.4 Symbolic Logic

It has been explained that logic is concerned with arguments and that arguments contain propositions or statements as their premisses and conclusions. These premisses and conclusions are not linguistic entities, such as declarative sentences, but are, rather, what declarative sentences are typically uttered to assert. The communication of propositions and arguments, however, requires the use of language, and this complicates our problem. Arguments formulated in English or any other natural language are often difficult to appraise because of the vague and equivocal nature of the words in which they are expressed, the ambiguity of their construction, the misleading idioms they may contain, and their pleasing but deceptive metaphorical style. The resolution of these difficulties is not the central problem for the logician, however, for even when they are resolved, the problem of deciding the validity or invalidity of the argument still remains.

To avoid the peripheral difficulties connected with ordinary language, workers in the various sciences have developed specialized technical vocabularies. The scientist economizes the space and time required for writing his reports and theories by adopting special symbols to express ideas that would otherwise require a long sequence of familiar words to formulate. This has the further advantage of reducing the amount of attention needed, for when a sentence or equation grows too long, its meaning is more difficult to grasp. The introduction of the exponent symbol in mathematics permits the expression of the equation

$$A \times A \times A \times A \times A \times A \times A \times A \times A \times A \times A \times A = B \times B \times B \times B \times B \times B \times B$$

more briefly and intelligibly as

$$A^{12} = B^7$$

A like advantage has been obtained by the use of graphic formulas in organic chemistry. The language of every advanced science has been enriched by similar symbolic innovations.

A special technical notation has been developed for logic as well. Aristotle made use of certain abbreviations to facilitate his own investigations. Modern symbolic logic augmented this base by the introduction of many more special symbols. The difference between the old and the new logic is one of degree rather than of kind, but the difference in degree is tremendous. Modern symbolic logic has become immeasurably more powerful a tool for analysis and deduction through the development of its own technical language. The special symbols of modern logic permit us to exhibit with greater clarity the logical structures of arguments that may be obscured by formulation in ordinary language. It is easier to divide arguments into the valid and the invalid when they are expressed in a special symbolic language, for with symbols the peripheral problems of vagueness, ambiguity, idiom, metaphor, and amphiboly do not arise. The introduction and use of special symbols serve not only to facilitate the appraisal of arguments, but also to clarify the nature of deductive inference.

The logician's special symbols are much better adapted to the actual drawing of inferences than is ordinary language. Their superiority in this respect is comparable to that of Arabic over the older Roman numerals for purposes of computation. It is easy to multiply 148 by 47, but very difficult to compute the product of CXLVIII and XLVII. Similarly, the drawing of inferences and the evaluation of arguments is greatly facilitated by the adoption of a special logical notation. To quote Alfred North Whitehead, an important contributor to the advance of symbolic logic:

> . . . by the aid of symbolism, we can make transitions in reasoning almost mechanically by the eye, which otherwise would call into play the higher faculties of the brain.[3]

Although this book treats symbolic logic systematically rather than historically, a few historical remarks may be appropriate at this point. Since the 1840s, symbolic logic has developed along two different historical paths. One of them began with the English mathematician George Boole (1815–1864). Boole applied algebraic notations and methods first to symbolize and then to validate arguments of the kind studied by Aristotle in the fourth century B.C. This route may be characterized as an effort to apply mathematical notations and methods to traditional, nonmathematical kinds of arguments. The other path began with the independent efforts of the English mathematician Augustus De Morgan (1806–1871) and the American scientist and philosopher Charles Peirce (1839–1914) to devise a very precise notation for relational arguments. The earlier logic had largely ignored this type of argument, which, nevertheless, plays a central role in mathematics. This historical path may be characterized as an effort to create a new quasi-mathematical kind of logical notation and analytical technique for use in mathematical derivations and demonstrations.

These two historical paths coalesced in the brilliant works of the German mathematician and philosopher Gottlob Frege (1848–1925), the Italian mathematician Guiseppe Peano (1858–1932), and the English philosophers Alfred North Whitehead (1861–1947) and Bertrand Russell (1872–1970), whose *Principia Mathematica* was an important landmark in the history of symbolic logic. Some of Boole's contributions are reported in the first two sections of Chapter 7 and in Appendix B. The contributions of the others have become so thoroughly incorporated into modern symbolic logic that only occasional references to their more distinctive ideas are appropriate.

[3] A. N. Whitehead, *An Introduction to Mathematics*, Oxford University Press, Oxford, England, 1911.

2

Arguments Containing Compound Statements

2.1 Simple and Compound Statements

All statements can be divided into two kinds, simple and compound. A *simple* statement is one that does not contain any other statement as a component part, whereas every *compound* statement does contain another statement as a component part. For example, 'Atmospheric testing of nuclear weapons will be discontinued or this planet will become uninhabitable' is a compound statement that contains, as its components, the two simple statements 'Atmospheric testing of nuclear weapons will be discontinued' and 'this planet will become uninhabitable'. The component parts of a compound statement may themselves be compound, of course.

As the term 'component' is used in logic, not every statement that is part of a larger statement is a component of the larger statement. For example, the last six words of the statement, 'The third wife of Bertrand Russell was a beautiful girl', can be regarded as a statement in its own right. But it is not a *component* part or a *component* of the larger statement containing it. For a part of a statement to be a component of a larger statement, two conditions must be satisfied. First, the part must be a statement in its own right; and second, if the part is replaced in the larger statement by any other statement, the result of that replacement must be meaningful. Although the first condition is satisfied in the example given, the second is not. For if the part 'Bertrand Russell was a beautiful girl' is replaced by the statement 'Where there's smoke, there's fire', the result of that replacement is arrant nonsense.[1]

The statement 'Roses are red and violets are blue' is a *conjunction*, a compound statement formed by inserting the word 'and' between two statements. Two statements so combined are called *conjuncts*. The word 'and' has other uses, however, as in the statement 'Castor and Pollux were twins', which is not compound, but a simple statement asserting a relationship. We introduce the dot '·' as a special symbol for combining statements conjunctively. Using

[1]This account of compound and component statements has been suggested by Professor C. Mason Myers of Northern Illinois University, Professor Alex Blum of Bar-Ilan University, and Professor James A. Martin of the University of Wyoming. See Martin's 'How Not to Define Truth-Functionality,' *Logique et Analyse*, 1970, 14^{e} Annee, N° 52, pp. 476–482.

this notation, the preceding conjunction is written 'Roses are red$\cdot$violets are blue'. Where p and q are any two statements whatever, their conjunction is written $p \cdot q$.

Every statement is either true or false, so we can speak of the *truth value* of a statement, where the truth value of a true statement is *true* and the truth value of a false statement is *false*. There are two broad categories into which compound statements can be divided, according to whether or not there is anything other than the truth values of its component statements that determines the truth value of the compound statement. The truth value of the conjunction of two statements is completely determined by the truth value of its conjuncts. A conjunction is true if both its conjuncts are true, but false otherwise. For this reason a conjunction is a *truth-functional* compound statement, and its conjuncts are *truth-functional* components of it.

Not every compound statement is truth-functional. For example, the truth-value of the compound statement 'Percival Lowell believed that Mars is inhabited' is not in any way determined by the truth value of its component simple statement 'Mars is inhabited'. This is so because it could be true that Percival Lowell believed that Mars is inhabited, regardless of whether it is inhabited or not. A person can believe one truth without believing another, and a person can believe one falsehood without believing all of them. So the component 'Mars is inhabited' is not a truth-functional component of the compound statement 'Percival Lowell believed that Mars is inhabited,' and the latter statement is not a truth-functional compound statement.

We define an occurrence of a component of a compound statement to be a *truth-functional component* of that compound statement provided that: If that occurrence of the component is replaced in the compound (or in any component of the compound statement which contains the component in question) by different statements that have the same truth value as each other, the different compound statements produced by these replacements will also have the same truth values as each other. And now a compound statement is defined to be a *truth-functional compound statement,* if all of its components are truth-functional components of the statement.[2]

The only compound statements we shall consider here will be truth-functionally compound statements. Therefore, in the rest of this book we shall use the term "simple statement" to refer to any statement that is not truth-functionally compound.

Since conjunctions are truth-functionally compound statements, our dot symbol is a truth-functional connective. Given any two statements p and q there are just four possible sets of truth values they can have, and in every case the truth value of their conjunction $p \cdot q$ is uniquely determined. The four possible cases can be exhibited as follows:

[2]Compare this definition with the one proposed by Professor David H. Sanford in his 'What Is a Truth Functional Component?' *Logique et Analyse,* 1970, 14^{e} Annee, N° 52, pp. 483–486.

in case p is true and q is true, $p \cdot q$ is true;
in case p is true and q is false, $p \cdot q$ is false;
in case p is false and q is true, $p \cdot q$ is false;
in case p is false and q is false, $p \cdot q$ is false.

Representing the truth values true and false by the capital letters 'T' and 'F', respectively, the way in which the truth value of a conjunction is determined by the truth values of its conjuncts can be displayed more briefly by means of a *truth table* as follows:

p	q	$p \cdot q$
T	T	T
T	F	F
F	T	F
F	F	F

Since it specifies the truth value of $p \cdot q$ in every possible case, this truth table can be taken as *defining* the dot symbol. Other English words such as 'moreover', 'furthermore', 'but', 'yet', 'still', 'however', 'also', 'nevertheless', 'although', and so forth, and even the comma and the semicolon, are also used to conjoin two statements into a single compound one, and all of them can be indifferently translated into the dot symbol, so far as truth values are concerned.

The statement 'It is not the case that lead is heavier than gold' is also compound, being the *negation* (or *denial* or *contradictory*) of its single component statement 'lead is heavier than gold'. We introduce the symbol '$\sim$', called a *curl* (or a *tilde*) to symbolize negation. There are often alternative formulations in English of the negation of a given statement. Thus where L symbolizes the statement 'lead is heavier than gold', the different statements 'it is not the case that lead is heavier than gold', 'it is false that lead is heavier than gold', 'it is not true that lead is heavier than gold', 'lead is not heavier than gold' are all indifferently symbolized as $\sim L$. More generally, where p is any statement whatever, its negation is written $\sim p$. Since the negation of a true statement is false and the negation of a false statement is true, we can take the following truth table as defining the curl symbol:

p	$\sim p$
T	F
F	T

When two statements are combined disjunctively by inserting the word 'or' between them, the resulting compound statement is a *disjunction* (or *alternation*), and the two statements so combined are called *disjuncts* (or *alternatives*). The word 'or' has two different senses, one of which is clearly intended in the statement 'Premiums will be waived in the event of sickness or unemployment'. The intention here is obviously that premiums are waived not only for sick persons and for unemployed persons, but also for persons who are both sick *and* unemployed. This sense of the word 'or' is called *weak* or *inclusive*. Where precision is at a premium, as in contracts and other legal documents, this sense is made explicit by use of the phrase 'and/or'.

A different sense of 'or' is intended when a restaurant lists 'tea or coffee' on its table d'hôte menu, meaning that for the stated price of the meal the customer can have one or the other, but *not both*. This second sense of 'or' is called *strong* or *exclusive*. Where precision is at a premium and the exclusive sense of 'or' is intended, the phrase 'but not both' is often added.

A disjunction which uses the inclusive 'or' asserts that *at least one disjunct is true*, whereas one which uses the exclusive 'or' asserts that *at least one disjunct is true but not both are true*. The *partial common meaning* that at least one disjunct is true, is the whole meaning of an inclusive disjunction and a part of the meaning of an exclusive disjunction.

In Latin, the word 'vel' expresses the inclusive sense of the word 'or' and the word 'aut' expresses the exclusive sense. It is customary to use the first letter of 'vel' to symbolize 'or' in its inclusive sense. Where p and q are any two statements whatever, their weak or inclusive disjunction is written $p \vee q$. The symbol 'v', called a *wedge* (or a *vee*), is a truth-functional connective, and is defined by the following truth table:

p	q	$p \vee q$
T	T	T
T	F	T
F	T	T
F	F	F

An obviously valid argument containing a disjunction is the following Disjunctive Syllogism:

The United Nations will become more responsible or there will be a third world war.
The United Nations will not become more responsible.
Therefore, there will be a third world war.

It is evident that a Disjunctive Syllogism is valid on *either* interpretation of the word 'or', that is, regardless of whether its first premiss asserts an inclusive or exclusive disjunction. It is usually difficult, and sometimes impossible, to discover which sense of the word 'or' is intended in a disjunction. But the typical valid argument that has a disjunction for a premiss is, like the Disjunctive Syllogism, valid on either interpretation of the word 'or'. Hence we effect a simplification by translating any occurrence of the word 'or' into the logical symbol 'v'—*regardless of which sense of 'or' is intended.* Of course, where it is explicitly stated that the disjunction is exclusive, by use of the added phrase 'but not both', for example, we do have the symbolic apparatus for expressing that sense, as will be explained below.

The use of parentheses, brackets, and braces for punctuating mathematical expressions is familiar. In the absence of a special convention, no number is uniquely denoted by the expression '$6 + 9 \div 3$', although when punctuation makes clear how its constituents are to be grouped, it denotes either 5 or 9. Punctuation is needed to resolve ambiguity in the language of symbolic logic too, since compound statements may themselves be combined to produce more complicated compounds. Ambiguity is present in $p \cdot q \vee r$, which could be either the conjunction of p with $q \vee r$, or else the disjunction of $p \cdot q$ with r. These two different senses are unambiguously given by different punctuations: $p \cdot (q \vee r)$ and $(p \cdot q) \vee r$. In case p and q are both false and r is true, the first punctuated expression is false (since its first conjunct is false), but the second punctuated expression is true (since its second disjunct is true). Here, a difference in punctuation makes all the difference between truth and falsehood. In symbolic logic, as in mathematics, we use parentheses, brackets, and braces for punctuation. To cut down on the number of punctuation marks required, however, we establish the symbolic convention that in any expression the curl will apply to the smallest component that the punctuation permits. Thus the ambiguity of $\sim p \vee q$, which might mean either $(\sim p) \vee q$ or $\sim(p \vee q)$, is resolved, by our convention, to mean the first of these. The curl can, and therefore by our convention *does*, apply to the first component p rather than to the larger expression $p \vee q$.

The word 'either' has a variety of different uses in English. It has conjunctive force in 'The Disjunctive Syllogism is valid on either interpretation of the word 'or'.' It frequently serves merely to introduce the first disjunct in a disjunction, as in 'Either the United Nations will become more responsible or there will be a third world war'. Perhaps the most useful function of the word 'either' is to punctuate some compound statements. Thus the sentence

> More stringent antipollution measures will be enacted and the laws will be strictly enforced or the quality of life will be degraded still further.

can have its ambiguity resolved in one direction by placing the word 'either' at its beginning, or in the other direction by inserting the word 'either' right after the word 'and'. Such punctuation is effected in our symbolic language by parentheses. The ambiguous formula $p \cdot q \vee r$ discussed in the preceding paragraph corresponds to the ambiguous sentence considered in this one. The two different punctuations of the formula correspond to the two different punctuations of the sentence effected by the two different insertions of the word 'either'.

Not all conjunctions are formulated by explicitly placing the word 'and' between complete sentences, as in 'Charlie's neat and Charlie's sweet'. Indeed, the latter would more naturally be expressed as 'Charlie's neat and sweet'. The familiar 'Jack and Jill went up the hill' is the more natural way of expressing the conjunction 'Jack went up the hill and Jill went up the hill'. It is the same with disjunctions: 'Either Alice or Betty will be elected' expresses more briefly the proposition alternatively formulated as 'Either Alice will be elected or Betty will be elected'; and 'Charlene will be either secretary or treasurer' expresses somewhat more briefly the same proposition as 'Either Charlene will be secretary or Charlene will be treasurer'.

The negation of a disjunction is often expressed by using the phrase 'neither-nor'. Thus the disjunction 'Either Alice or Betty will be elected' is denied by the statement 'Neither Alice nor Betty will be elected'. The disjunction would be symbolized as $A \vee B$ and its negation as either $\sim(A \vee B)$ or as $(\sim A)\cdot(\sim B)$. (The logical equivalence of these two formulas will be discussed in Section 2.4.) To deny that at least one of two statements is true is to assert that both of the two statements are false.

The word 'both' serves various functions. One is simply a matter of emphasis. To say 'Both Jack and Jill went up the hill' is only to emphasize that the two of them did what they are asserted to have done by saying 'Jack and Jill went up the hill'. A more useful function of the word 'both' is punctuational, like that of 'either'. 'Both . . . and -•- are not ---' is used to make the same statement as 'Neither . . . nor -•- is ---. In such sentences, the *order* of the words 'both' and 'not' is very significant. There is a great difference between

Alice and Betty will not both be elected.

and

Alice and Betty will both not be elected.

The former would be symbolized as $\sim(A\cdot B)$, the latter as $(\sim A)\cdot(\sim B)$.

Finally, it should be remarked that the word 'unless' can also be used in expressing the disjunction of two statements. Thus 'Our resources will soon be exhausted unless more recycling of materials is effected' and 'Unless more recycling of materials is effected our resources will soon be exhausted' can equally well be expressed as 'Either more recycling of materials is effected or our resources will soon be exhausted' and symbolized as $M \vee E$.

Since an exclusive disjunction asserts that at least one of its disjuncts is true but they are not both true, we can symbolize the exclusive disjunction of any two statements p and q quite simply as $(p \vee q)\cdot\sim(p\cdot q)$. Thus we are able to symbolize conjunctions, negations, and both inclusive and exclusive disjunctions. Any compound statement that is built up out of simple statements by repeated use of truth-functional connectives will have its truth value completely determined by the truth values of those simple statements. For example, if A and B are true statements and X and Y are false, the truth value of the compound statement $\sim[(\sim A \vee X) \vee \sim(B\cdot Y)]$ can be discovered as follows. Since A is true, $\sim A$ is false, and since X is false also, the disjunction $(\sim A \vee X)$

is false. Since Y is false, the conjunction $(B \cdot Y)$ is false, and so its negation $\sim(B \cdot Y)$ is true. Hence the disjunction $(\sim A \vee X) \vee \sim(B \cdot Y)$ is true, and its negation, which is the original statement, is false. Such a stepwise procedure, beginning with the inmost components, always permits us to determine the truth value of a truth-functionally compound statement from the truth values of its component simple statements.

EXERCISES[3]

I. If A and B are true statements and X and Y are false statements, which of the following compound statements are true?

*1. $\sim(A \vee X)$	11. $A \vee [X \cdot (B \vee Y)]$
2. $\sim A \vee \sim X$	12. $X \vee [A \cdot (Y \vee B)]$
3. $\sim B \cdot \sim Y$	13. $\sim\{\sim[\sim(A \cdot \sim X) \cdot \sim A] \cdot \sim X\}$
4. $\sim(B \cdot Y)$	14. $\sim\{\sim[\sim(A \cdot \sim B) \cdot \sim A] \cdot \sim A\}$
*5. $A \vee (X \cdot Y)$	*15. $[(A \cdot X) \vee \sim B] \cdot \sim[(A \cdot X) \vee \sim B]$
6. $(A \vee X) \cdot Y$	16. $[(X \cdot A) \vee \sim Y] \vee \sim[(X \cdot A) \vee \sim Y]$
7. $(A \vee B) \cdot (X \vee Y)$	17. $[A \cdot (X \vee Y)] \vee \sim[(A \cdot X) \vee (A \cdot Y)]$
8. $(A \cdot B) \vee (X \cdot Y)$	18. $[X \vee (A \cdot Y)] \vee \sim[(X \vee A) \cdot (X \vee Y)]$
9. $(A \cdot X) \vee (B \cdot Y)$	19. $[X \cdot (A \vee B)] \vee \sim[(X \vee A) \cdot (X \vee B)]$
*10. $A \cdot [X \vee (B \cdot Y)]$	20. $[X \vee (A \cdot Y)] \vee \sim[X \vee A) \vee (X \vee Y)]$

II. If A and B are known to be true, and X and Y are known to be false, but the truth values of P and Q are not known, of which of the following statements can you determine the truth value?

*1. $P \cdot X$	6. $P \cdot \sim P$
2. $B \vee Q$	7. $\sim P \vee (\sim A \vee P)$
3. $\sim A \cdot P$	8. $\sim P \vee (\sim X \vee P)$
4. $Q \vee \sim Q$	9. $\sim P \vee (\sim Q \vee P)$
*5. $P \vee \sim Y$	*10. $Q \vee \sim(P \cdot Q)$

11. $(P \cdot Q) \vee (\sim Q \vee \sim P)$
12. $(P \vee Q) \vee (\sim P \cdot \sim Q)$
13. $(P \vee Q) \cdot (\sim B \vee Y)$
14. $(P \cdot Q) \vee \sim(Q \cdot P)$
*15. $(P \cdot Q) \cdot (\sim A \vee X)$
16. $\sim(P \cdot Q) \vee P$
17. $(\sim P \vee Q) \vee (P \vee \sim Q)$
18. $[P \vee (Q \vee X)] \cdot \sim[(P \vee Q) \vee X]$
19. $[P \vee (Q \cdot A)] \cdot \sim[(P \vee Q) \cdot (P \vee A)]$
20. $[P \cdot (Q \vee X)] \vee \sim[(P \cdot Q) \vee (P \cdot X)]$

[3]Solutions to starred exercises will be found on pages 355–380.

III. Using the letters *A*, *B*, *C*, and *D* to abbreviate the simple statements: 'Atlanta wins their conference championship', 'Baltimore wins their conference championship', 'Chicago wins the superbowl', and 'Dallas wins the superbowl', symbolize the following:

*1. Either Atlanta wins their conference championship and Baltimore wins their conference championship or Chicago wins the superbowl.
2. Atlanta wins their conference championship and either Baltimore wins their conference championship or Dallas does not win the superbowl.
3. Atlanta and Baltimore will not both win their conference championships but Chicago and Dallas will both not win the superbowl.
4. Either Atlanta or Baltimore will win their conference championships but neither Chicago nor Dallas will win the superbowl.
*5. Either Chicago or Dallas will win the superbowl but they will not both win the superbowl.
6. Chicago will win the superbowl unless Atlanta wins their conference championship.
7. It is not the case that neither Atlanta nor Baltimore wins their conference championship.
8. Either Chicago or Dallas will fail to win the superbowl.
9. Either Chicago or Dallas will win the superbowl unless both Atlanta and Baltimore win their conference championships.
10. Either Chicago will win the superbowl and Dallas will not win the superbowl or both Atlanta and Baltimore will win their conference championships.

IV. Using capital letters to abbreviate simple statements, symbolize the following:

*1. The words of his mouth were smoother than butter, but war was in his heart. (Psalm 55:21)
2. Promotion cometh neither from the east, nor from the west, nor yet from the south. (Psalm 75:6)
3. As for man, his days are as grass: as a flower of the field, so he flourisheth. (Psalm 103:15)
4. Wine is a mocker, strong drink is raging. (Proverbs 20:1)
*5. God hath made man upright; but they have sought out many inventions (Ecclesiastes 7:29)
6. The race is *not* to the swift, nor the battle to the strong . . . (Ecclesiastes 9:11)
7. Love is strong as death; jealousy is cruel as the grave. (The Song of Solomon 8:6)
8. A bruised reed shall he not break, and the smoking flax shall he not quench. (Isaiah 42:3)
9. Saul and Jonathan were lovely and pleasant in their lives . . . (2 Samuel 1:23)
10. His eye was not dim, nor his natural force abated. (Deuteronomy 34:7)
11. The voice is Jacob's voice, but the hands are the hands of Esau. (Genesis 27:22)
12. He shall return no more to his house, neither shall his place know him any more. (Job 7:10)

2.2 Conditional Statements

The compound statement 'If the train is late, then we shall miss our connection' is a *conditional* (or a *hypothetical*, an *implication*, or an *implicative statement*). The component between the 'if' and the 'then' is called the *antecedent* (or the *implicans* or *protasis*), and the component that follows the 'then' is the *consequent* (or the *implicate* or *apodosis*). A conditional does not state either that its antecedent is true or that its consequent is true; it says only that *if* its antecedent is true, then its consequent is true also, that is, its antecedent *implies* its consequent. The key to the meaning of a conditional is the relation of *implication* that is asserted to hold between its antecedent and its consequent, in that order.

If we examine a number of different conditionals, we can see that there are different kinds of implications which they may express. In the conditional 'If all cats like liver and Dinah is a cat, then Dinah likes liver', the consequent follows *logically* from its antecedent. On the other hand, in the conditional 'If the figure is a triangle, then the figure has three sides', the consequent follows from the antecedent by the very *definition* of the word 'triangle'. The truth of the conditional 'If this piece of gold is placed in *aqua regia*, then this piece of gold dissolves' is not a matter of either logic or definition. The connection asserted here is *causal* and must be discovered empirically. These examples show that there are different kinds of implications that constitute different senses of the 'if-then' phrase. Having noted these differences, we now seek to find some identifiable common meaning, some partial meaning that is common to these different types of conditionals.

Our discussion of 'if-then' will parallel our previous discussion of the word 'or'. First, we pointed out two different senses of that word. Second, we noted that there was a common partial meaning: That *at least one disjunct is true* was seen to be involved in both the inclusive and the exclusive 'or'. Third, we introduced the special symbol 'v' to represent this common partial meaning (which was the whole meaning of 'or' in its inclusive sense). Fourth, we observed that because arguments like the Disjunctive Syllogism are valid on either interpretation of the word 'or', symbolizing *any* occurrence of the word 'or' by the wedge symbol, preserves the validity of such arguments. And because we are interested in arguments from the point of view of determining their validity, this translation of the word 'or' into 'v', which may abstract from or ignore part of its meaning in some cases, is wholly adequate for our present purposes.

A common partial meaning of these different kinds of conditional statements emerges when we ask what circumstances would suffice to establish the *falsehood* of a conditional. Under what circumstances would we agree that the conditional 'If this piece of gold is placed in this solution, then this piece of gold dissolves' is false? Clearly, the statement is false in the case that this piece of gold is actually placed in this solution and *does not dissolve*. Any conditional with a true antecedent and a false consequent must be false. Hence any conditional *if* p, *then* q is known to be false in case the conjunction $p \cdot \sim q$ is known to be true, that is, in case the antecedent of the conditional is true and its consequent false. For the conditional to be true, the indicated conjunction must be false, which means that the negation of that conjunction must be true. In other words, for any conditional *if* p, *then* q to be true, $\sim(p \cdot \sim q)$, the negation of the conjunction of its antecedent with the negation of its consequent, must be true also. We may, then, regard the latter as a *part* of the meaning of the former.

We introduce the new symbol '$\supset$', called a *horseshoe,* to represent the partial meaning that is common to all conditional statements. We define '$p \supset q$' as an abbreviation for '$\sim(p \cdot \sim q)$'. The horseshoe is a truth-functional connective, whose exact significance is indicated by the following truth table:

p	q	$\sim q$	$p \cdot \sim q$	$\sim(p \cdot \sim q)$	$p \supset q$
T	T	F	F	T	T
T	F	T	T	F	F
F	T	F	F	T	T
F	F	T	F	T	T

Here, the first two columns represent all possible truth values for the component statements p and q, and the third, fourth, and fifth represent successive stages in determining the truth value of the compound statement $\sim(p \cdot \sim q)$ in each case. The sixth column is identically the same as the fifth, since the formulas that head them are defined to express the same proposition. The horseshoe symbol must not be thought of as representing *the* meaning of 'if-then', or *the* relation of implication, but rather symbolizes a common partial factor of the various different kinds of implications signified by the 'if-then' phrase.

We can regard the horseshoe as symbolizing a special, extremely weak kind of implication. It is expedient for us to do so, since convenient ways to read '$p \supset q$' are 'if p, then q', 'p implies q', or 'p only if q'. The weak implication symbolized by '$\supset$' is called a *material implication.* Its special name indicates that it is a special concept and not to be confused with the other, more usual kinds of implication. Some conditional statements in English do express merely material implications, as for example 'If Russia is a democracy, then I'm a Dutchman'. It is clear that the implication expressed here is neither logical, definitional, nor causal. No 'real connection' is alleged to hold between what the antecedent states and what is stated by the consequent. This sort of conditional is ordinarily intended as an emphatic or humorous method of denying the truth of its antecedent, for it typically contains a notoriously or ridiculously false statement as its consequent. Any such assertion about truth values is adequately symbolized by using the truth-functional connective '$\supset$'.

Although most conditional statements express more than a merely material implication between antecedent and consequent, we now propose to symbolize *any* occurrence of 'if-then' by the truth-functional connective '$\supset$'. Such symbolizing abstracts from or ignores part of the meaning of most conditional statements. But the proposal can be justified on the grounds that the validity of valid arguments involving conditionals is preserved when the conditionals are regarded as expressing material implications only, as will be established in the following section.

Conditional statements can be expressed in a variety of ways. A statement of the form 'if p, then q' could equally well be expressed as 'if p, q', 'q if p', 'that p implies that q', 'that p entails that q' 'p only if q', 'that p is a sufficient condition that q', or as 'that q is a necessary condition that p', and any of these formulations will be symbolized as $p \supset q$.

EXERCISES

I. If A and B are true statements and X and Y are false statements, which of the following compound statements are true?

*1. $X \supset (X \supset Y)$
2. $(X \supset X) \supset Y$
3. $(A \supset X) \supset Y$
4. $(X \supset A) \supset Y$
*5. $A \supset (B \supset Y)$
6. $A \supset (X \supset B)$
7. $(X \supset A) \supset (B \supset Y)$
8. $(A \supset X) \supset (Y \supset B)$
9. $(A \supset B) \supset (\sim A \supset \sim B)$
*10. $(X \supset Y) \supset (\sim X \supset \sim Y)$
11. $(X \supset A) \supset (\sim X \supset \sim A)$
12. $(X \supset \sim Y) \supset (\sim X \supset Y)$
13. $[(A \cdot X) \supset Y] \supset (A \supset Y)$
14. $[(A \cdot B) \supset X] \supset [A \supset (B \supset X)]$
*15. $[(X \cdot Y) \supset A] \supset [X \supset (Y \supset A)]$
16. $[(A \cdot X) \supset B] \supset [A \supset (B \supset X)]$
17. $[X \supset (A \supset Y)] \supset [(X \supset A) \supset Y]$
18. $[X \supset (X \supset Y)] \supset [(X \supset X) \supset X]$
19. $[(A \supset B) \supset A] \supset A$
20. $[(X \supset Y) \supset X] \supset X$

II. If A and B are known to be true, and X and Y are known to be false, but the truth values of P and Q are not known, of which of the following statements can you determine the truth values?

*1. $X \supset (\sim Y \supset Q)$
2. $\sim Y \supset (P \supset \sim X)$
3. $(P \supset B) \supset Y$
4. $(P \vee A) \supset (Q \cdot X)$
*5. $(P \supset A) \supset (B \supset Y)$
6. $(B \supset P) \supset (Q \supset A)$
7. $A \supset (P \supset A)$
8. $(X \supset Q) \supset X$
9. $\sim(P \cdot X) \supset Y$
*10. $(Q \vee B) \supset X$
11. $[(P \supset A) \supset P] \supset P$
12. $[(P \supset X) \supset P] \supset P$
13. $[(P \supset Q) \supset P] \supset P$
14. $(P \supset X) \supset (X \supset P)$
*15. $(Q \supset B) \supset (A \supset Y)$
16. $(P \supset Q) \supset (\sim Q \supset \sim P)$
17. $(P \supset Q) \supset \{[P \supset (Q \supset A)] \supset (P \supset A)\}$
18. $(P \supset Q) \supset \{[P \supset (Q \supset X)] \supset (P \supset X)\}$
19. $(A \supset P) \supset \{[A \supset (P \supset Q)] \supset (A \supset Q)\}$
20. $(P \supset X) \supset \{[P \supset (X \supset Q)] \supset (P \supset Q)\}$

III. Symbolizing 'Amherst wins its first game' as A, 'Colgate wins its first game' as C, and 'Dartmouth wins its first game' as D, symbolize the following compound statements:

*1. Both Amherst and Colgate win their first games only if Dartmouth does not win its first game.
2. Amherst wins its first game if either Colgate wins its first game or Dartmouth wins its first game.
3. If Amherst wins its first game, then both Colgate and Dartmouth win their first games.
4. If Amherst wins its first game, then either Colgate or Dartmouth wins its first game.
*5. If Amherst does not win its first game, then it is not the case that either Colgate or Dartmouth wins its first game.
6. If it is not the case that both Amherst and Colgate win their first game, then both Colgate and Dartmouth win their first games.
7. If Amherst wins its first game, then not both Colgate and Dartmouth win their first games.
8. If Amherst does not win its first game, then both Colgate and Dartmouth do not win their first games.
9. Either Amherst wins its first game and Colgate does not win its first game or, if Colgate wins its first game, then Dartmouth does not win its first game.
*10. If Amherst wins its first game, then Colgate does not win its first game, but if Colgate does not win its first game, then Dartmouth wins its first game.
11. If Amherst wins its first game, then, if Colgate does not win its first game, then Dartmouth wins its first game.
12. Either Amherst and Colgate win their first games or it is not the case that if Colgate wins its first game, then Dartmouth wins it first game.
13. Amherst wins its first game only if either Colgate or Dartmouth does not win its first game.
14. If Amherst wins its first game only if Colgate wins its first game, then Dartmouth does not win it its first game.
15. If Amherst and Colgate both do not win their first games, then Amherst and Colgate do not both win their first games.

2.3 Argument Forms and Truth Tables

In this section, we develop a purely mechanical method for testing the validity of arguments containing truth-functionally compound statements. That method is closely related to the familiar technique of *refutation by logical analogy* that was used in the first chapter to show the invalidity of the argument

> If I am President, then I am famous.
> I am not President.
> Therefore, I am not famous.

That argument was shown to be invalid by constructing another argument of the same form:

> If Rockefeller is President, then Rockefeller is famous.
> Rockefeller is not President.
> Therefore, Rockefeller is not famous.

which is obviously invalid, since its premisses are true but its conclusion false. Any argument is proved to be invalid if another argument *of exactly the same form* can be constructed with true premisses and a false conclusion. This reflects the fact that validity and invalidity are purely *formal* characteristics of arguments: any two arguments having the same form are either both valid or both invalid, regardless of any differences in their subject matter.[4] The concept of two arguments having *exactly the same form* is one that deserves further examination.

It is convenient, in discussing forms of arguments, to use small letters from the middle part of the alphabet, '*p*', '*q*', '*r*', '*s*', . . . as *statement variables*, which are defined simply to be letters for which, or in place of which, statements may be substituted. Now, we define an *argument form* to be any array of symbols that contains statement variables, such that when statements are substituted for the statement variables—the same statement being substituted for every occurrence of the same statement variable throughout—the result is an argument. For definiteness, we establish the convention that in any argument form, '*p*' shall be the first statement variable that occurs in it, '*q*' shall be the second, '*r*' the third, and so on.

Any argument which results from the substitution of statements for the statement variables of an argument form is said to *have* that form, or to be a *substitution instance* of that argument form. If we symbolize the simple statement 'The United Nations will become more responsible' as *U*, and the simple statement 'There will be a third world war' as *W*, then the Disjunctive Syllogism presented earlier can be symbolized as

$$\begin{array}{l} U \vee W \\ \sim U \\ \therefore W \end{array} \qquad (1)$$

It has the form

$$\begin{array}{l} p \vee q \\ \sim p \\ \therefore q \end{array} \qquad (2)$$

[4] Here we assume that the simple statements involved are neither logically true (e.g., 'All equilateral triangles are triangles') nor logically false (e.g., 'Some triangles are nontriangles'). We assume also that the only logical relations among the simple statements involved are those asserted or entailed by the premisses. The point of these restrictions is to limit our considerations in Chapters 2 and 3 to truth-functional arguments alone, and to exclude other kinds of arguments whose validity turns on more complex logical considerations to be introduced in Chapters 4 and 5.

from which it results by substituting the statements *U* and *W* for the statement variables *p* and *q*, respectively. But that is not the only form of which it is a substitution instance. The same argument is obtained by substituting the statements $U \vee W$, $\sim U$, and *W* for the statement variables *p*, *q*, and *r*, respectively, in the argument form

$$\begin{array}{l} p \\ q \\ \therefore r \end{array} \tag{3}$$

We define *the specific form* of a given argument as that argument form from which the argument results by substituting a different *simple* statement for each distinct statement variable. Thus the specific form of the argument (1) is the argument form (2). Although the argument form (3) is *a* form of the argument (1), it is not *the specific form* of it. The technique of refutation by logical analogy can now be described more precisely. If the specific form of a given argument can be shown to have any substitution instance with true premisses and false conclusion, then the given argument is invalid.

The terms 'valid' and 'invalid' can be extended to apply to argument forms as well as arguments. An *invalid* argument form is one that has at least one substitution instance with true premisses and a false conclusion. The technique of refutation by logical analogy presupposes that any argument of which the specific form is an invalid argument form, is an invalid argument. Any argument form is *valid* that is not invalid: So a *valid* argument form is one that has *no* substitution instance with true premisses and false conclusion. Any given argument can be proved valid if it can be shown that the specific form of the given argument is a valid argument form.

To determine the validity or invalidity of an argument form, we must examine all possible substitution instances of it to see if any of them have true premisses and false conclusions. The arguments with which we are here concerned contain only simple statements and truth-functional compounds of them. We are interested only in the truth values of their premisses and conclusions. We can obtain all possible substitution instances whose premisses and conclusions have different truth values, by considering all possible arrangements of truth values for the statements that are substituted for the distinct statement variables in the argument form to be tested. Possible substitution instances can be set forth most conveniently in a truth table, with an initial or guide column for each distinct statement variable that appears in the argument form. Thus to prove the validity of the Disjunctive Syllogism form

$$\begin{array}{l} p \vee q \\ \sim p \\ \therefore q \end{array}$$

we construct the following truth table:

p	q	$p \vee q$	$\sim p$
T	**T**	**T**	**F**
T	**F**	**T**	**F**
F	**T**	**T**	**T**
F	**F**	**F**	**T**

Each row of this table represents a whole class of substitution instances. The **T**'s and **F**'s in the two initial columns represent the truth values of statements that can be substituted for the variables p and q in the argument form. These determine the truth values in the other columns, the third of which is headed by the first 'premiss' of the argument form, and the fourth by the second 'premiss'. The second column's heading is the 'conclusion' of the argument form. An examination of this truth table reveals that whatever statements are substituted for the variables p and q, the resulting argument cannot have true premisses and a false conclusion. This is so because the third row represents the only possible case in which both premisses are true, and there the conclusion is true also.

Because truth tables provide a purely mechanical or *effective* method of deciding the validity or invalidity of any argument of the general type here considered, we can now justify our proposal to symbolize all conditional statements by means of the truth-functional connective '$\supset$'. The justification for treating all implications as though they were mere material implications is that valid arguments containing conditional statements remain valid when those conditionals are interpreted as expressing material implications only. The three simplest and most intuitively valid forms of argument involving conditional statements are

Modus Ponens If p, then q
p
$\therefore q$

Modus Tollens If p, then q
$\sim q$
$\therefore \sim p$

and the

Hypothetical Syllogism If p, then q
If q, then r
$\therefore$ If p, then r

That they all remain valid when their conditionals are interpreted as expressing material implications is easily established by truth tables. The validity of

Modus Ponens is shown by the same truth table that defines the horseshoe symbol:

p	q	$p \supset q$
T	T	T
T	F	F
F	T	T
F	F	T

Here the two premisses are represented by the third and first columns, and the conclusion, by the second. Only the first row represents substitution instances in which both premisses are true, and in that row the conclusion is true also. The validity of *Modus Tollens* is shown by the truth table:

p	q	$p \supset q$	$\sim q$	$\sim p$
T	T	T	F	F
T	F	F	T	F
F	T	T	F	T
F	F	T	T	T

Here only the fourth row represents substitution instances in which both premisses (the third and fourth columns) are true, and there the conclusion (the fifth column) is true also. Since the Hypothetical Syllogism form contains three distinct statement variables, the truth table for it must have three initial columns and will require eight rows for listing all possible substitution instances:

p	q	r	$p \supset q$	$q \supset r$	$p \supset r$
T	T	T	T	T	T
T	T	F	T	F	F
T	F	T	F	T	T
T	F	F	F	T	F
F	T	T	T	T	T
F	T	F	T	F	T
F	F	T	T	T	T
F	F	F	T	T	T

In constructing it, the three initial columns represent all possible arrangements of truth values for the statements substituted for the statement variables p, q, and r. The fourth column is filled in by reference to the first and second, the fifth by reference to the second and third, and the sixth by reference to the first and third. The premisses are both true only in the first, fifth, seventh, and eighth rows, and in these rows the conclusion is true also. This suffices to demonstrate that the Hypothetical Syllogism remains valid when its conditionals are symbolized by means of the horseshoe symbol. Any doubts that remain about the claim that valid arguments containing conditionals remain valid when their conditionals are interpreted as expressing merely material implication, can be allayed by the reader's providing, symbolizing, and testing his own examples by means of truth tables.

To test the validity of an argument form by a truth table requires a table with a separate initial or guide column for each different statement variable, and a separate row for every possible assignment of truth values to the statement variables involved. Hence, testing an argument form containing n distinct statement variables, requires a truth table having 2^n rows. In constructing truth tables, it is convenient to fix upon some uniform pattern for inscribing the **T**'s and **F**'s in their initial or guide columns. In this book we shall follow the practice of simply alternating **T**'s and **F**'s down the extreme right-hand initial column and alternating pairs of **T**'s with pairs of **F**'s down the column directly to its left. Next, we shall alternate quadruples of **T**'s with quadruples of **F**'s, . . . , and finally, we shall fill in the top half of the extreme left-hand initial column with **T**'s and its bottom half with **F**'s.

There are two invalid argument forms that bear a superficial resemblance to the valid argument forms *Modus Ponens* and *Modus Tollens*. These are

$$\begin{array}{lcl} p \supset q & & p \supset q \\ q & \text{and} & \sim p \\ \therefore p & & \therefore \sim q \end{array}$$

and are known as the Fallacies of Affirming the Consequent and of Denying the Antecedent, respectively. The invalidity of both can be shown by a single truth table:

p	q	$p \supset q$	$\sim p$	$\sim q$
T	**T**	**T**	**F**	**F**
T	**F**	**F**	**F**	**T**
F	**T**	**T**	**T**	**F**
F	**F**	**T**	**T**	**T**

The two premisses in the Fallacy of Affirming the Consequent head the second and third columns, and are true in both the first and third rows. But the conclusion, which heads the first column, is false in the third row—which shows that the argument form does have a substitution instance with true premisses and a false conclusion, and is therefore invalid. Columns three and four are headed by the two premisses in the Fallacy of Denying the Antecedent, which are true in both the third and fourth rows. Its conclusion heads the fifth column and is false in the third row—which shows that the second argument form is invalid also.

It must be emphasized that although a valid argument form has only valid arguments as substitution instances, an invalid argument form can have both valid and invalid substitution instances. So to prove that a given argument is invalid, we must prove that *the specific form* of that argument is invalid.

EXERCISES

I. For each of the following arguments indicate which, if any, of the argument forms in Exercise II below have the given argument as a substitution instance, and indicate which, if any, is the specific form of the given argument:

*a. A
$\therefore A \vee B$

b. $C \cdot D$
$\therefore C$

c. $E \supset (F \cdot G)$
$\therefore \sim(F \cdot G) \supset \sim E$

d. H
I
$\therefore H \cdot I$

*e. $J \supset (K \cdot L)$
$J \vee (K \cdot L)$
$\therefore K \cdot L$

f. $M \supset (N \supset O)$
$O \supset \sim M$
$\therefore O \supset \sim N$

g. $(P \supset Q) \cdot (R \supset S)$
$\therefore P \supset Q$

h. $T \supset U$
$\therefore (T \supset U) \vee (V \cdot T)$

i. $W \supset X$
$\therefore X \supset (W \supset X)$

*j. $Y \vee (Z \cdot \sim Y)$
Y
$\therefore \sim(Z \cdot \sim Y)$

k. $(A \supset B) \cdot (C \supset D)$
$A \vee C$
$\therefore B \vee D$

l. $(E \supset F) \cdot (G \supset H)$
$\sim F \vee \sim G$
$\therefore \sim E \vee \sim H$

m. $I \supset J$
$\therefore (I \supset J) \supset (I \supset J)$

n. $K \supset (L \supset M)$
$K \supset L$
$\therefore K \supset M$

o. $N \supset (N \supset O)$
$N \supset N$
$\therefore N \supset O$

II. Use truth tables to determine the validity or invalidity of each of the following argument forms:

*1. $p \cdot q$
$\therefore p$

2. p
$\therefore p \cdot q$

3. $p \vee q$
$\therefore p$

4. p
$\therefore p \vee q$

*5. p
$\therefore p \supset q$

6. p
$\therefore q \supset p$

7. $p \supset q$
$\therefore \sim q \supset \sim p$

8. $p \supset q$
$\therefore \sim p \supset \sim q$

9. $p \supset (q \cdot r)$
$\therefore \sim(q \cdot r) \supset \sim p$

*10. $p \vee q$
p
$\therefore \sim q$

11. p
q
$\therefore p \cdot q$

12. $p \supset q$
$q \supset p$
$\therefore p \vee q$

13. $p \supset q$
$p \vee q$
$\therefore q$

14. $p \supset (q \supset r)$
$p \supset q$
$\therefore p \supset r$

*15. $(p \supset q) \cdot (p \supset r)$
p
$\therefore q \vee r$

16. $p \supset (q \vee r)$
$p \supset \sim q$
$\therefore p \vee r$

17. $(p \supset q) \cdot (r \supset s)$
$p \vee r$
$\therefore q \vee s$

18. $(p \supset q) \cdot (r \supset s)$
$\sim q \vee \sim s$
$\therefore \sim p \vee \sim r$

19. $(p \vee q) \supset (p \cdot q)$
$p \cdot q$
$\therefore p \vee q$

20. $p \vee (q \cdot \sim p)$
p
$\therefore \sim(q \cdot \sim p)$

21. $(p \vee q) \supset (p \cdot q)$
$\sim(p \vee q)$
$\therefore \sim(p \cdot q)$

III. Use truth tables to determine the validity or invalidity of each of the following arguments:

*1. If Alice is elected class president, then either Betty is elected vice-president or Carol is elected treasurer. Betty is elected vice-president. Therefore, if Alice is elected class-president, then Carol is not elected treasurer.

2. If Alice is elected class president, then either Betty is elected vice-president or Carol is elected treasurer. Carol is not elected treasurer. Therefore, if Betty is not elected vice-president, then Alice is not elected class president.

3. If Alice is elected class president, then Betty is elected vice-president and Carol is elected treasurer. Betty is not elected vice-president. Therefore, Alice is not elected class president.

4. If Alice is elected class president, then if Betty is elected vice-president, then Carol is elected treasurer. Betty is not elected vice-president. Therefore, either Alice is elected class president or Carol is elected treasurer.

*5. If the seed catalog is correct, then if the seeds are planted in April, then the flowers bloom in July. The flowers do not bloom in July. Therefore, if the seeds are planted in April, then the seed catalog is not correct.

6. If the seed catalog is correct, then if the seeds are planted in April, then the flowers bloom in July. The flowers bloom in July. Therefore, if the seed catalog is correct, then the seeds are planted in April.

7. If the seed catalog is correct, then if the seeds are planted in April, then the flowers bloom in July. The seeds are planted in April. Therefore, if the flowers do not bloom in July, then the seed catalog is not correct.

8. If the seed catalog is correct, then if the seeds are planted in April, then the flowers bloom in July. The flowers do not bloom in July. Therefore, if the seeds are not planted in April, then the seed catalog is not correct.

9. If Ed wins first prize, then Fred wins second prize, and if Fred wins second prize, then George is disappointed. Either Ed wins first prize or George is disappointed. Therefore, Fred does not win second prize.

*10. If Ed wins first prize, then either Fred wins second prize or George is disappointed. Fred does not win second prize. Therefore, if George is disappointed, then Ed does not win first prize.

11. If Ed wins first prize, then Fred wins second prize, and if Fred wins second prize, then George is disappointed. Either Fred does not win second prize or George is not disappointed. Therefore, Ed does not win first prize.

12. If Ed wins first prize, then Fred wins second prize, and if Fred wins second prize, then George is disappointed. Either Ed does not win first prize or Fred does not win second prize. Therefore, either Fred does not win second prize or George is not disappointed.

13. If the weather is warm and the sky is clear, then we go swimming and we go boating. It is not the case that if the sky is clear, then we go swimming. Therefore, the weather is not warm.

14. If the weather is warm and the sky is clear, then either we go swimming or we go boating. It is not the case that if the sky is clear, then we go swimming. Therefore, if we do not go boating, then the weather is not warm.

*15. If the weather is warm and the sky is clear, then either we go swimming or we go boating. It is not the case that if we do not go swimming, then the sky is not clear. Therefore, either the weather is warm or we go boating.

2.4 Statement Forms

The introduction of statement variables in the preceding section enabled us to define both argument forms in general and the specific form of a given argument. Now, we define a *statement form* to be any sequence of symbols containing statement variables, such that when statements are substituted for the statement variables—the same statement being substituted for every occurrence of the same statement variable throughout—the result is a statement. Again for definiteness, we establish the convention that in any statement form, 'p' shall be the first statement variable that occurs in it. The statement variable 'q' shall be the second, 'r' the third, and so on. Any statement which results from substituting statements for the statement variables of a statement form is said to *have* that form, or to be a *substitution instance* of it. Just as we distinguished the specific form of a given argument, so we distinguish *the specific form* of a given statement as that statement form from which the given statement results by substituting a different simple statement for each distinct statement variable. For example, where A, B, and C are different simple statements, the compound statement $A \supset (B \vee C)$ is a substitution instance of the statement form $p \supset q$, and also of the statement form $p \supset (q \vee r)$, but only the latter is the specific form of the given statement.

Although the statements 'Balboa discovered the Pacific Ocean' (B) and 'Balboa discovered the Pacific Ocean or else he didn't' ($B \vee \sim B$) are both true, we discover their truth in quite different ways. The truth of B is a matter of history, and must be learned through empirical investigation. Moreover, events might possibly have been such as to make B false; there is nothing *necessary* about the truth of B. But the truth of the statement $B \vee \sim B$ *can* be known independently of empirical investigation, and no events could possibly have made it false, for it is a necessary truth. The statement $B \vee \sim B$ is a formal truth, a substitution instance of a statement form *all* of whose substitution instances are true. A statement form that has only true substitution instances is said to be *tautologous*, or a *tautology*. The specific form of $B \vee \sim B$ is $p \vee \sim p$, and is proved a tautology by the following truth table:

p	$\sim p$	$p \vee \sim p$
T	F	T
F	T	T

That there are only **T**'s in the column headed by the statement form in question shows that all of its substitution instances are true. Any statement that is a substitution instance of a tautologous statement form is formally true and is, itself, said to be tautologous or a tautology.

Similarly, although the statements 'Cortez discovered the Pacific' (C) and 'Cortez discovered the Pacific and Cortez did not discover the Pacific' ($C \cdot \sim C$) are both false, we discover their falsehood in quite different ways. The first simply *happens* to be false, and that must be learned empirically; the second is necessarily false, and that can be known independently of empirical investigation. The statement $C \cdot \sim C$ is formally false, a substitution instance of a statement form, *all* of whose substitution instances are false. One statement is said to contradict, or to be a contradiction of, another statement when it is logically impossible for them both to be true. In this sense, *contradiction* is a relation between statements. But there is another, related sense of that term. When it is logically impossible for a particular statement to be true, that statement itself is said to be self-contradictory, or a self-contradiction. Such statements are also said more simply to be contradictory or contradictions. We shall follow the latter usage here. A statement form that has only false substitution instances is said to be *contradictory* or a *contradiction*, and the same terms are applied to its substitution instances. The statement form $p \cdot \sim p$ is proved a contradiction by the fact that in its truth table only **F**'s occur in the column that it heads.

Statements and statement forms that are neither tautologous nor contradictory are said to be *contingent* or *contingencies*. For example, p, $\sim p$, $p \vee q$, $p \cdot q$, and $p \supset q$ are contingent statement forms; and B, C, $\sim B$, $\sim C$, $B \cdot C$, $B \vee C$ are contingent *statements*. The term is appropriate, since their truth values are not formally determined but are dependent or contingent upon what happens to be the case.

It is easily proved that $p \supset (q \supset p)$ and $\sim p \supset (p \supset q)$ are tautologies. When expressed in English as 'A true statement is implied by any statement whatever', and as 'A false statement implies any statement whatever', they seem rather strange. They have been called by some writers the *paradoxes of material implication*. When it is kept in mind that the horseshoe symbol is a truth-functional connective that stands for *material* implication rather than either 'implication in general' or more usual kinds of implications such as logical or causal, then the tautologous statement forms in question are not at all surprising. And when the misleading English formulations are corrected by inserting the word 'materially' before 'implied' and 'implies', then the air of paradox vanishes. Material implication is a special, technical concept that the logician introduces and uses because it simplifies his task of discriminating valid from invalid arguments.

Two statements are said to be *materially equivalent* when they have the same truth value, and we symbolize the statement *that they are* materially equivalent by inserting the symbol '$\equiv$' between them. Being a truth-functional connective, the three-bar symbol is defined by the following truth table:

p	q	$p \equiv q$
T	**T**	**T**
T	**F**	**F**
F	**T**	**F**
F	**F**	**T**

To say that two statements are materially equivalent, is to say that they materially imply each other, as is easily verified by a truth table. Hence the three-bar symbol may be read 'if and only if'. A statement or statement form of the pattern $p \equiv q$ is called a *biconditional.* Two statements are said to be *logically equivalent* when the biconditional that expresses their material equivalence is a tautology. The 'principle of Double Negation', expressed as $p \equiv \sim\sim p$, is proved to be tautologous by a truth table.

There are two logical equivalences that express important interrelations of conjunctions, disjunctions, and negations. Because a conjunction asserts that both its conjuncts are true, its negation need assert only that at least one is false. Thus negating the conjunction $p \cdot q$ amounts to asserting the disjunction of the negations of p and q. This statement of equivalence is symbolized as $\sim(p \cdot q) \equiv (\sim p \vee \sim q)$, and proved to be a tautology by the following truth table:

p	q	$p \cdot q$	$\sim(p \cdot q)$	$\sim p$	$\sim q$	$\sim p \vee \sim q$	$\sim(p \cdot q) \equiv (\sim p \vee \sim q)$
T	T	T	F	F	F	F	T
T	F	F	T	F	T	T	T
F	T	F	T	T	F	T	T
F	F	F	T	T	T	T	T

Similarly, because a disjunction asserts merely that at least one disjunct is true, to negate it is to assert that both are false. Negating the disjunction $p \vee q$ amounts to asserting the conjunction of the negations of p and q. It is symbolized as $\sim(p \vee q) \equiv (\sim p \cdot \sim q)$ and is easily proved tautologous by a truth table. These two equivalences are known as De Morgan's Theorems, after the English mathematician-logician Augustus De Morgan (1806–1871) and can be stated compendiously in English as:

> The negation of the $\begin{Bmatrix}\text{conjunction}\\ \text{disjunction}\end{Bmatrix}$ of two statements is logically equivalent to the $\begin{Bmatrix}\text{disjunction}\\ \text{conjunction}\end{Bmatrix}$ of their negations.

Two statement forms are said to be *logically* equivalent if no matter what statements are substituted for their statement variables—the same statement being substituted for the same statement variable in both statement forms—the resulting pairs of statements are materially equivalent. Thus De Morgan's Theorem asserts that $\sim(p \vee q)$ and $\sim p \cdot \sim q$ are logically equivalent statement forms. By De Morgan's Theorem and the principle of Double Negation $\sim(p \cdot \sim q)$ and $\sim p \vee q$ are logically equivalent, hence either can be taken as defining $p \supset q$. The second alternative is the more usual choice.

To every argument corresponds a conditional statement whose antecedent is the conjunction of the argument's premisses and whose consequent is the argument's conclusion. That corresponding conditional is a tautology if and only if the argument is valid. Thus to the valid argument form

$$p \vee q$$
$$\sim p$$
$$\therefore q$$

corresponds the tautologous statement form $[(p \vee q) \cdot \sim p] \supset q$; and to the invalid argument form

$$p \supset q$$
$$q$$
$$\therefore p$$

corresponds the nontautologous statement form $[(p \supset q) \cdot q] \supset p$. An argument form is valid if and only if its truth table has a **T** under its conclusion in every row in which there are **T**'s under all of its premisses. Since an **F** can occur in the column headed by its corresponding conditional statement only where there are **T**'s under all of those premisses and an **F** under the conclusion, it is clear that there can be only **T**'s under a conditional that corresponds to a valid argument form. If an argument is valid, the statement that the conjunction of its premisses implies its conclusion is a tautology.

An alternative version of the truth table test of a statement form is the following, which corresponds to the preceding truth table.

$\sim$	$(p$	$\cdot$	$q)$	$\equiv$	$(\sim$	p	$\vee$	$\sim$	$q)$
F	**T**	**T**	**T**	**T**	**F**	**T**	**F**	**F**	**T**
T	**T**	**F**	**F**	**T**	**F**	**T**	**T**	**T**	**F**
T	**F**	**F**	**T**	**T**	**T**	**F**	**T**	**F**	**T**
T	**F**	**F**	**F**	**T**	**T**	**F**	**T**	**T**	**F**
(1)	(2)	(3)	(4)	(5)	(6)	(7)	(8)	(9)	(10)

Here columns (2), (4), (7), (10) are the initial or guide columns. Column (3) is filled in by reference to columns (2) and (4), and column (1) by reference to column (3). Column (6) is filled in by reference to column (7), column (9) is filled in by reference to column (10), and then column (8) by reference to columns (6) and (9). Finally, column (5) is filled in by reference to columns (1) and (8). That its main connective has only **T**'s under it in the truth table establishes that the statement form being tested is a tautology.

EXERCISES

I. Use truth tables to characterize the following statement forms as tautologous, contradictory, or contingent:

*****1.** $p \supset \sim p$

2. $(p \supset \sim p) \cdot (\sim p \supset p)$

3. $p \supset (p \supset p)$

4. $(p \supset p) \supset p$

*5. $p \supset (p \cdot p)$

6. $(p \cdot q) \supset p$

7. $(p \supset q) \supset [\sim(q \cdot r) \supset \sim(r \cdot p)]$

8. $(\sim p \cdot q) \cdot (q \supset p)$

9. $[(p \supset q) \supset q] \supset q$

10. $[(p \supset q) \supset p] \supset p$

II. Use truth tables to decide which of the following are logical equivalences:

*1. $(p \supset q) \equiv (\sim p \supset \sim q)$

2. $(p \supset q) \equiv (\sim q \supset \sim p)$

3. $[(p \cdot q) \supset r] \equiv [p \supset (q \supset r)]$

4. $[p \supset (q \supset r)] \equiv [(p \supset q) \supset r]$

*5. $[p \cdot (q \vee r)] \equiv [(p \cdot q) \vee (p \cdot r)]$

6. $[p \vee (q \cdot r)] \equiv [(p \vee q) \cdot r]$

7. $[p \vee (q \cdot r)] \equiv [(p \vee q) \cdot (p \vee r)]$

8. $(p \equiv q) \equiv [(p \cdot q) \vee (\sim p \cdot \sim q)]$

9. $p \equiv [p \cdot (p \supset q)]$

10. $p \equiv [p \cdot (q \supset p)]$

3

The Method of Deduction

3.1 Formal Proof of Validity

If arguments contain more than two or three different simple statements as components, it is cumbersome and tedious to use truth tables to test their validity. A more convenient method of establishing the validity of some arguments is to *deduce* their conclusions from their premisses by a sequence of shorter, more elementary arguments that are already known to be valid. Consider, for example, the following argument, in which five different simple statements occur:

> Either the Attorney General has imposed a strict censorship or if Black mailed the letter then Davis received a warning.
> If our lines of communication have not broken down completely, then if Davis received a warning, then Emory was informed about the matter.
> If the Attorney General has imposed a strict censorship, then our lines of communication have broken down completely.
> Our lines of communication have not broken down completely.
> Therefore, if Black mailed the letter, then Emory was informed about the matter.

It may be translated into our symbolism as

$$A \vee (B \supset D)$$
$$\sim C \supset (D \supset E)$$
$$A \supset C$$
$$\sim C$$
$$\therefore B \supset E$$

To establish the validity of this argument by means of a truth table would require a table with thirty-two rows. We can prove the given argument valid, however, by deducing its conclusion from its premisses by a sequence of just four arguments whose validity has already been established. From the third and fourth premisses, $A \supset C$ and $\sim C$, we validly infer $\sim A$ by *Modus Tollens*. From

$\sim A$ and the first premiss, $A \vee (B \supset D)$, we validly infer $B \supset D$ by a Disjunctive Syllogism. From the second and fourth premisses, $\sim C \supset (D \supset E)$ and $\sim C$, we validly infer $D \supset E$ by *Modus Ponens*. And finally, from these last two conclusions (or subconclusions), $B \supset D$ and $D \supset E$, we validly infer $B \supset E$ by a Hypothetical Syllogism. That its conclusion can be deduced from its premisses using valid arguments exclusively, *proves* the original argument to be valid. Here the elementary valid argument forms *Modus Ponens* (M.P.), *Modus Tollens* (M.T.), Disjunctive Syllogism (D.S.), and Hypothetical Syllogism (H.S.) are used as *Rules of Inference* by which conclusions are validly deduced from premisses.

A more formal and more concise way of writing out this proof of validity is to list the premisses and the statements deduced from them in one column, with "justifications" for the latter written beside them. In each case the "justification" for a statement specifies the preceding statements from which, and the Rule of Inference by which, the statement in question was deduced. It is convenient to put the conclusion to the right of the last premiss, separated from it by a slanting line which automatically marks all of the statements above it to be premisses. The formal proof of validity for the given argument can be written as

1.	$A \vee (B \supset D)$	
2.	$\sim C \supset (D \supset E)$	
3.	$A \supset C$	
4.	$\sim C$	$/ \therefore B \supset E$
5.	$\sim A$	3, 4, M.T.
6.	$B \supset D$	1, 5, D.S.
7.	$D \supset E$	2, 4, M.P.
8.	$B \supset E$	6, 7, H.S.

A *formal proof of validity* for a given argument is defined to be a sequence of statements, each of which is either a premiss of that argument or follows from preceding statements by an elementary valid argument, and such that the last statement in the sequence is the conclusion of the argument whose validity is being proved. This definition must be completed and made definite by specifying what is to count as an 'elementary valid argument'. We first define an *elementary valid argument* as any argument that is a substitution instance of an elementary valid argument form. Then, we present a list of just nine argument forms that are sufficiently obvious to be regarded as elementary valid argument forms and accepted as Rules of Inference.

One matter to be emphasized is that *any* substitution instance of an elementary valid argument form is an elementary valid argument. Thus the argument

$$\sim C \supset (D \supset E)$$
$$\sim C$$
$$\therefore D \supset E$$

is an elementary valid argument because it is a substitution instance of the elementary valid argument form *Modus Ponens* (M.P.). It results from

$$p \supset q$$
$$p$$
$$\therefore q$$

by substituting $\sim C$ for p and $D \supset E$ for q; therefore, it is of that form even though *Modus Ponens* is not *the specific form* of the given argument.

We begin our development of the method of deduction by presenting a list of just nine elementary valid argument forms that can be used in constructing formal proofs of validity:

RULES OF INFERENCE

1. *Modus Ponens* (M.P.)
   $p \supset q$
   p
   $\therefore q$
2. *Modus Tollens* (M.T.)
   $p \supset q$
   $\sim q$
   $\therefore \sim p$
3. Hypothetical Syllogism (H.S.)
   $p \supset q$
   $q \supset r$
   $\therefore p \supset r$
4. Disjunctive Syllogism (D.S.)
   $p \vee q$
   $\sim p$
   $\therefore q$
5. Constructive Dilemma (C.D.)
   $(p \supset q) \cdot (r \supset s)$
   $p \vee r$
   $\therefore q \vee s$
6. Destructive Dilemma (D.D.)
   $(p \supset q) \cdot (r \supset s)$
   $\sim q \vee \sim s$
   $\therefore \sim p \vee \sim r$
7. Simplification (Simp.)
   $p \cdot q$
   $\therefore p$
8. Conjunction (Conj.)
   p
   q
   $\therefore p \cdot q$
9. Addition (Add.)
   p
   $\therefore p \vee q$

These nine Rules of Inference are elementary valid argument forms, whose validity is easily established by truth tables. They can be used to construct formal proofs of validity for a wide range of more complicated arguments. The names listed are standard for the most part, and the use of their abbreviations permits formal proofs to be set down with a minimum of writing.

EXERCISES

I. For each of the following arguments, state the Rule of Inference by which its conclusion follows from its premiss or premisses:

*1. $(A \supset \sim B) \cdot (\sim C \supset D)$
$\therefore A \supset \sim B$

2. $E \supset \sim F$
$\therefore (E \supset \sim F) \vee (\sim G \supset H)$

3. $(I \equiv \sim J) \cdot (I \equiv \sim J)$
$\therefore I \equiv \sim J$

4. $K \vee (L \vee M)$
$\therefore [K \vee (L \vee M)] \vee [K \vee (L \vee M)]$

*5. $N \supset (O \equiv \sim P)$
$(O \equiv \sim P) \supset Q$
$\therefore N \supset Q$

6. $(R \equiv \sim S) \supset (T \supset U)$
$R \equiv \sim S$
$\therefore T \supset U$

7. $(V \supset W) \vee (X \supset Y)$
$\sim (V \supset W)$
$\therefore X \supset Y$

8. $(A \supset \sim B) \cdot [C \supset (D \cdot E)]$
$\sim \sim B \vee \sim (D \cdot E)$
$\therefore \sim A \vee \sim C$

9. $(F \supset \sim G) \supset (\sim H \vee \sim I)$
$F \supset \sim G$
$\therefore \sim H \vee \sim I$

*10. $[\sim (J \cdot K) \supset \sim L] \cdot (M \supset \sim N)$
$\sim (J \cdot K) \vee M$
$\therefore \sim L \vee \sim N$

11. $O \supset \sim P$
$\sim P \supset Q$
$\therefore (O \supset \sim P) \cdot (\sim P \supset Q)$

12. $(\sim R \equiv S) \vee (T \vee U)$
$\sim (\sim R \equiv S)$
$\therefore T \vee U$

13. $[(V \cdot \sim W) \supset X] \cdot [(W \cdot \sim Y) \supset Z]$
$(V \cdot \sim W) \vee (W \cdot \sim Y)$
$\therefore X \vee Z$

14. $[A \supset (B \vee C)] \supset [(D \cdot E) \equiv \sim F]$
$\sim [(D \cdot E) \equiv \sim F]$
$\therefore \sim [A \supset (B \vee C)]$

15. $\sim [G \supset (H \vee I)] \cdot \sim [(J \cdot K) \supset L]$
$\therefore \sim [G \supset (H \vee I)]$

II. Each of the following is a formal proof of validity for the indicated argument. State the 'justification' for each line that is not a premiss:

*1.
1. $(A \cdot B) \supset [A \supset (D \cdot E)]$
2. $(A \cdot B) \cdot C \quad / \therefore D \vee E$
3. $A \cdot B$
4. $A \supset (D \cdot E)$
5. A
6. $D \cdot E$
7. D
8. $D \vee E$

2.
1. $F \vee (G \vee H)$
2. $(G \supset I) \cdot (H \supset J)$
3. $(I \vee J) \supset (F \vee H)$
4. $\sim F \quad / \therefore H$
5. $G \vee H$
6. $I \vee J$
7. $F \vee H$
8. H

3.
1. $K \supset L$
2. $M \supset N$
3. $(O \supset N) \cdot (P \supset L)$
4. $(\sim N \vee \sim L) \cdot (\sim M \vee \sim O)$
$/ \therefore (\sim O \vee \sim P) \cdot (\sim M \vee \sim K)$
5. $(M \supset N) \cdot (K \supset L)$
6. $\sim N \vee \sim L$
7. $\sim M \vee \sim K$
8. $\sim O \vee \sim P$
9. $(\sim O \vee \sim P) \cdot (\sim M \vee \sim K)$

4.
1. $Q \supset (R \supset S)$
2. $(R \supset S) \supset T$
3. $(S \cdot U) \supset \sim V$
4. $\sim V \supset (R \equiv \sim W)$
5. $\sim T \vee \sim (R \equiv \sim W)$
$/ \therefore \sim Q \vee \sim (S \cdot U)$
6. $Q \supset T$
7. $(S \cdot U) \supset (R \equiv \sim W)$
8. $(Q \supset T) \cdot [(S \cdot U) \supset (R \equiv \sim W)]$
9. $\sim Q \vee \sim (S \cdot U)$

*5.
1. $(\sim X \vee \sim Y) \supset [A \supset (P \cdot \sim Q)]$
2. $(\sim X \cdot \sim R) \supset [(P \cdot \sim Q) \supset Z]$

3. $(\sim X \cdot \sim R) \cdot (\sim Z \vee A)$
   $/ \therefore A \supset Z$
4. $\sim X \cdot \sim R$
5. $(P \cdot \sim Q) \supset Z$
6. $\sim X$
7. $\sim X \vee \sim Y$
8. $A \supset (P \cdot \sim Q)$
9. $A \supset Z$

6.
1. $A \supset B$
2. $C \supset D$
3. $\sim B \vee \sim D$
4. $\sim\sim A$
5. $(E \cdot F) \supset C$
   $/ \therefore \sim(E \cdot F)$
6. $(A \supset B) \cdot (C \supset D)$
7. $\sim A \vee \sim C$
8. $\sim C$
9. $\sim(E \cdot F)$

7.
1. $(G \supset H) \supset (I \equiv J)$
2. $K \vee \sim(L \supset M)$
3. $(G \supset H) \vee \sim K$
4. $N \supset (L \supset M)$
5. $\sim(I \equiv J) \quad / \therefore \sim N$
6. $\sim(G \supset H)$
7. $\sim K$
8. $\sim(L \supset M)$
9. $\sim N$

8.
1. $(O \supset \sim P) \cdot (\sim Q \supset R)$
2. $(S \supset T) \cdot (\sim U \supset \sim V)$
3. $(\sim P \supset S) \cdot (R \supset \sim U)$
4. $(T \vee \sim V) \supset (W \cdot X)$
5. $O \vee \sim Q \quad / \therefore W \cdot X$
6. $\sim P \vee R$
7. $S \vee \sim U$
8. $T \vee \sim V$
9. $W \cdot X$

9.
1. $[(A \vee \sim B) \vee C] \supset [D \supset (E \equiv F)]$
2. $(A \vee \sim B) \supset [(F \equiv G) \supset H]$
3. $A \supset [(E \equiv F) \supset (F \equiv G)]$
4. $A \quad / \therefore D \supset H$
5. $A \vee \sim B$
6. $(A \vee \sim B) \vee C$
7. $D \supset (E \equiv F)$
8. $(E \equiv F) \supset (F \equiv G)$
9. $D \supset (F \equiv G)$
10. $(F \equiv G) \supset H$
11. $D \supset H$

***10.**
1. $H \supset (I \supset J)$
2. $K \supset (I \supset J)$
3. $(\sim H \cdot \sim K) \supset (\sim L \vee \sim M)$
4. $(\sim L \supset \sim N) \cdot (\sim M \supset \sim O)$
5. $(P \supset N) \cdot (Q \supset O)$
6. $\sim(I \supset J) \quad / \therefore \sim P \vee \sim Q$
7. $\sim H$
8. $\sim K$
9. $\sim H \cdot \sim K$
10. $\sim L \vee \sim M$
11. $\sim N \vee \sim O$
12. $\sim P \vee \sim Q$

III. Construct a formal proof of validity for each of the following arguments:

***1.** $A \supset B$
$C \supset D$
$(\sim B \vee \sim D) \cdot (\sim A \vee \sim B)$
$\therefore \sim A \vee \sim C$

2. $E \supset (F \cdot \sim G)$
$(F \vee G) \supset H$
E
$\therefore H$

3. $J \supset K$
$J \vee (K \vee \sim L)$
$\sim K$
$\therefore \sim L \cdot \sim K$

4. $M \supset N$
$N \supset O$
$(M \supset O) \supset (N \supset P)$
$(M \supset P) \supset Q$
$\therefore Q$

***5.** $(R \supset \sim S) \cdot (T \supset \sim U)$
$(V \supset \sim W) \cdot (X \supset \sim Y)$
$(T \supset W) \cdot (U \supset S)$
$V \vee R$
$\therefore \sim T \vee \sim U$

6. $A \supset (B \cdot C)$
$\sim A \supset [(D \supset E) \cdot (F \supset G)]$
$(B \cdot C) \vee [(\sim A \supset D) \cdot (\sim A \supset F)]$
$\sim(B \cdot C) \cdot \sim(G \cdot D)$
$\therefore E \vee G$

7. $(\sim H \vee I) \supset (J \supset K)$
$(\sim L \cdot \sim M) \supset (K \supset N)$
$(H \supset L) \cdot (L \supset H)$
$(\sim L \cdot \sim M) \cdot \sim O$
$\therefore J \supset N$

8. $(P \supset Q) \cdot (R \supset S)$
$(Q \supset T) \cdot (S \supset U)$
$(\sim P \supset T) \cdot (\sim Q \supset S)$
$\sim T$
$\therefore \sim R \vee \sim Q$

9. $V \supset W$
$X \supset Y$
$Z \supset W$
$X \supset A$
$W \supset X$
$[(V \supset Y) \cdot (Z \supset A)] \supset (V \vee Z)$
$\therefore Y \vee A$

10. $(B \vee C) \supset (D \vee E)$
$[(D \vee E) \vee F] \supset (G \vee H)$
$(G \vee H) \supset \sim D$
$E \supset \sim G$
B
$\therefore H$

IV. Construct a formal proof of validity for each of the following arguments, using the abbreviations suggested:

*1. If either algebra is required or geometry is required, then all students will study mathematics. Algebra is required and trigonometry is required. Therefore, all students will study mathematics. (*A*: Algebra is required. *G*: Geometry is required. *S*: All students will study mathematics. *T*: Trigonometry is required.)

2. Either Smith attended the meeting or Smith was not invited to the meeting. If the directors wanted Smith at the meeting, then Smith was invited to the meeting. Smith did not attend the meeting. If the directors did not want Smith at the meeting and Smith was not invited to the meeting, then Smith is on his way out of the company. Therefore, Smith is on his way out of the company. (*A*: Smith attended the meeting. *I*: Smith was invited to the meeting. *D*: The directors wanted Smith at the meeting. *W*: Smith is on his way out of the company.)

3. If a scarcity of commodities develops, then prices rise. If there is a change of administration, then fiscal controls will not be continued. If the threat of inflation persists, then fiscal controls will be continued. If there is overproduction, then prices do not rise. Either there is overproduction or there is a change of administration. Therefore, either a scarcity of commodities does not develop or the threat of inflation does not persist. (*S*: A scarcity of commodities develops. *P*: Prices rise. *C*: There is a change of administration. *F*: Fiscal controls will be continued. *I*: The threat of inflation persists. *O*: There is overproduction.)

4. If the investigation continues, then new evidence is brought to light. If new evidence is brought to light, then several leading citizens are implicated. If several leading citizens are implicated, then the newspapers stop publicizing the case. If continuation of the investigation implies that the newspapers stop publicizing the case, then the bringing to light of new evidence implies that the investigation continues. The investigation does not continue. Therefore, new evidence is not brought to light. (*C*: The investigation continues. *N*: New evidence is brought to light. *I*: Several leading citizens are implicated. *S*: The newspapers stop publicizing the case.)

*5. If the king does not castle and the pawn advances, then either the bishop is blocked or the rook is pinned. If the king does not castle, then if the bishop is blocked, then the game is a draw. Either the king castles or if the rook is pinned, then the exchange is lost. The king does not castle and the pawn advances. Therefore, either the game is a draw or the exchange is lost. (*K*: The king castles. *P*: The pawn advances. *B*: The bishop is blocked. *R*: The rook is pinned. *D*: The game is a draw. *E*: The exchange is lost.)

6. If Andrews is present, then Brown is present, and if Brown is present, then Cohen is not present. If Cohen is present, then Davis is not present. If Brown is present, then Emerson is present. If Davis is not present, then Farley is present. Either Emerson is not present or Farley is not present. Therefore, either Andrews is not present or Cohen is not present. (*A*: Andrews is present. *B*: Brown is present. *C*: Cohen is present. *D*: Davis is present. *E*: Emerson is present. *F*: Farley is present.)

7. If either George enrolls or Harry enrolls, then Ira does not enroll. Either Ira enrolls or Harry enrolls. If either Harry enrolls or George does not enroll, then Jim enrolls. George enrolls. Therefore, either Jim enrolls or Harry does not enroll. (*G*: George enrolls. *H*: Harry enrolls. *I*: Ira enrolls. *J*: Jim enrolls.)

8. If Tom received the message, then Tom took the plane, but if Tom did not take the plane, then Tom missed the meeting. If Tom missed the meeting, then Dave was elected to the board, but if Dave was elected to the board, then Tom received the message. If either Tom did not miss the meeting or Tom did not receive the message, then either Tom did not take the plane or Dave was not elected to the board. Tom did not miss the meeting. Therefore, either Tom did not receive the message or Tom did not miss the meeting. (*R*: Tom received the message. *P*: Tom took the plane. *M*: Tom missed the meeting. *D*: Dave was elected to the board.)

9. If Dick was recently vaccinated, then he has a fever. Either Dick was recently vaccinated, or if pocks begin to appear then Dick must be quarantined. Either Dick has measles or if a rash develops, then there are complications. If Dick has measles, then he has a fever. If Dick was not recently vaccinated and Dick does not have measles, then either a rash develops or pocks begin to appear. Dick does not have a fever. Therefore, either there are complications or Dick must be quarantined. (*V*: Dick was recently vaccinated. *F*: Dick has a fever. *P*: Pocks begin to appear. *Q*: Dick must be quarantined. *M*: Dick has measles. *R*: A rash develops. *C*: There are complications.)

*10. Either taxes are increased or if expenditures rise, then the debt ceiling is raised. If taxes are increased, then the cost of collecting taxes increases. If a rise in expenditures implies that the government borrows more money, then if the debt ceiling is raised, then interest rates increase. If taxes are not increased and the cost of collecting taxes does not increase, then if the debt ceiling is raised, then the government borrows more money. The cost of collecting taxes does not increase. Either interest rates do not increase or the government does not borrow more money. Therefore, either the debt ceiling is not raised or expenditures do not rise. (*T*: Taxes are increased. *E*: Expenditures rise. *D*: The debt ceiling is raised. *C*: The cost of collecting taxes increases. *G*: The government borrows more money. *I*: Interest rates increase.)

3.2 The Rule of Replacement

There are many valid truth-functional arguments that cannot be proved valid using only the nine Rules of Inference that have been given thus far. For example, a formal proof of validity for the obviously valid argument

$$A \cdot B$$
$$\therefore B$$

requires additional Rules of Inference.

Now the only compound statements that concern us here are truth-functional compound statements. Hence if any part of a compound statement is replaced by an expression that is logically equivalent to the part replaced, the truth value of the resulting statement is the same as that of the original statement. This is sometimes called the Rule of Replacement and sometimes the Principle of Extensionality.[1] We adopt the Rule of Replacement as an additional principle of inference. It permits us to infer from any statement the result of replacing all or part of that statement by any other statement logically equivalent to the part replaced. Thus, using the Principle of Double Negation (D.N.), which asserts the logical equivalence of p and $\sim\sim p$, we can infer from $A \supset \sim\sim B$ any of the statements

$$A \supset B, \quad \sim\sim A \supset \sim\sim B, \quad A \supset \sim\sim\sim\sim B, \quad \text{or} \quad \sim\sim(A \supset \sim\sim B)$$

by the Rule of Replacement.

To make this new rule definite, we list a number of logical equivalences with which it can be used. These equivalences constitute additional Rules of Inference that can be used in proving the validity of arguments. We number them consecutively after the first nine Rules already stated.

Rule of Replacement: Any of the following logically equivalent expressions can replace each other wherever they occur:

10. De Morgan's Theorems (De M.): $\sim(p \cdot q) \equiv (\sim p \vee \sim q)$.
$\sim(p \vee q) \equiv (\sim p \cdot \sim q)$.

11. Commutation (Com.): $(p \vee q) \equiv (q \vee p)$.
$(p \cdot q) \equiv (q \cdot p)$.

12. Association (Assoc.): $[p \vee (q \vee r)] \equiv [(p \vee q) \vee r]$.
$[p \cdot (q \cdot r)] \equiv [(p \cdot q) \cdot r]$.

13. Distribution (Dist.): $[p \cdot (q \vee r)] \equiv [(p \cdot q) \vee (p \cdot r)]$.
$[p \vee (q \cdot r)] \equiv [(p \vee q) \cdot (p \vee r)]$.

[1] It will be stated more formally in an appropriate context and demonstrated in Chapter 8.

14. Double Negation (D.N.): $p \equiv \sim\sim p$.
15. Transposition (Trans.): $(p \supset q) \equiv (\sim q \supset \sim p)$.
16. Material Implication (Impl.): $(p \supset q) \equiv (\sim p \vee q)$.
17. Material Equivalence (Equiv.): $(p \equiv q) \equiv [(p \supset q) \cdot (q \supset p)]$.
$(p \equiv q) \equiv [(p \cdot q) \vee (\sim p \cdot \sim q)]$
18. Exportation (Exp.): $[(p \cdot q) \supset r] \equiv [p \supset (q \supset r)]$.
19. Tautology (Taut.): $p \equiv (p \vee p)$.
$p \equiv (p \cdot p)$.

A formal proof of validity for the argument given on page 39 can now be written:

1. $A \cdot B$ $/ \therefore B$
2. $B \cdot A$ 1, Com.
3. B 2, Simp.

Some argument forms, although very elementary and perfectly valid, are not included in our list of nineteen Rules of Inference. Although the argument

$$A \cdot B$$
$$\therefore B$$

is obviously valid, its form

$$p \cdot q$$
$$\therefore q$$

is not included in our list. Hence B does not follow from $A \cdot B$ by any single elementary valid argument form, *as defined by our list*. It can be deduced, however, by using *two* elementary valid arguments, as shown above. We could add the intuitively valid argument form

$$p \cdot q$$
$$\therefore q$$

to our list, of course, but if we expanded our list in this way we might end up with a list that was too long and, therefore, unmanageable.

The list of Rules of Inference contains several redundancies. For example, *Modus Tollens* could be dropped from our list without any real weakening of the machinery. This is so because any step deduced by its use could be deduced by using other Rules of the list instead. Thus in our first proof, on page 33, line 5, $\sim A$, which was deduced from lines 3 and 4, $A \supset C$ and $\sim C$, by *Modus Tollens*, could have been deduced without it, since $\sim C \supset \sim A$ follows from $A \supset C$ by Transposition, and $\sim A$ from $\sim C \supset \sim A$ and $\sim C$ by *Modus Ponens*.

Modus Tollens is so common and intuitive a principle of inference, however, that it has been included anyway. Others have also been included for convenience, despite their logical redundancy.

The test of whether or not a given sequence of statements is a formal proof, is *effective*. That is, direct observation will suffice to decide, of every line beyond the premisses, whether or not it actually does follow from preceding lines by one of the given Rules of Inference. No 'thinking' is required: neither in thinking about what the statements mean, nor in using logical intuition to check the validity of any line's deduction. Even where the 'justification' of a statement is not written beside it, there is a finite, mechanical procedure for deciding whether or not the deduction is legitimate. Each line is preceded by only a finite number of other lines and only a finite number of Rules of Inference have been adopted. Although time consuming, it can be verified by inspection whether the line in question follows from any single preceding line or from any pair of preceding lines by any one of the Rules of Inference listed. For example, in the foregoing proof, line 2, $B \cdot A$, is preceded only by line 1, $A \cdot B$. Its legitimacy can be decided by observing that although it does not follow from $A \cdot B$ by *Modus Ponens*, nor by *Modus Tollens*, nor by a Hypothetical Syllogism, and so on through number 10, when we come to number 11, we can *see*, simply by looking at their forms, that line 2 follows from line 1 by the principle of Commutation. Similarly, the legitimacy of any line can be decided by a finite number of observations, none of which involves anything more than comparing shapes or patterns. To preserve this effectiveness, we lay down the rule that only one Rule of Inference should be applied at a time. The explanatory notation beside each statement is not, strictly speaking, part of the proof, but it is helpful and should always be included.

Although the test of whether or not a given sequence of statements is a formal proof is effective, *constructing* such a formal proof is *not* an effective procedure. In this respect, the present method differs from the method of the preceding chapter. The use of truth tables is *completely* mechanical: Given any argument of the general sort with which we are now concerned, its validity can always be tested by following the simple rules presented in Chapter 2. But in constructing a formal proof of validity on the basis of the nineteen Rules of Inference listed, it is necessary to *think* or 'figure out' where to begin and how to proceed. Although we have no effective or purely mechanical method of procedure, it is usually much easier to construct a formal proof of validity than to write out a truth table with perhaps dozens or hundreds or even thousands of rows.

There is an important difference between the first nine and the last ten Rules of Inference. The first nine can be applied only to whole lines of a proof. Thus A can be inferred from $A \cdot B$ by Simplification only if $A \cdot B$ constitutes a whole line. But neither A nor $A \supset C$ follows from $(A \cdot B) \supset C$ by Simplification or by any other Rule of Inference. A does not follow because A can be false while $(A \cdot B) \supset C$ is true. $A \supset C$ does not follow because, if A is true and B and C are both false, $(A \cdot B) \supset C$ is true, whereas $A \supset C$ is false. On the other hand, any of the last ten Rules of Inference can be applied either to whole lines or to parts of lines. Not only can the statement $A \supset (B \supset C)$ be inferred from the whole line $(A \cdot B) \supset C$ by Exportation, but from the line $[(A \cdot B) \supset C] \vee D$ we can infer $[A \supset (B \supset C)] \vee D$ by Exportation. The Rule of Replacement authorizes specified logically equivalent expressions to replace each other wherever they occur, even where they do not constitute whole lines of a proof. But the first nine Rules of Inference can be used only with whole lines of a proof serving as premisses.

There are two very important differences between *substitution* and *replacement*. First, in substituting statements for statement variables in an argument form, *any* statement variable can have *any* statement substituted for it. In replacement, on the other hand, a statement can be replaced only by a statement logically equivalent to it, where the logical equivalence of the two statements is given in Rules 10 through 19. Second, in substitution, every occurrence of a statement variable in an argument form must have the same statement substituted for it. In replacement, however, one (or several) occurrences of a statement may be replaced by a logically equivalent statement, without other occurrences of that statement having to be replaced also.

In the absence of mechanical rules for the construction of formal proofs of validity, some rules of thumb or hints on procedure may be suggested. The first is simply to begin deducing conclusions from the premisses by the given Rules of Inference. As more and more of these subconclusions become available as premisses for further deductions, one is more likely to see how to deduce the conclusion of the argument that is to be proved valid.

Another hint is to try to eliminate statements that occur in the premisses, but not in the conclusion. Such elimination can proceed only in accordance with the Rules of Inference. The Rules contain many techniques for eliminating statements, however. Simplification is one such rule: By its application, the right-hand conjunct of a conjunction can be dropped—provided that the conjunction is a whole line in the proof. By Commutation, the left-hand conjunct can be switched over to the right-hand side, from which location it can be dropped by Simplification. The 'middle' term q can be eliminated by a Hypothetical Syllogism, given two premisses or subconclusions of the patterns $p \supset q$ and $q \supset r$. Distribution is a useful rule for transforming a disjunction of the form $p \vee (q \cdot r)$ into the conjunction $(p \vee q) \cdot (p \vee r)$, the conjuncts of which can be eliminated, either by Simplification alone, or by Commutation followed by Simplification. Another rule of thumb is to introduce, by Addition, a statement that occurs in the conclusion but not in the premisses. Another method is to work backward from the conclusion by looking for some statement or pair of statements from which the conclusion could be deduced by one of the Rules of Inference. Then those intermediate statements are deduced either from the premisses or from other intermediate statements, and so on, until some that are derivable from the premisses are found. A judicious combination of these methods is often the best way to proceed. Practice, of course, is the best way to acquire facility in using the method of deduction.

EXERCISES

I. For each of the following arguments, state the Rule of Inference by which its conclusion follows from its premiss:

*1. $(\sim A \supset B) \cdot (C \vee \sim D)$
$\therefore (\sim A \supset B) \cdot (\sim D \vee C)$

2. $(\sim E \vee F) \cdot (G \vee \sim H)$
$\therefore (E \supset F) \cdot (G \vee \sim H)$

3. $(I \supset \sim J) \vee (\sim K \supset \sim L)$
$\therefore (I \supset \sim J) \vee (L \supset K)$

4. $M \supset \sim(N \vee \sim O)$
$\therefore M \supset (\sim N \cdot \sim\sim O)$

*5. $[P \supset (Q \vee R)] \vee [P \supset (Q \vee R)]$
$\therefore P \supset (Q \vee R)$

6. $[S \cdot (T \cdot U)] \supset (V \equiv \sim W)$
$\therefore [(S \cdot T) \cdot U] \supset (V \equiv \sim W)$

7. $[X \cdot (Y \cdot Z)] \supset (A \equiv \sim B)$
$\therefore X \supset [(Y \cdot Z) \supset (A \equiv \sim B)]$

8. $(C \cdot \sim D) \supset (E \equiv \sim F)$
$\therefore (C \cdot \sim D) \supset [(E \cdot \sim F) \vee (\sim E \cdot \sim\sim F)]$

9. $(G \vee H) \cdot (I \vee J)$
$\therefore [(G \vee H) \cdot I] \vee [(G \vee H) \cdot J]$

*10. $(K \cdot L) \supset \{M \cdot [(N \cdot O) \cdot P]\}$
$\therefore (K \cdot L) \supset \{M \cdot [(O \cdot N) \cdot P]\}$

11. $\sim\{Q \vee \sim[(R \cdot \sim S) \cdot (T \vee \sim U)]\}$
$\therefore \sim\{Q \vee [\sim(R \cdot \sim S) \vee \sim(T \vee \sim U)]\}$

12. $\sim V \supset \{W \supset [\sim(X \cdot Y) \supset \sim Z]\}$
$\therefore \sim V \supset \{[W \cdot \sim(X \cdot Y)] \supset \sim Z\}$

13. $[A \vee (B \vee C)] \vee [(D \vee D) \vee E]$
$\therefore [A \vee (B \vee C)] \vee [D \vee (D \vee E)]$

14. $(F \supset G) \cdot \{[(G \supset H) \cdot (H \supset G)] \supset (H \supset I)\}$
$\therefore (F \supset G) \cdot \{(G \equiv H) \supset (H \supset I)\}$

*15. $J \equiv \sim\{[(K \cdot \sim L) \vee \sim M] \cdot [(K \cdot \sim L) \vee \sim N]\}$
$\therefore J \equiv \sim\{(K \cdot \sim L) \vee (\sim M \cdot \sim N)\}$

16. $O \supset [(P \cdot \sim Q) \equiv (P \cdot \sim\sim R)]$
$\therefore O \supset [(P \cdot \sim Q) \equiv (\sim\sim P \cdot \sim\sim R)]$

17. $\sim S \equiv \{\sim\sim T \supset [\sim\sim\sim U \vee (\sim T \cdot S)]\}$
$\therefore \sim S \equiv \{\sim\sim\sim T \vee [\sim\sim\sim U \vee (\sim T \cdot S)]\}$

18. $V \supset \{(\sim W \supset \sim\sim X) \vee [(\sim Y \supset Z) \vee (\sim Z \supset \sim Y)]\}$
$\therefore V \supset \{(\sim X \supset W) \vee [(\sim Y \supset Z) \vee (\sim Z \supset \sim Y)]\}$

19. $(A \cdot \sim B) \supset [(C \cdot C) \supset (C \supset D)]$
$\therefore (A \cdot \sim B) \supset [C \supset (C \supset D)]$

20. $(E \cdot \sim F) \supset [G \supset (G \supset H)]$
$\therefore (E \cdot \sim F) \supset [(G \cdot G) \supset H)]$

II. Each of the following is a formal proof of validity for the indicated argument. State the 'justification' for each line that is not a premiss:

*1.
1. $(A \vee B) \supset (C \cdot D)$
2. $\sim C$ $/\therefore \sim B$
3. $\sim C \vee \sim D$
4. $\sim(C \cdot D)$
5. $\sim(A \vee B)$
6. $\sim A \cdot \sim B$
7. $\sim B \cdot \sim A$
8. $\sim B$

2.
1. $(E \cdot F) \cdot G$
2. $(F \equiv G) \supset (H \vee I)$ $/\therefore I \vee H$
3. $E \cdot (F \cdot G)$
4. $(F \cdot G) \cdot E$

5. $F \cdot G$
6. $(F \cdot G) \vee (\sim F \cdot \sim G)$
7. $F \equiv G$
8. $H \vee I$
9. $I \vee H$

3. 1. $(J \cdot K) \supset L$
2. $(J \supset L) \supset M$
3. $\sim K \vee N \quad /\therefore K \supset (M \cdot N)$
4. $(K \cdot J) \supset L$
5. $K \supset (J \supset L)$
6. $K \supset M$
7. $\sim K \vee M$
8. $(\sim K \vee M) \cdot (\sim K \vee N)$
9. $\sim K \vee (M \cdot N)$
10. $K \supset (M \cdot N)$

4. 1. $(O \supset \sim P) \cdot (P \supset Q)$
2. $Q \supset O$
3. $\sim R \supset P \quad /\therefore R$
4. $\sim Q \vee O$
5. $O \vee \sim Q$
6. $(O \supset \sim P) \cdot (\sim Q \supset \sim P)$
7. $\sim P \vee \sim P$
8. $\sim P$
9. $\sim \sim R$
10. R

***5.** 1. $S \supset (T \supset U)$
2. $U \supset \sim U$
3. $(V \supset S) \cdot (W \supset T) \quad /\therefore V \supset \sim W$
4. $(S \cdot T) \supset U$
5. $\sim U \vee \sim U$
6. $\sim U$
7. $\sim (S \cdot T)$
8. $\sim S \vee \sim T$
9. $\sim V \vee \sim W$
10. $V \supset \sim W$

6. 1. $X \supset (Y \supset Z)$
2. $X \supset (A \supset B)$
3. $X \cdot (Y \vee A)$
4. $\sim Z \quad /\therefore B$
5. $(X \cdot Y) \supset Z$
6. $(X \cdot A) \supset B$
7. $(X \cdot Y) \vee (X \cdot A)$
8. $[(X \cdot Y) \supset Z] \cdot [(X \cdot A) \supset B]$
9. $Z \vee B$
10. B

7. 1. $C \supset (D \supset \sim C)$
2. $C \equiv D \quad /\therefore \sim C \cdot \sim D$
3. $C \supset (\sim \sim C \supset \sim D)$
4. $C \supset (C \supset \sim D)$
5. $(C \cdot C) \supset \sim D$
6. $C \supset \sim D$
7. $\sim C \vee \sim D$
8. $\sim (C \cdot D)$
9. $(C \cdot D) \vee (\sim C \cdot \sim D)$
10. $\sim C \cdot \sim D$

8. 1. $E \cdot (F \vee G)$
2. $(E \cdot G) \supset \sim (H \vee I)$
3. $(\sim H \vee \sim I) \supset \sim (E \cdot F)$
$/\therefore H \equiv I$
4. $(E \cdot G) \supset (\sim H \cdot \sim I)$
5. $\sim (H \cdot I) \supset \sim (E \cdot F)$
6. $(E \cdot F) \supset (H \cdot I)$
7. $[(E \cdot F) \supset (H \cdot I)] \cdot [(E \cdot G)$
$\supset (\sim H \cdot \sim I)]$
8. $(E \cdot F) \vee (E \cdot G)$
9. $(H \cdot I) \vee (\sim H \cdot \sim I)$
10. $H \equiv I$

9. 1. $J \vee (\sim K \vee J)$
2. $K \vee (\sim J \vee K)$
$/\therefore (J \cdot K) \vee (\sim J \cdot \sim K)$
3. $(\sim K \vee J) \vee J$
4. $\sim K \vee (J \vee J)$
5. $\sim K \vee J$
6. $K \supset J$
7. $(\sim J \vee K) \vee K$
8. $\sim J \vee (K \vee K)$
9. $\sim J \vee K$
10. $J \supset K$
11. $(J \supset K) \cdot (K \supset J)$
12. $J \equiv K$
13. $(J \cdot K) \vee (\sim J \cdot \sim K)$

10. 1. $(L \vee M) \vee (N \cdot O)$
2. $(\sim L \cdot O) \cdot \sim (\sim L \cdot M)$
$/\therefore \sim L \cdot N$
3. $\sim L \cdot [O \cdot \sim (\sim L \cdot M)]$
4. $\sim L$
5. $L \vee [M \vee (N \cdot O)]$
6. $M \vee (N \cdot O)$
7. $(M \vee N) \cdot (M \vee O)$
8. $M \vee N$
9. $\sim L \cdot (M \vee N)$
10. $(\sim L \cdot M) \vee (\sim L \cdot N)$
11. $\sim (\sim L \cdot M) \cdot (\sim L \cdot O)$
12. $\sim (\sim L \cdot M)$
13. $\sim L \cdot N$

III. Construct a formal proof of validity for each of the following arguments:

*1. $\sim A$
$\therefore A \supset B$

2. C
$\therefore D \supset C$

3. $E \supset (F \supset G)$
$\therefore F \supset (E \supset G)$

4. $H \supset (I \cdot J)$
$\therefore H \supset I$

*5. $K \supset L$
$\therefore K \supset (L \vee M)$

6. $N \supset O$
$\therefore (N \cdot P) \supset O$

7. $(Q \vee R) \supset S$
$\therefore Q \supset S$

8. $T \supset \sim(U \supset V)$
$\therefore T \supset U$

9. $W \supset (X \cdot \sim Y)$
$\therefore W \supset (Y \supset X)$

*10. $A \supset \sim(B \supset C)$
$(D \cdot B) \supset C$
D
$\therefore \sim A$

11. $E \supset F$
$E \supset G$
$\therefore E \supset (F \cdot G)$

12. $H \supset (I \vee J)$
$\sim I$
$\therefore H \supset J$

13. $(K \vee L) \supset \sim(M \cdot N)$
$(\sim M \vee \sim N) \supset (O \equiv P)$
$(O \equiv P) \supset (Q \cdot R)$
$\therefore (L \vee K) \supset (R \cdot Q)$

14. $S \supset T$
$S \vee T$
$\therefore T$

*15. $(\sim U \vee V) \cdot (U \vee W)$
$\sim X \supset \sim W$
$\therefore V \vee X$

16. $A \supset (B \supset C)$
$C \supset (D \cdot E)$
$\therefore A \supset (B \supset D)$

17. $E \supset F$
$G \supset F$
$\therefore (E \vee G) \supset F$

18. $[(H \cdot I) \supset J] \cdot [\sim K \supset (I \cdot \sim J)]$
$\therefore H \supset K$

19. $[L \cdot (M \vee N)] \supset (M \cdot N)$
$\therefore L \supset (M \supset N)$

*20. $O \supset (P \supset Q)$
$P \supset (Q \supset R)$
$\therefore O \supset (P \supset R)$

21. $S \supset (T \cdot U)$
$(T \vee U) \supset V$
$\therefore S \supset V$

22. $\sim W \vee [(X \supset Y) \cdot (Z \supset Y)]$
$W \cdot (X \vee Z)$
$\therefore Y$

23. $(A \vee B) \supset (C \cdot D)$
$\sim A \supset (E \supset \sim E)$
$\sim C$
$\therefore \sim E$

24. $(F \supset G) \cdot (H \supset I)$
$F \vee H$
$(F \supset \sim I) \cdot (H \supset \sim G)$
$\therefore G \equiv \sim I$

*25. $J \cdot (K \cdot L)$
$(K \cdot J) \supset [M \vee (N \vee O)]$
$\sim M \cdot (\sim P \cdot \sim O)$
$\therefore N$

26. $Q \vee (R \cdot S)$
$(Q \supset T) \cdot (T \supset S)$
$\therefore S$

27. $(U \supset V) \cdot (W \supset X)$
$\therefore (U \vee W) \supset (V \vee X)$

28. $(Y \supset Z) \cdot (A \supset B)$
$\therefore (Y \cdot A) \supset (Z \cdot B)$

29. $(C \supset D) \cdot (E \supset F)$
$G \supset (C \vee E)$
$\therefore G \supset (D \vee F)$

30. $(H \supset I) \cdot (J \supset K)$
$H \vee J$
$(H \supset \sim K) \cdot (J \supset \sim I)$
$(I \cdot \sim K) \supset L$
$K \supset (I \vee M)$
$\therefore L \vee M$

IV. Construct a formal proof of validity for each of the following arguments, using the suggested notation in each case.

*1. If I study, I make good grades. If I do not study, I enjoy myself. Therefore, either I make good grades or I enjoy myself. (*S*, *G*, *E*)

2. If the supply of silver remains constant and the use of silver increases, then the price of silver rises. If an increase in the use of silver implies that the price of silver rises, then there will be a windfall for speculators. The supply of silver remains constant. Therefore, there will be a windfall for speculators. (*S*, *U*, *P*, *W*)

3. Either Adams is elected chairman or both Brown and Clark are elected to the board. If either Adams is elected chairman or Brown is elected to the board, then Davis will lodge a protest. Therefore, either Clark is elected to the board or Davis will lodge a protest. (*A*, *B*, *C*, *D*)

4. If he uses good bait, then if the fish are biting, then he catches the legal limit. He uses good bait, but he does not catch the legal limit. Therefore, the fish are not biting. (*G*, *B*, *C*)

*5. Either the governor and the lieutenant-governor will both run for reelection or the primary race will be wide open and the party will be torn by dissension. The governor will not run for reelection. Therefore, the party will be torn by dissension. (*G*, *L*, *W*, *T*)

6. If the Dodgers win the pennant, then they will win the Series. Therefore, if the Dodgers win the pennant, then if they continue to hit, then they will win the Series. (*P*, *S*, *H*)

7. If he attracts the farm vote, then he will carry the rural areas, and if he attracts the labor vote, then he will carry the urban centers. If he carries both the urban centers and the rural areas, then he is certain to be elected. He is not certain to be elected. Therefore, either he does not attract the farm vote or he does not attract the labor vote. (*F*, *R*, *L*, *U*, *C*)

8. Either Argentina does not join the alliance or Brazil boycotts it, but if Argentina joins the alliance, then Chile boycotts it. If Brazil boycotts the alliance, then if Chile boycotts it, then Ecuador will boycott it. Therefore, if Argentina joins the alliance then Ecuador will boycott it. (*A*, *B*, *C*, *E*)

9. If Argentina joins the alliance, then both Brazil and Chile will boycott it. If either Brazil or Chile boycotts the alliance, then the alliance will be ineffective. Therefore, if Argentina joins the alliance, then the alliance will be ineffective. (*A*, *B*, *C*, *I*)

*10. Steve took either the bus or the train. If he took the bus or drove his own car, then he arrived late and missed the first session. He did not arrive late. Therefore, he took the train. (*B*, *T*, *C*, *L*, *M*)

11. If you enroll in the course and study hard, then you will pass, but if you enroll in the course and do not study hard, then you will not pass. Therefore, if you enroll in the course, then either you study hard and pass or you do not study hard and do not pass. (*E*, *S*, *P*)

12. If Argentina joins the alliance, then either Brazil or Chile will boycott it. If Brazil boycotts the alliance, then Chile will boycott it, also. Therefore, if Argentina joins the alliance, then Chile will boycott it. (*A*, *B*, *C*)

13. If either Argentina or Brazil joins the alliance, then both Chile and Ecuador will boycott it. Therefore, if Argentina joins the alliance, then Chile will boycott it. (*A*, *B*, *C*, *E*)

14. If prices fall or wages rise, then both retail sales and advertising activities

increase. If retail sales increase, then jobbers make more money, but jobbers do not make more money. Therefore, prices do not fall. (*P, W, R, A, J*)

*15. If I work, then I earn money, but if I am idle, then I enjoy myself. Either I work or I am idle. However, if I work, then I do not enjoy myself, while if I am idle, then I do not earn money. Therefore, I enjoy myself if and only if I do not earn money. (*W, M, I, E*)

16. If you enter the primary, then if you campaign vigorously, then you win the nomination. If you win the nomination and receive the support of the party regulars, then you will be elected. If you take the party platform seriously, then you will receive the support of the party regulars but will not be elected. Therefore, if you enter the primary, then if you campaign vigorously, then you do not take the party platform seriously. (*P, C, N, R, E, T*)

17. Either the tariff is lowered or imports continue to decrease and our own industries prosper. If the tariff is lowered, then our own industries prosper. Therefore, our own industries prosper. (*T, I, O*)

18. Either he has his old car repaired or he buys a new car. If he has his old car repaired, then he will owe a lot of money to the garage. If he owes a lot of money to the garage, then he will not soon be out of debt. If he buys a new car, then he must borrow money from the bank, and if he must borrow money from the bank, then he will not soon be out of debt. Either he will soon be out of debt or his creditors will force him into bankruptcy. Therefore, his creditors will force him into bankruptcy. (*R, N, G, S, B, C*)

19. If she goes on a picnic, then she wears sport clothes. If she wears sport clothes, then she does not attend both the banquet and the dance. If she does not attend the banquet, then she still has her ticket, but she does not still have her ticket. She does attend the dance. Therefore, she does not go on a picnic. (*P, S, B, D, T*)

20. If she studies the sciences, then she prepares to earn a good living, and if she studies the humanities, then she prepares to live a good life. If she prepares to earn a good living or she prepares to live a good life, then her college years are well spent. But her college years are not well spent. Therefore, she does not study either the sciences or the humanities. (*S, E, H, L, C*)

*21. If you plant tulips, then your garden will bloom early, and if you plant asters, then your garden will bloom late. So if you plant either tulips or asters, then your garden will bloom either early or late. (*T, E, A, L*)

22. If you plant tulips, then your garden will bloom early, and if you plant asters, then your garden will bloom late. So if you plant both tulips and asters, then your garden will bloom both early and late. (*T, E, A, L*)

23. If we go to Europe, then we tour Scandinavia. If we go to Europe, then if we tour Scandinavia, then we visit Norway. If we tour Scandinavia, then if we visit Norway, then we shall take a trip on a fiord. Therefore, if we go to Europe, then we shall take a trip on a fiord. (*E, S, N, F*)

24. If Argentina joins the alliance, then either Brazil or Chile boycotts it. If Ecuador joins the alliance, then either Chile or Peru boycotts it. Chile does not boycott it. Therefore, if neither Brazil nor Peru boycotts it, then neither Argentina nor Ecuador joins the alliance. (*A, B, C, E, P*)

25. If either Argentina or Brazil joins the alliance, then if either Chile or Ecuador boycotts it, then although Peru does not boycott it, Venezuela

boycotts it. If either Peru or Nicaragua does not boycott it, the Uruguay will join the alliance. Therefore, if Argentina joins the alliance, then if Chile boycotts it, then Uruguay will join the alliance. (*A*, *B*, *C*, *E*, *P*, *V*, *N*, *U*)

26. If either Saudi Arabia raises the price of oil or Saudi Arabia does not raise the price of oil, then America will be in grave difficulties. Therefore, America will be in grave difficulties. (S, *A*)

3.3 Proving Invalidity

We can establish the invalidity of an argument by using a truth table to show that the specific form of that argument is invalid. The truth table proves invalidity if it contains at least one row in which truth values are assigned to the statement variables in such a way that the premisses are made true and the conclusion false. If we can devise such a truth value assignment without constructing the entire truth table, we shall have a shorter method of proving invalidity.

Consider the invalid argument

> If the Senator votes against this bill, then he is opposed to more severe penalties against tax evaders.
> If the Senator is a tax evader himself, then he is opposed to more severe penalties against tax evaders.
> Therefore, if the Senator votes against this bill, he is a tax evader himself.

which may be symbolized as

$$V \supset O$$
$$H \supset O$$
$$\therefore V \supset H$$

Instead of constructing a truth table for the specific form of this argument, we can prove its invalidity by making an assignment of truth values to the component simple statements *V*, *O*, and *H*, which will make the premisses true and the conclusion false. The conclusion is made false by assigning **T** to *V* and **F** to *H*; and both premisses are made true by assigning **T** to *O*. This method of proving invalidity is closely related to the truth table method. In effect, making the indicated truth value assignment amounts to describing one row of the relevant truth table—a row that suffices to establish the invalidity of the argument being tested. The relationship appears more clearly, perhaps, when the truth value assignments are written out horizontally:

V	O	H	$V \supset O$	$H \supset O$	$V \supset H$
T	**T**	**F**	**T**	**T**	**F**

This new method of proving invalidity is shorter than writing out a complete truth table. The amount of time and work saved is proportionally greater for more complicated arguments. In proving the invalidity of more extended arguments, a certain amount of trial and error may be needed to discover a truth value assignment that works. But even so, this method is quicker and easier than writing out the entire truth table. It is obvious that the present method will suffice to prove the invalidity of any argument that can be shown to be invalid by a truth table.

EXERCISES

Prove the invalidity of each of the following arguments by the method of assigning truth values:

*1. $A \supset B$
$C \supset D$
$B \vee C$
$\therefore A \vee D$

2. $E \supset (F \vee G)$
$G \supset (H \cdot I)$
$\sim H$
$\therefore E \supset I$

3. $J \supset (K \supset L)$
$K \supset (\sim L \supset M)$
$(L \vee M) \supset N$
$\therefore J \supset N$

4. $(O \vee P) \supset Q$
$Q \supset (P \vee R)$
$O \supset (\sim S \supset P)$
$(S \supset O) \supset \sim R$
$\therefore P \equiv Q$

*5. $T \equiv U$
$U \equiv (V \cdot W)$
$V \equiv (T \vee X)$
$T \vee X$
$\therefore T \cdot X$

6. $X \equiv (Y \supset Z)$
$Y \equiv (\sim X \cdot \sim Z)$
$Z \equiv (X \vee \sim Y)$
Y
$\therefore X \vee Z$

7. $(A \supset B) \cdot (C \supset D)$
$A \vee C$
$(B \vee D) \supset (E \cdot F)$
$E \supset (F \supset G)$
$G \supset (A \supset H)$
$\therefore H$

8. $I \vee (J \cdot K)$
$(I \vee J) \supset (L \equiv \sim M)$
$(L \supset \sim M) \supset (M \cdot \sim N)$
$(N \supset O) \cdot (O \supset M)$
$(J \supset K) \supset O$
$\therefore O$

9. $P \equiv (Q \equiv \sim R)$
$Q \supset (\sim R \vee \sim S)$
$[R \supset (Q \vee \sim T)] \cdot (P \supset Q)$
$[U \supset (S \cdot T)] \cdot (T \supset \sim V)$
$[(Q \cdot R) \supset \sim U] \cdot [U \supset (Q \vee R)]$
$(Q \vee V) \cdot \sim V$
$\therefore \sim U \cdot \sim V$

10. $W \equiv (X \vee Y)$
$X \equiv (Z \supset Y)$
$Y \equiv (Z \equiv \sim A)$
$Z \equiv (A \supset B)$
$A \equiv (B \equiv Z)$
$B \vee \sim W$
$\therefore W \equiv B$

3.4 The Rule of Conditional Proof

Next, we introduce a new rule to use in the method of deduction: the rule of Conditional Proof. In this section, the new rule will be applied only to arguments whose conclusions are conditional statements. The new rule can best be explained and justified by reference to the principle of Exportation and the correspondence, noted in Chapter 2, between valid argument forms and tautologies.

To every argument there corresponds a conditional statement whose antecedent is the conjunction of the argument's premisses and whose consequent is the argument's conclusion. As has been remarked, an argument is valid if and only if its corresponding conditional is a tautology. If an argument has a conditional statement for its conclusion, which we may symbolize as $A \supset C$, then if we symbolize the conjunction of its premisses as P,[2] the argument is valid if and only if the conditional

$$P \supset (A \supset C) \tag{1}$$

is a tautology. If we can deduce the conclusion $A \supset C$ by a sequence of elementary valid arguments, from the premisses conjoined in P, we thereby prove the argument to be valid and the associated conditional (1) to be a tautology. By the principle of Exportation, (1) is logically equivalent to

$$(P \cdot A) \supset C \tag{2}$$

But (2) is the conditional associated with a somewhat different argument. This second argument has, as its premisses, all of the premisses of the first argument *plus* an additional premiss that is the antecedent of the conclusion of the first argument. The conclusion of the second argument is the consequent of the conclusion of the first argument. Now, if we deduce the conclusion of the second argument, C, from the premisses conjoined in $P \cdot A$ by a sequence of elementary valid arguments, we thereby prove that its associated conditional statement (2) is a tautology. But since (1) and (2) are logically equivalent, this fact proves that (1) is a tautology also, from which it follows that the original argument—with one less premiss and the conditional conclusion $A \supset C$—is valid also. Now, the rule of Conditional Proof permits us to infer the validity of any argument

$$\begin{array}{l} P \\ \therefore A \supset C \end{array}$$

from a formal proof of validity for the argument

$$\begin{array}{l} P \\ A \\ \therefore C \end{array}$$

Given any argument whose conclusion is a conditional statement, a proof of its validity, using the rule of Conditional Proof—that is, a conditional proof of its validity—is constructed by assuming the antecedent of its conclusion as an

[2] For definiteness, we specify that the conjunction has its conjuncts associated to the left, that is, if the premisses are P_1, P_2, and P_3, the conjunction is $(P_1 \cdot P_2) \cdot P_3$. If there is a fourth premiss, P_4, the conjunction is $[(P_1 \cdot P_2) \cdot P_3] \cdot P_4$, and so on.

additional premiss, and then deducing the consequent of its conclusion by a sequence of elementary valid arguments. Thus a conditional proof of validity for the argument

$$(A \vee B) \supset (C \cdot D)$$
$$(D \vee E) \supset F$$
$$\therefore A \supset F$$

may be written as

1.	$(A \vee B) \supset (C \cdot D)$	
2.	$(D \vee E) \supset F$	$/\therefore A \supset F$
3.	A	$/\therefore F$ (C.P.)
4.	$A \vee B$	3, Add.
5.	$C \cdot D$	1, 4, M.P.
6.	$D \cdot C$	5, Com.
7.	D	6, Simp.
8.	$D \vee E$	7, Add.
9.	F	2, 8, M.P.

Here, the second slant line and three-dot 'therefore' symbol, as well as the parenthesized 'C.P.', indicate that the rule of Conditional Proof is being used.

Since the rule of Conditional Proof can be used in dealing with any valid argument having a conditional statement as conclusion, it can be applied more than once in the course of the same deduction. Thus a conditional proof of validity for

$$A \supset (B \supset C)$$
$$B \supset (C \supset D)$$
$$\therefore A \supset (B \supset D)$$

will be a proof of validity for

$$A \supset (B \supset C)$$
$$B \supset (C \supset D)$$
$$A$$
$$\therefore B \supset D$$

and since the latter, itself, has a conditional conclusion, it can be given a conditional proof by proving the validity of

$$A \supset (B \supset C)$$
$$B \supset (C \supset D)$$
$$A$$
$$B$$
$$\therefore D$$

Each use of the conditional method should be signaled by an additional slant line and 'therefore' sign, in addition to the notation '(C.P.)'. The suggested proof would be written:

1. $A \supset (B \supset C)$
2. $B \supset (C \supset D)$ $/\therefore A \supset (B \supset D)$
3. A $/\therefore B \supset D$ (C.P.)
4. B $/\therefore D$ (C.P.)
5. $B \supset C$ 1, 3, M.P.
6. C 5, 4, M.P.
7. $C \supset D$ 2, 4, M.P.
8. D 7, 6, M.P.

The rule of Conditional Proof is a genuine addition to the proof apparatus of Sections 3.1 and 3.2. Not only does it permit the construction of *shorter* proofs of validity for arguments that could be proved valid by appealing to the original list of nineteen Rules of Inference alone, but it permits us to establish the validity of valid arguments whose validity could *not* be proved by reference to the original list alone.

It is proved in Appendix A[3] that the obviously valid argument

$$A \supset B$$
$$\therefore A \supset (A \cdot B)$$

cannot be proved valid using only the original list of nineteen Rules of Inference. But it is easily proved valid using the new rule of Conditional Proof. Its conditional proof of validity is

1. $A \supset B$ $/\therefore A \supset (A \cdot B)$
2. A $/\therefore A \cdot B$ (C·P.)
3. B 1, 2, M.P.
4. $A \cdot B$ 2, 3, Conj.

EXERCISES

Give conditional proofs of validity for Exercises *21, 22, 23, 24, and 25 on pages 47–48.

3.5 The Rule of Indirect Proof

The method of *indirect proof,* often called the method of proof by *reductio ad absurdum,* is familiar to all who have studied elementary geometry. In deriving his theorems, Euclid often begins by assuming the opposite of what

[3]See pages 333–336.

he wants to prove. If that assumption leads to a contradiction, or 'reduces to an absurdity', then that assumption must be false, and so its negation—the theorem to be proved—must be true.

An indirect proof of validity for a given argument is constructed by assuming, as an additional premiss, the negation of its conclusion, and then deriving an explicit contradiction from the augmented set of premisses. Thus an indirect proof of validity for the argument

$$A \supset (B \cdot C)$$
$$(B \vee D) \supset E$$
$$D \vee A$$
$$\therefore E$$

may be set down as follows:

1.	$A \supset (B \cdot C)$	
2.	$(B \vee D) \supset E$	
3.	$D \vee A$ $/\therefore E$	
4.	$\sim E$	I.P. (Indirect Proof)
5.	$\sim(B \vee D)$	2, 4, M.T.
6.	$\sim B \cdot \sim D$	5, De M.
7.	$\sim D \cdot \sim B$	6, Com.
8.	$\sim D$	7, Simp.
9.	A	3, 8, D.S.
10.	$B \cdot C$	1, 9, M.P.
11.	B	10, Simp.
12.	$\sim B$	6, Simp.
13.	$B \cdot \sim B$	11, 12, Conj.

Here line 13 is an explicit contradiction, so the demonstration is complete. The validity of the original argument follows by the rule of Indirect Proof.

Adding the rule of Indirect Proof serves to strengthen our proof apparatus still further. Any argument whose conclusion is a tautology can be shown to be valid, by the method of truth tables, regardless of what its premisses may be. But if the tautologous conclusion of an argument is not a conditional statement, and the premisses are consistent with each other and quite irrelevant to that conclusion,[4] then the argument cannot be proved valid using only the methods of Sections 3.1, 3.2, and 3.4. The rule of Indirect Proof must be used. Although the argument

$$A$$
$$\therefore B \vee (B \supset C)$$

[4]In this context, one statement is irrelevant to another if there is no simple statement that is a component of both.

cannot be proved valid by the methods set forth in the preceding sections, its validity is easily established by using the rule of Indirect Proof. One proof of its validity is this:

1.	A	$/\therefore B \vee (B \supset C)$
2.	$\sim[B \vee (B \supset C)]$	I.P.
3.	$\sim[B \vee (\sim B \vee C)]$	2, Impl.
4.	$\sim[(B \vee \sim B) \vee C]$	3, Assoc.
5.	$\sim(B \vee \sim B) \cdot \sim C$	4, De M.
6.	$\sim(B \vee \sim B)$	5, Simp.
7.	$\sim B \cdot \sim\sim B$	6, De M.

Our nineteen Rules of Inference, plus the rules of Conditional and Indirect Proof, provide us with a method of deduction that is complete. Any argument whose validity can be established by the use of truth tables can be proved valid by the method of deduction as set forth in Sections 3.1, 3.2, 3.4 and 3.5. This will not be proved, however, until the end of Chapter 8.

EXERCISES

For each of the following arguments, construct both a formal proof of validity and an indirect proof and compare their lengths:

1. $A \vee (B \cdot C)$
$A \supset C$
$\therefore C$

2. $(D \vee E) \supset (F \supset G)$
$(\sim G \vee H) \supset (D \cdot F)$
$\therefore G$

***3.** $(H \supset I) \cdot (J \supset K)$
$(I \vee K) \supset L$
$\sim L$
$\therefore \sim(H \vee J)$

4. $(M \vee N) \supset (O \cdot P)$
$(O \vee Q) \supset (\sim R \cdot S)$
$(R \vee T) \supset (M \cdot U)$
$\therefore \sim R$

***5.** $(V \supset \sim W) \cdot (X \supset Y)$
$(\sim W \supset Z) \cdot (Y \supset \sim A)$
$(Z \supset \sim B) \cdot (\sim A \supset C)$
$V \cdot X$
$\therefore \sim B \cdot C$

3.6 Proofs of Tautologies

The conditional and indirect methods of proof can be used not only to establish the validity of arguments, but also to prove that certain statements and statement forms are tautologies. Any conditional statement corresponds, in a sense, to an argument whose single premiss is the antecedent of the conditional, and whose conclusion is the conditional's consequent. The conditional is a tautology if and only if that argument is valid. Hence a conditional is proved tautologous by deducing its consequent from its antecedent by a sequence of elementary valid arguments. Thus the statement $(A \cdot B) \supset A$ is proved tautologous by the same sequence of lines that proves the validity of the argument

$$A \cdot B$$
$$\therefore A$$

It has already been noted that the conditional method can be used repeatedly in a single proof. Thus the conditional statement

$$(Q \supset R) \supset [(P \supset Q) \supset (P \supset R)]$$

is proved tautologous by the following:

1. $Q \supset R$ $/\therefore (P \supset Q) \supset (P \supset R)$ (C.P.)
2. $P \supset Q$ $/\therefore P \supset R$ (C.P.)
3. $P \supset R$ 2, 1, H.S.

For some complicated conditional statements, this method of proving them tautologous is shorter and easier than constructing truth tables.

There are many tautologies that are not conditional in form. To these, the preceding method cannot be applied. But any tautology can be established as tautologous by the indirect method. As applied to an *argument,* the indirect method of proving validity proceeds by adding the negation of its conclusion to the argument's premisses and then deducing a contradiction by a sequence of elementary valid arguments. As applied to a *statement,* the indirect method of proving it tautologous proceeds by taking its negation as premiss and then deducing an explicit contradiction by a sequence of elementary valid arguments. Thus the statement $B \vee \sim B$ is proved to be a tautology by the following:

1. $\sim(B \vee \sim B)$ $/\therefore B \vee \sim B$ (I.P.)
2. $\sim B \cdot \sim\sim B$ 1, De M.

To say that a statement is a tautology is to assert that its truth is unconditional, so it can be established without appealing to any other statements as premisses. Another, perhaps not too misleading, way of saying the same thing is to assert the validity of the 'argument' that has the statement in question as 'conclusion', but has no premisses at all. If the 'conclusion' *is* a tautology, then the method of deduction permits us to prove that the 'argument' is valid, even though it has no premisses—using either the rule of Conditional Proof or the rule of Indirect Proof. Any tautology can be established by the method of deduction, although this will not be proved until the end of Chapter 8.

EXERCISES

I. Use the method of conditional proof to verify that the following are tautologies:

*1. $P \supset (Q \supset P)$

2. $[P \supset (Q \supset R)] \supset [(P \supset Q) \supset (P \supset R)]$

3. $[P \supset (Q \supset R)] \supset [Q \supset (P \supset R)]$

4. $(P \supset Q) \supset (\sim Q \supset \sim P)$

*5. $\sim\sim P \supset P$

6. $P \supset \sim\sim P$

7. $(A \supset B) \supset [(B \supset C) \supset (A \supset C)]$

8. $[(A \supset B) \cdot (A \supset C)] \supset [A \supset (B \vee C)]$

9. $[(A \supset B) \cdot (A \supset C)] \supset [A \supset (B \cdot C)]$

*10. $(A \supset B) \supset [A \supset (A \cdot B)]$

11. $(A \supset B) \supset [(\sim A \supset B) \supset B]$

12. $(A \supset B) \supset [(A \cdot C) \supset (B \cdot C)]$

13. $[(A \supset B) \supset B] \supset (A \vee B)$

14. $(B \supset C) \supset [(A \vee B) \supset (C \vee A)]$

*15. $[A \supset (B \cdot C)] \supset \{[B \supset (D \cdot E)] \supset (A \supset D)\}$

16. $[(A \vee B) \supset C] \supset \{[(C \vee D) \supset E] \supset (A \supset E)\}$

17. $[(A \supset B) \supset A] \supset A$

18. $P \supset (P \cdot P)$

19. $(P \cdot Q) \supset P$

20. $(P \supset Q) \supset [\sim(Q \cdot R) \supset \sim(R \cdot P)]$

II. Use the method of indirect proof to verify that the following are tautologies:

*1. $(A \supset B) \vee (A \supset \sim B)$

2. $(A \supset B) \vee (\sim A \supset B)$

3. $(A \supset B) \vee (B \supset A)$

4. $(A \supset B) \vee (B \supset C)$

*5. $(A \supset B) \vee (\sim A \supset C)$

6. $A \vee (A \supset B)$

7. $P \equiv \sim\sim P$

8. $A \equiv [A \cdot (A \vee B)]$

9. $A \equiv [A \vee (A \cdot B)]$

10. $\sim[(A \supset \sim A) \cdot (\sim A \supset A)]$

3.7 The Strengthened Rule of Conditional Proof

In the preceding sections, the method of Conditional Proof was applied only to arguments whose conclusions were conditional in form. But in the next chapter, it will be convenient to use something like the method of Conditional Proof for arguments whose conclusions are not explicit conditional statements. To accomplish this purpose, we strengthen our rule of Conditional Proof and thereby give it wider applicability.

To formulate our Strengthened Rule of Conditional Proof, it is useful to adopt a new method of writing out proofs that make use of the Conditional Method. As explained in Section 3.5, we used the method of Conditional Proof to establish the validity of an argument that has a conditional as conclusion by adding the antecedent of that conditional to the argument's premisses as an assumption, and then deducing the conditional's consequent. The notation in Section 3.5 involved the use of an additional slant line and an extra *therefore* sign, as in proving the validity of the argument

$$A \supset B$$
$$\therefore A \supset (A \cdot B)$$

by the four-line proof on page 52.

A Conditional Proof of validity for that same argument is set down in our new notation as the following sequence of five lines:

1. $A \supset B$ $/\therefore A \supset (A \cdot B)$
2. A assumption
3. B 1, 2, M.P.
4. $A \cdot B$ 2, 3, Conj.
5. $A \supset (A \cdot B)$ 2–4, C.P.

Here, the fifth line is inferred not from any one or two of the preceding lines, but from the *sequence* of lines 2, 3, 4, which constitutes a valid deduction of line 4 from lines 1 and 2. In line 5, we infer the validity of the argument

$$A \supset B$$
$$\therefore A \supset (A \cdot B)$$

from the demonstrated validity of the argument

$$A \supset B$$
$$A$$
$$\therefore A \cdot B$$

That inference is 'justified' by noting the sequence of lines to which appeal is made, using the letters 'C.P.' to show that the principle of Conditional Proof is being used.

In the second of the preceding proofs, line 2, the assumption, has lines 3 and 4 dependent upon it. Line 5, however, is *not* dependent upon line 2, but only upon line 1. Line 5 is, therefore, *outside* or *beyond the scope* of the assumption made as line 2. When an assumption is made in a Conditional Proof of validity, its 'scope' is always *limited,* never extending all the way to the last line of the demonstration.

A notation is now introduced that is very helpful in keeping track of assumptions and their *scopes*. A bent arrow is used for this purpose, with its head pointing at the assumption from the left, its shaft bent down to run along all lines within the scope of the assumption, and then bent inward to mark the end of the scope of that assumption. The scope of the assumption in the preceding proof is indicated thus:

1.	$A \supset B$	$/\therefore A \supset (A \cdot B)$
→ 2.	A	assumption
3.	B	1, 2, M.P.
4.	$A \cdot B$	2, 3, Conj.
5.	$A \supset (A \cdot B)$	2–4, C.P.

It should be observed that *only* a line inferred by the principle of Conditional Proof ends the scope of an assumption, and that *every* use of the rule of Conditional Proof serves to end the scope of an assumption. When the scope of an assumption has been ended, the assumption is said to have been *discharged.* No subsequent line can be justified by reference to it or to any line lying between it and the line inferred by the rule of Conditional Proof that discharges it. Only lines lying between an assumption of limited scope and the line that discharges it, can be justified by reference to that assumption. After one assumption of limited scope has been discharged, another such assumption may be made and then discharged. Or a second assumption of limited scope may be written within the scope of a first. Scopes of different assumptions may follow each other, or one scope may be contained entirely within another.

If the scope of an assumption does *not* extend all the way to the end of a proof, then the final line of the proof does not *depend* on that assumption, but has been proved to follow from the original premisses alone. Hence we need not restrict ourselves to using as assumptions only the antecedents of conditional conclusions. *Any* proposition can be taken as an assumption of limited scope, for the final line that is the conclusion will always be beyond its scope and independent of it.

A more complex demonstration that involves making *two* assumptions is the following (incidentally, when our bent arrow notation is used, the word 'assumption' need not be written, since each assumption is sufficiently identified by the arrowhead on its left):

1.	$(A \vee B) \supset [(C \vee D) \supset E]$ $/\therefore A \supset [(C \cdot D) \supset E]$	
→ 2.	A	
3.	$A \vee B$	2, Add.
4.	$(C \vee D) \supset E$	1, 3, M.P.
→ 5.	$C \cdot D$	
6.	C	5, Simp.
7.	$C \vee D$	6, Add.
8.	E	4, 7, M.P.
9.	$(C \cdot D) \supset E$	5–8, C.P.
10.	$A \supset [(C \cdot D) \supset E]$	2–9, C.P.

In this proof, lines 2 through 9 lie within the scope of the first assumption, while lines 5, 6, 7, and 8 lie within the scope of the second assumption. From these examples, it is clear that the scope of an assumption α in a proof consists of all lines α through φ, where the line following φ is of the form $\alpha \supset \varphi$ and is inferred by C.P. from that sequence of lines. In the preceding proof, the second assumption lies within the scope of the first because it lies between the first assumption and line 10, which is inferred by C.P. from the sequence of lines 2 through 9.

When we use this new method of writing out a Conditional Proof of validity, the scope of every original premiss extends all the way to the end of the proof. The original premisses may be supplemented by additional assumptions, provided that the latter's scopes are limited and do not extend to the end of the proof. Each line of a formal proof of validity must be either a premiss, or an assumption of limited scope, or must follow validly from one or two preceding lines by a Rule of Inference, or must follow from a *sequence* of preceding lines by the principle of Conditional Proof.

Some arguments can be proved valid by the Strengthened Rule of Conditional Proof, although the original Rule of Conditional Proof, as stated in Section 3.4, cannot be applied to them. Consider the following argument and its proof:

1.	$A \supset B$	
2.	$[A \supset (A \cdot B)] \supset C$ $/\therefore C$	
→ 3.	A	
4.	B	1, 3, M.P.
5.	$A \cdot B$	3, 4, Conj.
6.	$A \supset (A \cdot B)$	3–5, C.P.
7.	C	2, 6, M.P.

Here the conclusion, C, is not a conditional statement, so the original Rule of Conditional Proof as stated in Section 3.4 is simply not applicable. But the Strengthened Rule of Conditional Proof, which permits *any* statement to be taken as an assumption of limited scope, here permits a quick and easy proof of validity.

Of course, the Rule of Indirect Proof could be used here, but in the present case, it would result in a longer proof. Because the Strengthened Rule of Conditional Proof permits *any* assumption of limited scope to be made, we can, if we wish, assume the negation of the argument's conclusion. Thus the Strengthened Rule of Conditional Proof *subsumes* the Rule of Indirect Proof, although when C.P. rather than I.P. is used and a contradiction is deduced, we are required to go on *through* or *past* the contradiction, instead of stopping with it. We must do that in order to discharge the assumption and end its scope. We go on *through* the contradiction by deducing the conclusion of the argument from that contradiction.

It is easy to show that, from a contradiction, *any* conclusion can validly be deduced. In other words, any argument of the form

$$
\begin{array}{l}
p \\
\sim p \\
\therefore q
\end{array}
$$

is valid, no matter what statements are substituted for the variables p and q. Thus from lines 11 and 12 in the proof on page 53, the conclusion E can be derived in just two more lines. Such a continuation would proceed:

14.	$B \vee E$	11, Add.
15.	E	14, 12, D.S.

Hence it is possible to regard an indirect proof of the validity of a given argument, not as the deduction of its validity *from the fact that* a contradiction was obtained, but rather as the deduction of the conclusion of that argument *from the contradiction itself*. To transform the proof of validity on page 53 into a proof using the Strengthened Rule of Conditional Proof, simply make the following changes. Insert an arrow head pointing to line 4 to label it an assumption. Then delete line 13 altogether, because the contradiction is already explicit in lines 11 and 12. Then deduce $B \vee E$ (as line 13) from B (line 11) by Addition. Then deduce E (as line 14) from line 13 and line 12 by a Disjunctive Syllogism. We are still within the scope of our assumption. To discharge that assumption, we bend the arrow's shaft in underneath line 12 and write beneath it:

15.	$\sim E \supset E$	4–12, C.P.

We do not yet have the desired conclusion, but we can get it from line 15 in just three easy steps.

16.	$\sim\sim E \vee E$	15, Impl.
17.	$E \vee E$	16, D.M.
18.	E	17, Taut.

From now on, the Strengthened Rule of Conditional Proof will be referred to simply as the Rule of Conditional Proof.[5]

[5] For further discussion of the relations between the Rule of Indirect Proof and the (Strengthened) Rule of Conditional Proof, see Donald Scherer, 'The Form of *Reductio ad Absurdum*', *Mind,* vol. 80 (1971), pp. 247–252; Paul Foulkes, 'The Form of *Reductio ad Absurdum*', *Mind,* vol. 82 (1973), pp. 579–580; Charles H. Lambros, 'Scherer on *Reductio ad Absurdum*', *Mind,* vol. 82 (1973), pp. 581–585; J. M. Lee, 'The Form of *Reductio ad Absurdum*', *Notre Dame Journal of Formal Logic,* vol. 14 (1973), pp. 381–386; and Lyman C. D. K. Kulathungam, '*Reductio-ad-Absurdum:* A Family Feud between Copi and Scherer', *Notre Dame Journal of Formal Logic,* vol. 16 (1975), pp. 245–254.

EXERCISES

Use the strengthened method of conditional proof to prove the validity of the following arguments:

*1. $A \supset B$
$B \supset [(C \supset {\sim}{\sim}C) \supset D]$
$\therefore A \supset D$

2. $(E \vee F) \supset G$
$H \supset (I \cdot J)$
$\therefore (E \supset G) \cdot (H \supset I)$

3. $(K \supset L) \cdot (M \supset N)$
$(L \vee N) \supset \{[O \supset (O \vee P)] \supset (K \cdot M)\}$
$\therefore K \equiv M$

4. $Q \vee (R \supset S)$
$[R \supset (R \cdot S)] \supset (T \vee U)$
$(T \supset Q) \cdot (U \supset V)$
$\therefore Q \vee V$

5. $[W \supset ({\sim}X \cdot {\sim}Y)] \cdot [Z \supset {\sim}(X \vee Y)]$
$({\sim}A \supset W) \cdot ({\sim}B \supset Z)$
$(A \supset X) \cdot (B \supset Y)$
$\therefore X \equiv Y$

6. $(C \vee D) \supset (E \supset F)$
$[E \supset (E \cdot F)] \supset G$
$G \supset [({\sim}H \vee {\sim}{\sim}H) \supset (C \cdot H)]$
$\therefore C \equiv G$

3.8 Shorter Truth Table Technique—*Reductio Ad Absurdum* Method

There is still another method of testing the validity of arguments and of classifying statements as tautologous, contradictory, or contingent. In the preceding section, it was pointed out that an argument is invalid if and only if it is possible to assign truth values to its component simple statements in such a way as to make all its premisses true and its conclusion false. It is impossible to make such truth value assignments in the case of a valid argument. Hence to prove the validity of an argument, it suffices to prove that no such truth values can be assigned. We do so by showing that its premisses can be made true and its conclusion false only by assigning truth values *inconsistently*, so that some component statement is assigned both a **T** and an **F**. If the truth value **T** is assigned to each premiss of a valid argument and the truth value **F** is assigned to its conclusion, this will necessitate assigning *both* **T** and **F** to some component statement, which is, of course, a contradiction.

Thus to prove the validity of the argument

$$(A \vee B) \supset (C \cdot D)$$
$$(D \vee E) \supset F$$
$$\therefore A \supset F$$

we assign **T** to each premiss and **F** to the conclusion. Assigning **F** to the conclusion requires that **T** be assigned to *A* and **F** be assigned to *F*. Since **T** is assigned to *A*, the antecedent of the first premiss is true, and since the premiss has been assigned **T**, its consequent must be true also, so **T** must be assigned to both *C* and *D*. Since **T** is assigned to *D*, the antecedent of the second premiss is true, and since the premiss has been assigned **T**, its consequent must also be true, so **T** must be assigned to *F*. We have already been forced to assign **F** to

F, however, to make the conclusion false. Hence the argument is invalid only if the statement *F* is both true and false, which is impossible. This method of proving the validity of an argument is a version of the *reductio ad absurdum* technique that uses truth value assignments rather than Rules of Inference.

It is easy to extend the use of this method to the classification of statements (and statement forms). Thus to certify that Peirce's Law $[(p \supset q) \supset p] \supset p$ is a tautology, we assign it the truth value **F**, which requires us to assign **T** to its antecedent $[(p \supset q) \supset p]$ and **F** to its consequent p. For the conditional $[(p \supset q) \supset p]$ to be true while *its* consequent p is false, its antecedent $(p \supset q)$ must be assigned the truth value **F** also. But for the conditional $p \supset q$ to be false, its antecedent p must be assigned **T** and its consequent q assigned **F**. However, we were previously forced to assign **F** to p, so assuming Peirce's Law false leads to a contradiction, which proves it a tautology.

If it *is* possible to assign truth values consistently to its components on the assumption that it is false, then the expression in question is not a tautology, but must be either contradictory or contingent. In such a case, we attempt to assign truth values to make it true. If this attempt leads to a contradiction, the expression cannot possibly be true and must be a contradiction. But if truth values can be assigned to make it true and other truth values can be assigned to make it false, then it is neither a tautology nor a contradiction, but is contingent.

The *reductio ad absurdum* method of assigning truth values is by far the quickest and easiest method of testing arguments and classifying statements. It is, however, more readily applied in some cases than in others. If **F** is assigned to a disjunction, **F** must be assigned to both disjuncts, and where **T** is assigned to a conjunction, **T** must be assigned to both conjuncts. Here the sequence of assignments is forced. But where **T** is assigned to a disjunction or **F** to a conjunction, that assignment, by itself, does not determine *which* disjunct is true or *which* conjunct is false. At this point we shall have to experiment and make various 'trial assignments', which will tend to diminish the advantage of the method for such cases. Despite these complications, however, in the vast majority of cases, the *reductio ad absurdum* method is superior to any other method known.

EXERCISES

1. Use the *reductio ad absurdum* method of assigning truth values to decide the validity or invalidity of the arguments and argument forms in the exercises on pages 25–26.
2. Use the *reductio ad absurdum* method of assigning truth values to establish that the statements in Exercises I and II on page 56 are tautologies.
3. Use the *reductio ad absurdum* method of assigning truth values to classify the statement forms in Exercise I on pages 30–31 as tautologous, contradictory, or contingent.

4

Quantification Theory

4.1 Singular Propositions and General Propositions

The logical techniques developed thus far apply only to arguments whose validity depends upon the way in which simple statements are truth-functionally combined into compound statements. Those techniques cannot be applied to such arguments as the following:

> All humans are mortal.
> Socrates is human.
> Therefore, Socrates is mortal.

The validity of such an argument depends upon the inner logical structure of the noncompound statements it contains. To appraise such arguments, we must develop methods for analyzing noncompound statements and symbolizing their inner structures.

The second premiss of the preceding argument is a *singular proposition;* it says that the individual Socrates has the attribute of being human. We call 'Socrates' the *subject* term and 'human' the *predicate* term. Any (affirmative) singular proposition states that the individual referred to by its subject term has the attribute designated by its predicate term. We regard as individuals not only persons, but any *things,* such as animals, cities, nations, planets, or stars, of which attributes can significantly be predicated. Attributes can be designated not only by adjectives, but by nouns or even verbs. Thus 'Leslie is a gossip' and 'Leslie gossips' have the same meaning, which can also be expressed as 'Leslie is gossipy'.

In symbolizing singular propositions, we use the small letters '*a*' through '*w*' to denote individuals, ordinarily using the first letter of an individual's name to denote that individual. Because these symbols denote individuals, we call them 'individual constants'. To designate attributes, we use capital letters, being guided by the same principle in their selection. Thus in the context of the preceding argument, we denote Socrates by the small letter '*s*' and symbolize the attributes *human* and *mortal* by the capital letters '*H*' and '*M*'. To express a singular proposition in our symbolism, we write the symbol for its predicate term to the left of the symbol for its subject term. Thus we symbolize 'Socrates is human' as '*Hs*' and 'Socrates is mortal' as '*Ms*'.[1]

Examining the symbolic formulations of singular propositions having the same predicate term, we observe them to have a common pattern. The symbolic formulations of the singular propositions 'Aristotle is human', 'Boston is human', 'California is human', 'Descartes is human', . . . , which are '*Ha*', '*Hb*', '*Hc*', '*Hd*', . . . , each consists of the attribute symbol '*H*' followed by an individual constant. We use the expression '*Hx*', to symbolize the pattern common to all such singular propositions. The small letter '*x*'—called an 'individual variable'—is a mere *place marker* that serves to indicate where an individual constant can be written to produce a singular proposition. The singular propositions '*Ha*', '*Hb*', '*Hc*', '*Hd*', . . . are either true or false; but '*Hx*' is neither true nor false, since it is not a proposition. Such expressions as '*Hx*' are called 'propositional functions'. These are defined to be expressions that contain individual variables and become propositions when their individual variables are replaced by individual constants.[2] Any singular proposition can be regarded as a *substitution instance* of the propositional function from which it results by the substitution of an individual constant for the individual variable in the propositional function. The process of obtaining a proposition from a propositional function by substituting a constant for a variable is called 'instantiation'. The negative singular propositions 'Aristotle is not human' and 'Boston is not human', symbolized as '$\sim Ha$' and '$\sim Hb$', result by *instantiation* from the propositional function '$\sim Hx$', of which they are substitution instances. Thus we see that symbols other than attribute symbols and individual variables can occur in propositional functions.

General propositions such as 'Everything is mortal' and 'Something is mortal' differ from singular propositions in not containing the names of any individuals. However, they also can be regarded as resulting from propositional functions, not by instantiation, but by the process called 'generalization' or 'quantification'. The first example, 'Everything is mortal', can alternatively be expressed as

Given any individual thing whatever, it is mortal.

Here the relative pronoun 'it' refers back to the word 'thing', which precedes it in the statement. We can use the individual variable '*x*' in place of the pronoun 'it' and its antecedent to paraphrase the first general proposition as

Given any *x*, *x* is mortal.

Then we can use the notation already introduced to rewrite it as

Given any *x*, *Mx*.

[1] Some logicians enclose the individual constant in parentheses, symbolizing 'Socrates is human' as '*H*(*s*)', but we shall not follow that practice here.

[2] Some writers have defined 'propositional functions' to be the *meanings* of such expressions; but here we define them as the expressions themselves.

The phrase 'Given any *x*' is called a 'universal quantifier', and is symbolized by '(x)'. Using this new symbol, we can completely symbolize our first general proposition as

$$(x)Mx$$

We can, similarly, paraphrase the second general proposition, 'Something is mortal', successively as

> There is at least one thing that is mortal.
> There is at least one thing such that it is mortal.
> There is at least one *x* such that *x* is mortal.

and as

> There is at least one *x* such that *Mx*.

The phrase 'There is at least one *x* such that' is called an 'existential quantifier' and is symbolized as '$(\exists x)$'. Using the new symbol, we can completely symbolize our second general proposition as

$$(\exists x)Mx$$

A general proposition is formed from a propositional function by placing either a universal or an existential quantifier before it. It is obvious that the universal quantification of a propositional function is true if and only if all of its substitution instances are true. Also, the existential quantification of a propositional function is true if and only if it has at least one true substitution instance. If we grant that there is at least one individual, then every propositional function has at least one substitution instance (true or false). Under this assumption, if the universal quantification of a propositional function is true, then its existential quantification must be true as well.

A further relationship between universal and existential quantification can be shown by considering two additional general propositions, 'Something is not mortal' and 'Nothing is mortal', which are the respective negations of the first two general propositions just considered. 'Something is not mortal' is symbolized as '$(\exists x)\sim Mx$' and 'Nothing is mortal' is symbolized as '$(x)\sim Mx$'. These show that the negation of the universal (existential) quantification of a propositional function is logically equivalent to the existential (universal) quantification of the new propositional function that results from placing a negation symbol in front of the first propositional function. Where we use the Greek letter *phi* to represent any attribute symbol whatever, the general connections between universal and existential quantification can be described in terms of the following square array:

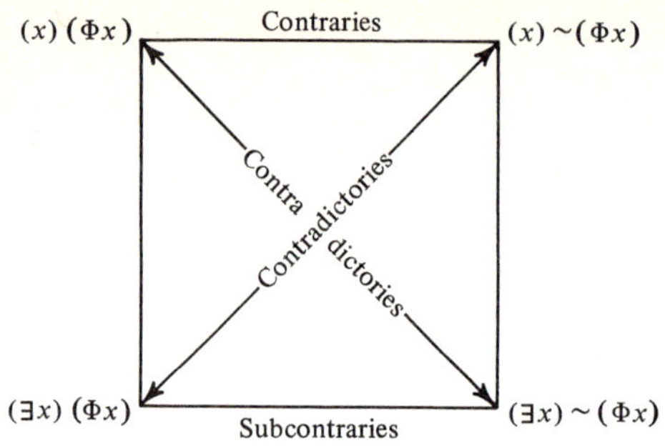

Assuming the existence of at least one individual, we can say that the two top propositions are *contraries*, that is, they might both be false but cannot both be true; the two bottom propositions are *subcontraries*, that is, they can both be true but cannot both be false. Propositions that are at opposite ends of the diagonals are *contradictories*, of which one must be true and the other false. Finally, on each side, the truth of the lower proposition is implied by the truth of the proposition directly above it.

Traditional logic emphasized four types of subject-predicate propositions illustrated by the following:

> All humans are mortal.
> No humans are mortal.
> Some humans are mortal.
> Some humans are not mortal.

These were classified as 'universal affirmative', 'universal negative', 'particular affirmative', and 'particular negative', respectively, and their types abbreviated as '**A**', '**E**', '**I**', '**O**', again respectively. (The letter names have been presumed to come from the Latin '***A****ff****I****rmo*' and '*n****E****g****O***', meaning *I affirm* and *I deny*.) These four special forms of subject-predicate propositions are easily symbolized by means of propositional functions and quantifiers.[3] The first of them, the **A** proposition, can successively be paraphrased as

> Given any individual thing whatever, if it is human, then it is mortal.
> Given any x, if x is human, then x is mortal.
> Given any x, x is human $\supset$ x is mortal.

and finally symbolized as

$$(x)(Hx \supset Mx)$$

[3]An alternative method of symbolizing them is presented in Section 7.1.

Our symbolic formulation of the **A** proposition is the universal quantification of the complex propositional function '$Hx \supset Mx$', which has as its substitution instances not singular propositions but conditionals whose antecedents and consequents are singular propositions that have the same subject terms. Among the substitution instances of the propositional function '$Hx \supset Mx$' are the conditionals '$Ha \supset Ma$', '$Hb \supset Mb$', '$Hc \supset Mc$', and so on. In symbolizing an **A** proposition the parentheses (or square brackets or braces) serve as punctuation marks to indicate that the universal quantifier '(x)' *applies to* or *has within its scope* the whole of the complex propositional function '$Hx \supset Mx$'. The notion of the *scope of a quantifier* is very important, for differences in scope correspond to differences in meaning. The expression '$(x)(Hx \supset Mx)$' is a proposition which asserts that all substitution instances of the propositional function '$Hx \supset Mx$' are true. On the other hand, the expression '$(x)Hx \supset Mx$' is a propositional function whose substitution instances are '$(x)Hx \supset Ma$', '$(x)Hx \supset Mb$', '$(x)Hx \supset Mc$', and so forth.[4]

The **E** proposition 'No humans are mortal' may similarly be paraphrased successively as

> Given any individual thing whatever, if it is human, then it is not mortal.
> Given any x, if x is human, then x is not mortal.
> Given any x, x is human $\supset$ x is not mortal.

and then symbolized as

$$(x)(Hx \supset \sim Mx)$$

Similarly, the **I** proposition 'Some humans are mortal', may be paraphrased as

> There is at least one thing that is human and mortal.
> There is at least one thing such that it is human and it is mortal.
> There is at least one x such that x is human and x is mortal.
> There is at least one x such that x is human $\cdot$ x is mortal.

and completely symbolized as

$$(\exists x)(Hx \cdot Mx)$$

Finally, the **O** proposition 'Some humans are not mortal', becomes

> There is at least one thing that is human but not mortal.
> There is at least one thing such that it is human and it is not mortal.
> There is at least one x such that x is human and x is not mortal.

[4] We have the same symbolic convention for quantifiers (both universal and existential) that we established for negation on page 12: a quantifier applies to or has for its scope the smallest component that the punctuation permits.

and then symbolized as the existential quantification of a complex function

$$(\exists x)(Hx \cdot \sim Mx)$$

Where the Greek letters *phi* and *psi* are used to represent any attribute symbols whatever, the four general subject-predicate propositions of traditional logic may be represented in a square array as

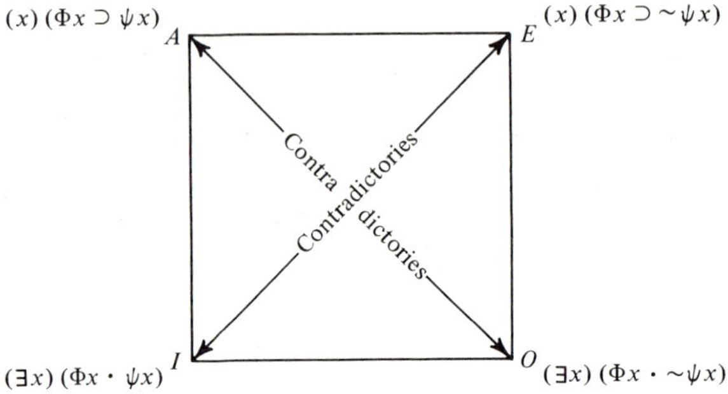

Of these, the **A** and the ***O*** are contradictories, and the ***E*** and the ***I*** are contradictories also. But none of the other relationships discussed in connection with the square array on page 66 hold for the traditional **A**, ***E***, ***I***, and ***O*** propositions, even if we assume that there is at least one individual in the universe. Where 'Φx' is a propositional function that has no true substitution instances, then regardless of what attribute is symbolized by 'Ψ', the propositional functions '$\Phi x \supset \Psi x$' and '$\Phi x \supset \sim\Psi x$' have only true substitution instances. This is so because all their substitution instances are conditional statements with false antecedents. In such cases both the **A** and ***E*** propositions that are the universal quantifications of these complex propositional functions are true, so **A** and ***E*** propositions are not contraries. Again, where 'Φx' is a propositional function that has no true substitution instances, then, regardless of what 'Ψx' might be, the propositional functions '$\Phi x \cdot \Psi x$' and '$\Phi x \cdot \sim\Psi x$' have only false substitution instances, for their substitution instances are conjunctions whose first conjuncts are false. In such cases the ***I*** and ***O*** propositions that are the existential quantifications of these complex propositional functions are false. Thus ***I*** and ***O*** propositions are not subcontraries. In all such cases, then, since the **A** and ***E*** propositions are true and the ***I*** and ***O*** propositions are false, the truth of a universal does *not* imply the truth of the corresponding particular; no implication relation holds between them.

If we make the assumption that there is at least one individual, then '$(x)(\Phi x \supset \Psi x)$' does imply '$(\exists x)(\Phi x \supset \Psi x)$'. The latter, however, is *not* an ***I*** proposition. An ***I*** proposition of the form 'Some Φ's are Ψ's is symbolized as '$(\exists x)(\Phi x \cdot \Psi x)$', which asserts that there is at least one thing having both the attribute Φ and the attribute Ψ. But the proposition '$(\exists x)(\Phi x \supset \Psi x)$' asserts only that there is at least one object that either has the attribute Ψ or does not have the attribute Φ, which is a very different and much weaker assertion.

The four traditional subject-predicate forms **A**, **E**, **I**, and **O** are not the only forms of general propositions. There are others that involve the quantification of more complicated propositional functions. Thus the general proposition 'All members are either parents or teachers', which does *not* mean the same as 'All members are parents or all members are teachers', is symbolized as '$(x)[Mx \supset (Px \vee Tx)]$'. And the general proposition 'Some Senators are either disloyal or misguided', is symbolized as '$(\exists x)[Sx \cdot (Dx \vee Mx)]$'. It should be observed that such a proposition as 'Apples and bananas are nourishing' can be symbolized either as the conjunction of two **A** propositions, '$[(x)(Ax \supset Nx)] \cdot [(x)(Bx \supset Nx)]$', or as a single noncompound general proposition, '$(x)[(Ax \vee Bx) \supset Nx]$'. It should *not*, however, be symbolized as '$(x)[(Ax \cdot Bx) \supset Nx]$', since to say that apples and bananas are nourishing is to say that anything is nourishing which is *either* an apple *or* a banana, *not* that anything is nourishing which is *both* an apple and a banana (whatever that might be). It must be emphasized that there are no mechanical rules for translating statements from English into our logical notation. In every case one must *understand the meaning* of the English sentence and then *re-express* that meaning in terms of propositional functions and quantifiers.

EXERCISES

I. Translate each of the following into the logical notation of propositional functions and quantifiers. In each case, use the abbreviations suggested and begin each formula with a quantifier, *not* with a negation symbol:

*1. Snakes are reptiles. (*Sx*: *x* is a snake. *Rx*: *x* is a reptile.)
2. Snakes are not all poisonous. (*Sx*: *x* is a snake. *Px*: *x* is poisonous.)
3. Children are present. (*Cx*: *x* is a child. *Px*: *x* is present.)
4. Executives all have secretaries. (*Ex*: *x* is an executive. *Sx*: *x* has a secretary.)
*5. Only executives have secretaries. (*Ex*: *x* is an executive. *Sx*: *x* has a secretary.)
6. Only property owners may vote in special municipal elections. (*Px*: *x* is a property owner. *Vx*: *x* may vote in special municipal elections.)
7. Employees may use only the service elevator. (*Ux*: *x* is an elevator that employees may use. *Sx*: *x* is a service elevator.)
8. Only employees may use the service elevator. (*Ex*: *x* is an employee. *Ux*: *x* may use the service elevator.)
9. All that glitters is not gold. (*Gx*: *x* glitters. *Ax*: *x* is gold.)
*10. None but the brave deserve the fair. (*Bx*: *x* is brave. *Dx*: *x* deserves the fair.)
11. Not every visitor stayed for dinner. (*Vx*: *x* is a visitor. *Sx*: *x* stayed for dinner.)
12. Not any visitor stayed for dinner. (*Vx*: *x* is a visitor. *Sx*: *x* stayed for dinner.)
13. Nothing in the house escaped destruction. (*Hx*: *x* was in the house. *Ex*: *x* escaped destruction.)
14. Some students are both intelligent and hard workers. (*Sx*: *x* is a student. *Ix*: *x* is intelligent. *Hx*: *x* is a hard worker.)
*15. No coat is waterproof, unless it has been specially treated. (*Cx*: *x* is a coat. *Wx*: *x* is waterproof. *Sx*: *x* has been specially treated.)

16. Some medicines are dangerous only if taken in excessive amounts. (*Mx*: *x* is a medicine. *Dx*: *x* is dangerous. *Ex*: *x* is taken in excessive amounts.)
17. All fruits and vegetables are wholesome and delicious. (*Fx*: *x* is a fruit. *Vx*: *x* is a vegetable. *Wx*: *x* is wholesome. *Dx*: *x* is delicious.)
18. Everything enjoyable is either immoral, illegal, or fattening. (*Ex*: *x* is enjoyable. *Mx*: *x* is moral. *Lx*: *x* is legal. *Fx*: *x* is fattening.)
19. A professor is a good lecturer if and only if he is both well informed and entertaining. (*Px*: *x* is a professor. *Gx*: *x* is a good lecturer. *Wx*: *x* is well informed. *Ex*: *x* is entertaining.)
*20. Only policemen and firemen are both indispensable and underpaid. (*Px*: *x* is a policeman. *Fx*: *x* is a fireman. *Ix*: *x* is indispensable. *Ux*: *x* is underpaid.)
21. Not every actor is talented who is famous. (*Ax*: *x* is an actor. *Tx*: *x* is talented. *Fx*: *x* is famous.)
22. Any girl is healthy if she is well nourished and exercises regularly. (*Gx*: *x* is a girl. *Hx*: *x* is healthy. *Wx*: *x* is well nourished. *Ex*: *x* exercises regularly.)
23. It is not true that every watch will keep good time if and only if it is wound regularly and not abused. (*Wx*: *x* is a watch. *Kx*: *x* keeps good time. *Rx*: *x* is wound regularly. *Ax*: *x* is abused.)
24. Not every person who talks a great deal has a great deal to say. (*Px*: *x* is a person. *Tx*: *x* talks a great deal. *Hx*: *x* has a great deal to say.)
*25. No automobile that is over ten years old will be repaired if it is severely damaged. (*Ax*: *x* is an automobile. *Ox*: *x* is over ten years old. *Rx*: *x* will be repaired. *Dx*: *x* is severely damaged.)

In symbolizing the following, use the abbreviations: *Hx*: *x* is a horse. *Gx*: *x* is gentle. *Tx*: *x* has been well trained.

26. Some horses are gentle and have been well trained.
27. Some horses are gentle only if they have been well trained.
28. Some horses are gentle if they have been well trained.
29. Any horse is gentle that has been well trained.
*30. Any horse that is gentle has been well trained.
31. No horse is gentle unless it has been well trained.
32. Any horse is gentle if it has been well trained.
33. Any horse has been well trained if it is gentle.
34. Any horse is gentle if and only if it has been well trained.
*35. Gentle horses have all been well trained.
36. Only well-trained horses are gentle.
37. Only gentle horses have been well trained.
38. Only horses are gentle if they have been well trained.
39. Some horses are gentle even though they have not been well trained.
*40. If something is a well-trained horse, then it must be gentle.
41. Some horses that are well trained are not gentle.
42. Some horses are neither gentle nor well trained.
43. No horse that is well trained fails to be gentle.
44. A horse is gentle only if it has been well trained.
45. If anything is a gentle horse, then it has been well trained.
46. If any horse is well trained, then it is gentle.

II. Symbolize the following, using propositional functions and quantifiers:

*1. Blessed is he that considereth the poor. (Psalm 41:1)
2. He that hath knowledge spareth his words. (Proverbs 17:27)
3. Whoso findeth a wife findeth a good thing. (Proverbs 18:22)
4. He that maketh haste to be rich shall not be innocent. (Proverbs 28:20)
*5. They shall sit every man under his vine and under his fig-tree. (Micah 4:4)
6. He that increaseth knowledge increaseth sorrow. (Ecclesiastes 1:18)
7. Nothing is secret which shall not be made manifest. (Luke 8:17)
8. Whom The Lord loveth He chasteneth. (Hebrews 12:6)
9. If a man desire the office of a bishop, he desireth a good work. (I Timothy 3:1)
10. He that hateth dissembleth with his lips, and layeth up deceit within him. (Proverbs 26:24)

4.2 Proving Validity: Preliminary Quantification Rules

To construct formal proofs of validity for arguments symbolized by means of quantifiers and propositional functions, we must augment our list of Rules of Inference. We shall add four rules governing quantification, offering an over-simplified preliminary statement of them in this section and giving a more adequate formulation in Section 4.5.

1. Universal Instantiation (Preliminary Version). Because the universal quantification of a propositional function is true if and only if all substitution instances of that propositional function are true, we can add to our list of Rules of Inference the principle that any substitution instance of a propositional function can validly be inferred from its universal quantification. We can express this rule symbolically as

$$\frac{(x)(\Phi x)}{\therefore \Phi \nu} \qquad \text{(where } \nu \text{ is any individual symbol)}$$

Since this rule permits substitution instances to be inferred from universal quantifications, we refer to it as the 'principle of Universal Instantiation', and abbreviate it as '**UI**.'[5] The addition of **UI** permits us to give a formal proof of validity for the argument: 'All humans are mortal; Socrates is human; therefore, Socrates is mortal'.

1. $(x)(Hx \supset Mx)$
2. $Hs \quad / \therefore Ms$
3. $Hs \supset Ms$ 1, **UI**
4. Ms 3, 2, M.P.

[5]This rule and the three that follow are variants of rules for 'natural deduction', which were devised independently by Gerhard Gentzen and Stanislaw Jaśkowski in 1934.

2. Universal Generalization (Preliminary Version). We can explain our next rule by analogy with fairly standard mathematical practice. A geometer may begin a proof by saying, 'Let *ABC* be any arbitrarily selected triangle'. Then he may go on to prove that the triangle *ABC* has some specified attribute, and concludes that *all* triangles have that attribute. Now, what justifies his final conclusion? Why does it follow from triangle *ABC*'s having a specified attribute that *all* triangles do? The answer is that if no assumption other than its triangularity is made about *ABC*, then the expression '*ABC*' can denote any triangle whatsoever. And if the argument has established that *any* triangle whatsoever must have the attribute in question, then it follows that *all* triangles do. We now introduce a notation analogous to that of the geometer in his reference to 'any arbitrarily selected triangle'. The hitherto unused small letter 'y' will be used to denote *any arbitrarily selected individual.* In this usage, the expression 'Φy' is a substitution instance of the propositional function 'Φx', and it asserts that *any arbitrarily selected individual* has the property Φ. Clearly, 'Φy' follows validly from '$(x)(\Phi x)$' by **UI**, since what is true of all individuals is true of any arbitrarily selected individual. The inference is equally valid in the other direction, since what is true of *any arbitrarily selected individual* must be true of *all* individuals. We augment our list of Rules of Inference further by adding the principle that the universal quantification of a propositional function can validly be inferred from its substitution instance with respect to the symbol 'y'. Since this rule permits the inference of general propositions that are universal quantifications, we refer to it as the 'principle of Universal Generalization', and abbreviate it as '**UG**'. Our symbolic expression for this second quantification rule is

$$\frac{\Phi y}{\therefore (x)(\Phi x)}$$

(where 'y' denotes any arbitrarily selected individual, and Φy is not within the scope of any assumption containing the special symbol 'y')

We can use the new notation and the additional rule **UG** to construct a formal proof of validity for the argument: 'No mortals are perfect; all humans are mortal; therefore, no humans are perfect'.

1. $(x)(Mx \supset \sim Px)$
2. $(x)(Hx \supset Mx)$ $/\therefore (x)(Hx \supset \sim Px)$
3. $Hy \supset My$ 2, **UI**
4. $My \supset \sim Py$ 1, **UI**
5. $Hy \supset \sim Py$ 3, 4, H.S.
6. $(x)(Hx \supset \sim Px)$ 5, **UG**

We can explain the need for the two restrictions in our statement of **UG** by considering two obviously invalid arguments. First, the silly argument 'Chicago is large, therefore, everything is large'. If we ignored the restriction on **UG** that the premiss must contain the special symbol 'y', we might be led to construct the following 'proof':

1. $Lc \quad /\therefore (x)Lx$
2. $(x)Lx$ 1, **UG** (*wrong*)

The mistake occurs at line 2. The individual symbol in line 1 is c, but **UG** can be used only on a line containing 'y'.

The second invalid argument to consider is the obviously invalid: 'Not everything is edible; therefore, nothing is edible'. If we ignored the restriction on **UG** that the Φy must not be within the scope of an assumption containing the special symbol 'y', we might be led to construct the following 'proof':

1. $\sim(x)Ex \quad /\therefore (x)\sim Ex$
2. Ey
3. $(x)Ex$ 2, **UG** (*wrong*)
4. $Ey \supset (x)Ex$ 2–3, C.P.
5. $\sim Ey$ 4, 1, M.T.
6. $(x)\sim Ex$ 5, **UG**

The mistake here occurs at line 3. The line 2 to which **UG** is (mis)applied, *is* within the scope of an assumption containing the special symbol 'y'. This shows, incidentally, that we regard every assumption as being (or lying) within its own scope. To avoid errors like these, we must obey the indicated restrictions in using **UG**.

3. Existential Generalization (Preliminary Version). Because the existential quantification of a propositional function is true if and only if that propositional function has at least one true substitution instance, we can add to our list of Rules of Inference the principle that the existential quantification of a propositional function can validly be inferred from any substitution instance of that propositional function. This rule permits the inference of general propositions that are existentially quantified, so we call it the 'principle of Existential Generalization' and abbreviate it as '**EG**'. Its symbolic formulation is

$$\frac{\Phi\nu}{\therefore (\exists x)(\Phi x)} \qquad \text{(where } \nu \text{ is any individual symbol)}$$

4. Existential Instantiation (Preliminary Version). One further quantification rule is required. The existential quantification of a propositional function asserts that there exists at least one individual the substitution of whose name for the variable 'x' in that propositional function will yield a true substitution instance of it. Of course, we may not know anything else about that individual.

But we can take any individual constant other than '*y*', say, '*w*', which has had no prior occurrence in that context and use it to denote the individual or one of the individuals whose existence has been asserted by the existential quantification. Knowing that there is such an individual, and having agreed to denote it by '*w*', we can infer from the existential quantification of a propositional function the substitution instance of that propositional function with respect to the individual symbol '*w*'. We add, as our final quantification rule, the principle that from the existential quantification of a propositional function we may validly infer the truth of its substitution instance with respect to an individual constant that has no prior occurrence in that context. The new rule may be written as

$$\frac{(\exists x)(\Phi x)}{\therefore \Phi \nu}$$ (where ν is an individual constant, other than '*y*', that has no prior occurrence in the context)

It is referred to as the 'principle of Existential Instantiation' and abbreviated as '**EI**'.

We make use of the last two quantification rules in constructing a formal proof of validity for the argument: 'All dogs are carnivorous; some animals are dogs; therefore, some animals are carnivorous'.

1.	$(x)(Dx \supset Cx)$	
2.	$(\exists x)(Ax \cdot Dx)$	$/ \therefore (\exists x)(Ax \cdot Cx)$
3.	$Aw \cdot Dw$	2, **EI**
4.	$Dw \supset Cw$	1, **UI**
5.	$Dw \cdot Aw$	3, Com.
6.	Dw	5, Simp.
7.	Cw	4, 6, M.P.
8.	Aw	3, Simp.
9.	$Aw \cdot Cw$	8, 7, Conj.
10.	$(\exists x)(Ax \cdot Cx)$	9, **EG**

We can show the need for the indicated restriction on the use of **EI** by considering the obviously invalid argument: 'Some cats are animals; some dogs are animals; therefore, some cats are dogs'. If we ignored the restriction on **EI** that the substitution instance inferred by it can contain only an individual constant that had no prior occurrence in the context, we might be led to construct the following 'proof':

1.	$(\exists x)(Cx \cdot Ax)$	
2.	$(\exists x)(Dx \cdot Ax)$	$/ \therefore (\exists x)(Cx \cdot Dx)$
3.	$Cw \cdot Aw$	1, **EI**
4.	$Dw \cdot Aw$	2, **EI** (*wrong*)

5.	Cw	3, Simp.
6.	Dw	4, Simp.
7.	$Cw \cdot Dw$	5, 6, Conj.
8.	$(\exists x)(Cx \cdot Dx)$	7, **EG**

The mistake here occurs at line 4. The second premiss assures us that there is at least one thing that is both a dog and an animal. We are not free to use the symbol '*w*' to denote that thing, however, because '*w*' has already been used to denote one of the things that was asserted by the first premiss to be both a cat and an animal. To avoid errors of this sort, we must obey the indicated restriction in using **EI**. It should be clear that whenever we use both **EI** and **UI** in a proof to instantiate with respect to the same individual constant, we must use **EI** first. (It has been suggested that we could avoid the necessity of using **EI** before **UI** by changing the restriction on **EI** to read 'where ν is an individual constant, other than '*y*', that was not introduced into the context by any previous use of **EI**.' Even apart from its apparent circularity, that formulation would not prevent the construction of an erroneous 'formal proof of validity' for such an invalid argument as 'Some men are handsome. Socrates is a man. Therefore, Socrates is handsome.')

Like the first nine Rules of Inference presented in Section 3.1, the four quantification rules **UI**, **UG**, **EG**, and **EI** can be applied only to whole lines in a proof.

Any assumption of limited scope may be made in a Conditional Proof of validity and, in particular, we are free to make an assumption of the form 'Φy'. Thus the argument 'All freshmen and sophomores are invited and will be welcome; therefore, all freshmen are invited' is proved valid by the following Conditional Proof:

1.	$(x)[(Fx \vee Sx) \supset (Ix \cdot Wx)]$	$/\therefore (x)(Fx \supset Ix)$
→ 2.	Fy	
3.	$(Fy \vee Sy) \supset (Iy \cdot Wy)$	1, **UI**
4.	$Fy \vee Sy$	2, Add.
5.	$Iy \cdot Wy$	3, 4, M.P.
6.	Iy	5, Simp.
7.	$Fy \supset Iy$	2–6, C.P.
8.	$(x)(Fx \supset Ix)$	7, **UG**

More than one assumption of limited scope can be made in proving the validity of arguments involving quantifiers, as in the following Conditional Proof:

1.	$(x)[(Ax \vee Bx) \supset (Cx \cdot Dx)]$	
2.	$(x)\{(Cx \vee Ex) \supset [(Fx \vee Gx) \supset Hx\}$	$/\therefore (x)[Ax \supset (Fx \supset Hx)]$
3.	$(Ay \vee By) \supset (Cy \cdot Dy)$	1, **UI**
4.	$(Cy \vee Ey) \supset [(Fy \vee Gy) \supset Hy]$	2, **UI**

→ 5.	Ay	
6.	$Ay \vee By$	5, Add
7.	$Cy \cdot Dy$	3, 6, M.P.
8.	Cy	7, Simp.
9.	$Cy \vee Ey$	8, Add.
10.	$(Fy \vee Gy) \supset Hy$	4, 9, M.P.
→ 11.	Fy	
12.	$Fy \vee Gy$	11, Add.
13.	Hy	10, 12, M.P.
14.	$Fy \supset Hy$	11–13, C.P.
15.	$Ay \supset (Fy \supset Hy)$	5–14, C.P.
16.	$(x)[Ax \supset (Fx \supset Hx)]$	15, **UG**

EXERCISES

I. Construct formal proofs of validity for the following arguments, using the rule of Conditional Proof wherever you wish:

***1.** $(x)(Ax \supset Bx)$
$\sim Bt$
$\therefore \sim At$

2. $(x)(Cx \supset Dx)$
$(x)(Ex \supset \sim Dx)$
$\therefore (x)(Ex \supset \sim Cx)$

3. $(x)(Fx \supset \sim Gx)$
$(\exists x)(Hx \cdot Gx)$
$\therefore (\exists x)(Hx \cdot \sim Fx)$

4. $(x)(Ix \supset Jx)$
$(\exists x)(Ix \cdot \sim Jx)$
$\therefore (x)(Jx \supset Ix)$

***5.** $(x)(Kx \supset Lx)$
$(x)[(Kx \cdot Lx) \supset Mx]$
$\therefore (x)(Kx \supset Mx)$

6. $(x)(Nx \supset Ox)$
$(x)(Px \supset Ox)$
$\therefore (x)[(Nx \vee Px) \supset Ox]$

7. $(x)(Qx \supset Rx)$
$(\exists x)(Qx \vee Rx)$
$\therefore (\exists x)Rx$

8. $(x)[Sx \supset (Tx \supset Ux)]$
$(x)[Ux \supset (Vx \cdot Wx)]$
$\therefore (x)[Sx \supset (Tx \supset Vx)]$

9. $(x)[(Xx \vee Yx) \supset (Zx \cdot Ax)]$
$(x)[(Zx \vee Ax) \supset (Xx \cdot Yx)]$
$\therefore (x)(Xx \equiv Zx)$

10. $(x)[(Bx \supset Cx) \cdot (Dx \supset Ex)]$
$(x)[(Cx \vee Ex) \supset \{[Fx \supset (Gx \supset Fx)] \supset (Bx \cdot Dx)\}]$
$\therefore (x)(Bx \equiv Dx)$

II. Construct formal proofs of validity for the following arguments, using the rule of Conditional Proof wherever you wish:

***1.** All athletes are brawny. Charles is not brawny. Therefore, Charles is not an athlete. (*Ax, Bx, c*)

2. No contractors are dependable. Some contractors are engineers. Therefore, some engineers are not dependable. (*Cx, Dx, Ex*)

3. All fiddlers are gay. Some hunters are not gay. Therefore, some hunters are not fiddlers. (*Fx, Gx, Hx*)

4. No judges are idiots. Kanter is an idiot. Therefore, Kanter is not a judge. (*Jx*, *Ix*, *k*)

*5. All liars are mendacious. Some liars are newspapermen. Therefore, some newspapermen are mendacious. (*Lx*, *Mx*, *Nx*)

6. No osteopaths are pediatricians. Some quacks are pediatricians. Therefore, some quacks are not osteopaths. (*Ox*, *Px*, *Qx*)

7. Only salesmen are retailers. Not all retailers are travelers. Therefore, some salesmen are not travelers. (*Sx*, *Rx*, *Tx*)

8. There are no uniforms that are not washable. There are no washable velvets. Therefore, there are no velvet uniforms. (*Ux*, *Wx*, *Vx*)

9. Only authoritarians are bureaucrats. Authoritarians are all churlish. Therefore, any bureaucrat is churlish. (*Ax*, *Bx*, *Cx*)

*10. Dates are edible. Only items of food are edible. All items of food are good. Therefore, all dates are good. (*Dx*, *Ex*, *Fx*, *Gx*)

11. All dancers are graceful. Mary is a student. Mary is a dancer. Therefore, some students are graceful. (*Dx*, *Gx*, *Sx*, *m*)

12. Tigers are fierce and dangerous. Some tigers are beautiful. Therefore, some dangerous things are beautiful. (*Tx*, *Fx*, *Dx*, *Bx*)

13. Bananas and grapes are fruits. Fruits and vegetables are nourishing. Therefore, bananas are nourishing. (*Bx*, *Gx*, *Fx*, *Vx*, *Nx*)

14. A communist is either a fool or a knave. Fools are naive. Not all communists are naive. Therefore, some communists are knaves. (*Cx*, *Fx*, *Kx*, *Nx*)

*15. All butlers and valets are both obsequious and dignified. Therefore, all butlers are dignified. (*Bx*, *Vx*, *Ox*, *Dx*)

16. All houses built of brick are warm and cozy. All houses in Englewood are built of brick. Therefore, all houses in Englewood are warm. (*Hx*, *Bx*, *Wx*, *Cx*, *Ex*)

17. All professors are learned. All learned professors are savants. Therefore, all professors are learned savants. (*Px*, *Lx*, *Sx*)

18. All diplomats are public servants. Some diplomats are eloquent. All eloquent public servants are orators. Therefore, some diplomats are orators. (*Dx*, *Px*, *Ex*, *Ox*)

19. Doctors and lawyers are college graduates. Any altruist is an idealist. Some lawyers are not idealists. Some doctors are altruists. Therefore, some college graduates are idealists. (*Dx*, *Lx*, *Cx*, *Ax*, *Ix*)

*20. Bees and wasps sting if they are either angry or frightened. Therefore, any bee stings if it is angry. (*Bx*, *Wx*, *Sx*, *Ax*, *Fx*)

21. Any authors are successful if and only if they are well read. All authors are intellectuals. Some authors are successful but not well read. Therefore, all intellectuals are authors. (*Ax*, *Sx*, *Wx*, *Ix*)

22. Every passenger is either in first class or in tourist class. Each passenger is in tourist class if and only if he is not wealthy. Some passengers are wealthy. Not all passengers are wealthy. Therefore, some passengers are in tourist class. (*Px*, *Fx*, *Tx*, *Wx*)

23. All members are both officers and gentlemen. All officers are fighters. Only a pacifist is either a gentleman or not a fighter. No pacifists are gentlemen if they are fighters. Some members are fighters if and only if they are officers. Therefore, not all members are fighters. (*Mx*, *Ox*, *Gx*, *Fx*, *Px*)

24. Wolfhounds and terriers are hunting dogs. Hunting dogs and lap dogs are domesticated animals. Domesticated animals are gentle and useful. Some wolfhounds are neither gentle nor small. Therefore, some terriers are small but not gentle. (Wx, Tx, Hx, Lx, Dx, Gx, Ux, Sx)
25. No man who is a candidate will be defeated if he is a good campaigner. Any man who runs for office is a candidate. Any candidate who is not defeated will be elected. Every man who is elected is a good campaigner. Therefore, any man who runs for office will be elected if and only if he is a good campaigner. (Mx, Cx, Dx, Gx, Rx, Ex)

4.3 Proving Invalidity

In the preceding chapter we proved the invalidity of invalid arguments containing truth-functional compound statements by assigning truth values to their component simple statements in such a way as to make their premisses true and their conclusions false. We use a closely related method to prove the invalidity of invalid arguments involving quantifiers. The method of proving invalidity about to be described is connected with our basic assumption that there is at least one individual.

The assumption that there is at least one individual could be satisfied in infinitely many different ways: if there is exactly one individual, or if there are exactly two individuals, or if there are exactly three individuals, and so forth. For any such case, there is a strict logical equivalence between noncompound general propositions and truth-functional compounds of singular propositions. If there is exactly one individual, say, a, then

$$(x)(\Phi x) \equiv \Phi a \quad \text{and} \quad (\exists x)(\Phi x) \equiv \Phi a$$

If there are exactly two individuals, say a and b, then

$$(x)(\Phi x) \equiv (\Phi a \cdot \Phi b) \quad \text{and} \quad (\exists x)(\Phi x) \equiv (\Phi a \vee \Phi b)$$

And for any number k, if there are exactly k individuals, say $a, b, c, \ldots$, k, then

$$(x)(\Phi x) \equiv (\Phi a \cdot \Phi b \cdot \Phi c \cdot \ldots \cdot \Phi k)$$

and

$$(\exists x)(\Phi x) \equiv (\Phi a \vee \Phi b \vee \Phi c \vee \ldots \vee \Phi k)$$

The truth of these biconditionals is an immediate consequence of our definitions of the universal and existential quantifiers. No use is made here of the four quantification rules presented in the preceding section. So for any possible nonempty universe or model containing any finite number of individuals, every general proposition is logically equivalent to some truth-functional compound of singular propositions. Hence, for any such model, every argument whose propositions contain quantifiers is logically equivalent to some argument containing only singular propositions and truth-functional compounds of them.

An argument whose propositions contain quantifiers is valid if and only if it is valid no matter how many individuals there are, so long as there is at least one individual. So an argument involving quantifiers is valid if and only if for every possible nonempty universe or model it is logically equivalent to a truth-functional argument that is valid. Hence we can prove the invalidity of a given argument by displaying or describing a model for which the given argument is logically equivalent to an *invalid* truth-functional argument. We can accomplish this purpose by translating the given argument, whose propositions contain quantifiers, into a logically equivalent argument containing only singular propositions and truth-functional compounds of them, and then using the method of assigning truth values to prove the truth-functional argument invalid. For example, given the argument

> All whales are heavy.
> All elephants are heavy.
> Therefore, all whales are elephants.

we first symbolize it as

$$(x)(Wx \supset Hx)$$
$$(x)(Ex \supset Hx)$$
$$\therefore (x)(Wx \supset Ex)$$

In the case of a model containing exactly one individual, say, *a*, the given argument is logically equivalent to

$$Wa \supset Ha$$
$$Ea \supset Ha$$
$$\therefore Wa \supset Ea$$

which is proved invalid by assigning the truth value **T** to '*Wa*' and '*Ha*' and **F** to '*Ea*'. (This assignment of truth values is a shorthand way of describing the model in question as one that contains only the one individual *a*, which is *W* (a whale), and *H* (heavy), but not *E* (an elephant).)[6] Hence the original argument is not valid for a model containing exactly one individual and is, therefore, *invalid*.

[6]Here we assume that the simple propositional functions '*Ax*', '*Bx*', '*Cx*', . . . are neither necessary, that is, logically true of all individuals (for example, *x* is identical with itself), nor impossible, that is, logically false of all individuals (for example, *x* is different from itself). We also assume that the only logical relations among the simple propositional functions are those asserted or logically implied by the premisses of the argument being proved invalid. The point of these restrictions is to permit the arbitrary assignment of truth values to substitution instances of these simple propositional functions without inconsistency—for our model-descriptions must, of course, be consistent.

It must be emphasized that, in proving the invalidity of arguments involving quantifiers, *no use is made of our quantification rules*. For a model containing just the one individual a, we do not *infer* the statement '$Wa \supset Ha$' from the statement '$(x)(Wx \supset Hx)$' by **UI**. Those two statements are logically equivalent for that model, because in it '$Wa \supset Ha$' is the *only* substitution instance of the propositional function '$Wx \supset Hx$'.

It *can* happen that an invalid argument involving quantifiers is logically equivalent, for any model containing exactly one individual, to a valid truth-functional argument, although it will be logically equivalent, for any model containing more than one individual, to an invalid truth-functional argument. For example, consider the argument

All whales are heavy.
Some elephants are heavy.
Therefore, all whales are elephants.

which is symbolized as

$$(x)(Wx \supset Hx)$$
$$(\exists x)(Ex \cdot Hx)$$
$$\therefore (x)(Wx \supset Ex)$$

For a model containing just the one individual a, this argument is logically equivalent to

$$Wa \supset Ha$$
$$Ea \cdot Ha$$
$$\therefore Wa \supset Ea$$

which is a valid argument. But for a model consisting of the two individuals a and b, the given argument is logically equivalent to

$$(Wa \supset Ha) \cdot (Wb \supset Hb)$$
$$(Ea \cdot Ha) \vee (Eb \cdot Hb)$$
$$\therefore (Wa \supset Ea) \cdot (Wb \supset Eb)$$

which is proved invalid by assigning the truth value **T** to 'Wa', 'Wb', 'Ha', 'Hb', 'Eb', and the truth value **F** to 'Ea'. Hence the original argument is invalid, because there is a model for which it is logically equivalent to an invalid truth-functional argument.

Another illustration is

Some dogs are pointers.
Some dogs are spaniels.
Therefore, some pointers are spaniels.

which we symbolize as

$$(\exists x)(Dx \cdot Px)$$
$$(\exists x)(Dx \cdot Sx)$$
$$\therefore (\exists x)(Px \cdot Sx)$$

For a model containing just the one individual *a*, it is logically equivalent to

$$Da \cdot Pa$$
$$Da \cdot Sa$$
$$\therefore Pa \cdot Sa$$

which is valid. But for a model consisting of the two individuals *a* and *b*, it is equivalent to

$$(Da \cdot Pa) \vee (Db \cdot Pb)$$
$$(Da \cdot Sa) \vee (Db \cdot Sb)$$
$$\therefore (Pa \cdot Sa) \vee (Pb \cdot Sb)$$

which is proved invalid by assigning the truth value **T** to '*Da*', '*Db*', '*Pa*', '*Sb*', and the truth value **F** to '*Pb*' and '*Sa*'. Here, too, the original argument is invalid, because there is a model for which it is logically equivalent to an invalid truth-functional argument.

An invalid argument involving quantifiers may be valid for any model containing fewer than *k* individuals, even though it must be invalid for every model containing *k* or more individuals. Hence in using this method to prove the invalidity of an argument involving quantifiers, it may be necessary to consider larger and larger models. The question naturally arises: How large a model must we consider in trying to prove the invalidity of an argument of this type? A theoretically satisfactory answer to this question has been found. If an argument contains *n* different predicate symbols, then if it is valid for a model containing 2^n individuals, then it is valid for every model or universally valid.[7] This result holds only for propositional functions of one variable and is not true of the relational predicates discussed in Chapter 5. Although theoretically satisfactory, this solution is not of much practical help. If we were to go straight to the theoretically crucial case for deciding the validity or invalidity of any of the arguments already considered in this section, we should have to consider models containing eight individuals. And for some of the following exercises, the theoretically crucial case would be a model containing $2^8 = 256$ individuals. In fact, however, none of the following exercises requires consideration of models containing more than three individuals to prove their invalidity.

[7] See Paul Bernays and Moses Schönfinkel. 'Zum Entscheidungsproblem der mathematischen Logik'. *Mathematische Annalen*, vol. 99 (1928), and Wilhelm Ackermann. *Solvable Cases of the Decision Problem*, North-Holland Publishing Company, Amsterdam, 1954, chap. 4.

EXERCISES

I. Prove that each of the following arguments is invalid:

*__1.__ $(\exists x)(Ax \cdot Bx)$
Ac
$\therefore Bc$

2. $(x)(Cx \supset \sim Dx)$
$\sim Cj$
$\therefore Dj$

3. $(x)(Ex \supset Fx)$
$(x)(Gx \supset Fx)$
$\therefore (x)(Ex \supset Gx)$

4. $(x)(Hx \supset \sim Ix)$
$(\exists x)(Ix \cdot \sim Ix)$
$\therefore (x)(Hx \supset Jx)$

*__5.__ $(\exists x)(Kx \cdot Lx)$
$(\exists x)(\sim Kx \cdot \sim Lx)$
$\therefore (\exists x)(Lx \cdot \sim Kx)$

6. $(x)[Mx \supset (Nx \cdot Ox)]$
$(\exists x)(Px \cdot Nx)$
$(\exists x)(Px \cdot \sim Ox)$
$\therefore (x)(Mx \supset \sim Px)$

7. $(x)[Qx \supset (Rx \cdot Sx)]$
$(\exists x)(Tx \cdot Rx)$
$(\exists x)(Tx \cdot \sim Sx)$
$\therefore (x)(Qx \supset Tx)$

8. $(x)[Ux \supset (Vx \supset Wx)]$
$(x)[Vx \supset (Ux \supset \sim Wx)]$
$(\exists x)(Ux \cdot Wx)$
$\therefore (\exists x)(Ux \cdot Vx)$

9. $(\exists x)(Xx \cdot Yx)$
$(x)(Xx \supset Zx)$
$(\exists x)(Zx \cdot \sim Xx)$
$\therefore (\exists x)(Zx \cdot \sim Yx)$

10. $(x)(Ax \supset Bx)$
$(\exists x)(Cx \cdot Bx)$
$(\exists x)(Cx \cdot \sim Bx)$
$\therefore (x)(Ax \supset Cx)$

II. Prove that each of the following arguments is invalid:

*__1.__ All astronauts are brave. Jim is brave. Therefore, Jim is an astronaut.
2. No cowboys are dudes. Bill is not a dude. Therefore, Bill is a cowboy.
3. All evergreens are fragrant. Some gum trees are not fragrant. Therefore, some evergreens are not gum trees.
4. All heathens are idolaters. No heathen is joyful. Therefore, no idolater is joyful.
*__5.__ No kittens are large. Some mammals are large. Therefore, no kittens are mammals.
6. All novelists are observant. Some poets are not observant. Therefore, no novelists are poets.
7. All statesmen are intelligent. Some politicians are intelligent. Not all politicians are intelligent. Therefore, no statesmen are politicians.
8. All statesmen are intelligent. Some politicians are intelligent. Not all politicians are intelligent. Therefore, all statesmen are politicians.
9. All statesmen are politicians. Some statesmen are intelligent. Some politicians are not statesmen. Therefore, some politicians are not intelligent.
10. Horses and cows are mammals. Some animals are mammals. Some animals are not mammals. Therefore, all horses are animals.

III. Prove the validity or prove the invalidity of each of the following arguments:

*__1.__ All aviators are brave. Jones is brave. Therefore, Jones is an aviator.
2. All collegians are debonair. Smith is a collegian. Therefore, Smith is debonair.

3. No educators are fools. All gamblers are fools. Therefore, no educators are gamblers.
4. No historians are illiterates. All illiterates are underprivileged. Therefore, no historians are underprivileged.
*5. Only citizens are voters. Not all residents are citizens. Therefore, some residents are not voters.
6. Only citizens are voters. Not all citizens are residents. Therefore, some voters are not residents.
7. Automobiles and wagons are vehicles. Some automobiles are Fords. Some automobiles are trucks. All trucks are vehicles. Therefore, some Fords are trucks.
8. Automobiles and wagons are vehicles. Some automobiles are Fords. Some automobiles are trucks. All vehicles are trucks. Therefore, some Fords are trucks.
9. Automobiles and wagons are vehicles. Some automobiles are Fords. Some automobiles are trucks. Some wagons are not vehicles. Therefore, some Fords are trucks.
*10. All tenors are either overweight or effeminate. No overweight tenor is effeminate. Some tenors are effeminate. Therefore, some tenors are overweight.
11. All tenors are either overweight or effeminate. No overweight tenor is effeminate. Some tenors are effeminate. Therefore, some tenors are not overweight.
12. No applicant will be either hired or considered who is either untrained or inexperienced. Some applicants are inexperienced beginners. All applicants who are women will be disappointed if they are not hired. Every applicant is a woman. Some women will be hired. Therefore, some applicants will be disappointed.
13. No candidate is either elected or appointed who is either a liberal or a radical. Some candidates are wealthy liberals. All candidates who are politicians are disappointed if they are not elected. Every candidate is a politician. Some politicians are elected. Therefore, some candidates are not disappointed.
14. Abbots and bishops are churchmen. No churchmen are either dowdy or elegant. Some bishops are elegant and fastidious. Some abbots are not fastidious. Therefore, some abbots are dowdy.
15. Abbots and bishops are churchmen. No churchmen are both dowdy and elegant. Some bishops are elegant and fastidious. Some abbots are not fastidious. Therefore, some abbots are not dowdy.

4.4 Multiply-General Propositions

Thus far, we have limited our attention to general propositions containing only a single quantifier. A general proposition that contains exactly one quantifier is said to be *singly* general. We turn next to *multiply*-general propositions, which contain two or more quantifiers. In our use of the term, any compound statement whose components are general propositions is to be counted as a multiply-general proposition. For example, the conditional 'If all dogs are carnivorous then some animals are carnivorous', symbolized as '$(x)(Dx \supset Cx) \supset (\exists x)(Ax \cdot Cx)$', is a multiply-general proposition. Other multiply-general propositions are more complex and require a more complicated notation. To develop the new notation, we must turn again to the notion of a propositional function.

All propositional functions considered up to now have had as substitution instances either singular propositions or truth-functional compounds of singular propositions having the same subject terms. If we consider a compound statement whose components are singular propositions having *different* subject terms, such as '$Fa \cdot Gb$', we can regard it as a substitution instance either of the propositional function '$Fx \cdot Gb$' or of the propositional function '$Fa \cdot Gx$'. Some propositional functions, we see, may contain singular propositions as parts. And if we consider a compound statement of which one component is a general proposition and the other component is a singular proposition, such as "If all dogs are carnivorous, then Rover is carnivorous', symbolized as '$(x)(Dx \supset Cx) \supset Cr$', we can regard it as a substitution instance of the propositional function '$(x)(Dx \supset Cx) \supset Cx$'. Thus we see that some propositional functions may contain general propositions as parts.

At this point, two new technical terms may properly be introduced. An occurrence of the variable 'x', which does not occur within or lie within the scope of a universal or existential quantifier[8] '(x)' or '$(\exists x)$', will be called a *free occurrence* of that variable. On the other hand, an occurrence of the variable 'x', which is either part of a quantifier or lies within the scope of a quantifier '(x)' or '$(\exists x)$', will be called a *bound occurrence* of that variable.[9] Thus in the expression '$(x)(Dx \supset Cx) \supset Cx$', the first occurrence of the variable 'x' is *part of a quantifier* and is, therefore, considered to be *bound*. The second and third occurrences are bound occurrences also. But the fourth occurrence is a free occurrence. Thus we see that propositional functions may contain both free and bound occurrences of variables. On the other hand, all occurrences of variables in propositions must be bound, since every proposition must be either true or false. A propositional function must contain at least one free occurrence of a variable, but no proposition can contain any free occurrences of any variable.

The proposition '$Fa \cdot Gb$' can also be regarded as a substitution instance of '$Fx \cdot Gy$', where the latter is a propositional function containing *two different variables*. Up to now, we have explicitly admitted only one individual variable, the letter 'x'. However, in our previous *use* of the letter 'y' to denote *any arbitrarily selected* individual, we were, in effect, using it as a variable without acknowledging this fact. In introducing a letter by **EI** to denote *some particular* individual having a specified attribute without knowing *which* individual was denoted by it, we were, in effect, using that letter as a variable, also. We

[8] As explained on page 67.

[9] An alternative, less common nomenclature refers to free variables as 'real' variables, and to bound variables as 'apparent' variables.

now proceed to acknowledge explicitly what was implicit in our former usage. Some propositional functions may contain two or more different individual variables. It will be convenient to have a larger supply of individual variables available, so we readjust our notational conventions to include the letters '*u*', '*v*', '*w*', '*x*', '*y*', and '*z*' as individual variables. Propositional functions now include such expressions as 'Fu', '$Fu \vee Gw$', '$(Fx \cdot Gy) \supset Hz$', '$Fx \vee (Gy \cdot Hx)$', and the like.

In substituting constants for variables to obtain a proposition from a propositional function, the same constant must be substituted for every free occurrence of the same variable. Thus among the substitution instances of the propositional function '$Fx \vee (Gy \cdot Hx)$' are

$$\begin{array}{lll} Fa \vee (Gb \cdot Ha), & Fa \vee (Gc \cdot Ha), & Fa \vee (Gd \cdot Ha), \ldots \\ Fb \vee (Ga \cdot Hb), & Fb \vee (Gc \cdot Hb), & Fb \vee (Gd \cdot Hb), \ldots \\ Fc \vee (Ga \cdot Hc), & Fc \vee (Gb \cdot Hc), & Fc \vee (Gd \cdot Hc), \ldots \\ \ldots\ldots\ldots & \ldots\ldots\ldots & \ldots\ldots\ldots \end{array}$$

but *not* such propositions as '$Fa \vee (Gb \cdot Hc)$'. On the other hand, the *same* constant can be substituted for free occurrences of *different* variables, provided, of course, that if it is substituted for any free occurrence of a variable, it must be substituted for all free occurrences of that variable. Thus additional substitution instances of the propositional function '$Fx \vee (Gy \cdot Hx)$' are '$Fa \vee (Ga \cdot Ha)$', '$Fb \vee (Gb \cdot Hb)$', '$Fc \vee (Gc \cdot Hc)$', . . .

Having admitted the letters '*u*', '*v*', '*w*', '*y*', and '*z*' as individual variables in addition to '*x*', we now adjust our notation for universal and existential quantification to conform to our expanded stock of variables. The proposition 'All *F*'s are *G*'s' may be alternatively symbolized as '$(u)(Fu \supset Gu)$', '$(v)(Fv \supset Gv)$', '$(w)(Fw \supset Gw)$', '$(x)(Fx \supset Gx)$', '$(y)(Fy \supset Gy)$', or '$(z)(Fz \supset Gz)$'. Similarly the proposition, 'There are some *H*'s', may be alternatively symbolized as '$(\exists u)Hu$', '$(\exists v)hv$', '$(\exists w)Hw$', '$(\exists x)Hx$', '$(\exists y)Hy$', or '$(\exists z)Hz$'. The difference between '$(x)Fx$' and '$(y)Fy$' [as between '$(\exists x)Gx$' and '$(\exists y)Gy$'] is purely notational and either may be written in place of the other. Of course, where a propositional function contains free occurrences of two or more different variables, such as '$Fx \cdot Gy$', the two propositional functions that result from quantifying it differently as

$$(x)(Fx \cdot Gy) \quad \text{and} \quad (y)(Fx \cdot Gy)$$

are very different indeed, and their difference is more than merely notational. The substitution instances of the first are

$$(x)(Fx \cdot Ga), \quad (x)(Fx \cdot Gb), \quad (x)(Fx \cdot Gc), \ldots$$

whereas the substitution instances of the second are

$$(y)(Fa \cdot Gy), \quad (y)(Fb \cdot Gy), \quad (y)(Fc \cdot Gy), \ldots$$

If every individual has the attribute F, and some, but not all, individuals have the attribute G, then some substitution instances of the first will be true propositions, whereas all substitution instances of the second will be false, a considerable difference. This example should serve to indicate the need for speaking, not of "*the* universal (or existential) quantification of a propositional function', but rather of "the universal (or existential) quantification of a propositional function *with respect to the variable* 'x'" or "the universal (or existential) quantification of a propositional function *with respect to the variable* 'y'", and so on.

It should be clear that since '$(x)(Fx \supset Gx)$' and '$(y)(Fy \supset Gy)$' are alternative translations of the proposition 'Everything that is an F is also a G', the universal quantification of '$Fx \supset Gx$' with respect to 'x' has the same meaning and is logically equivalent to the universal quantification with respect to 'y' of the propositional function that results from substituting 'y' for every free occurrence of 'x' in '$Fx \supset Gx$'—for the result of that substitution is '$Fy \supset Gy$'. In the early stages of our work, it will be desirable to have at most one quantification with respect to a given variable in a single proposition. This is not strictly *necessary*, but it is helpful in preventing confusion. Thus the first multiply-general proposition considered, 'If all dogs are carnivorous, then some animals are carnivorous', is more conveniently symbolized as '$(x)(Dx \supset Cx) \supset (\exists y)(Ay \cdot Cy)$' than as '$(x)(Dx \supset Cx) \supset (\exists x)(Ax \cdot Cx)$', although neither is *incorrect.*

It has been remarked that no proposition can contain a free occurrence of any variable. Hence in symbolizing any proposition, we must take care that every occurrence of every variable used lies within the scope of a quantifier with respect to that variable. Some examples will help to make the matter clear. The proposition

> If something is wrong with the house, then everyone in the house complains.

is properly symbolized as a conditional whose antecedent and consequent contain different quantifiers:

$$(\exists x)(x \text{ is wrong with the house}) \supset (y)\,[(y \text{ is a person in the house}) \supset (y \text{ complains})]$$

Here the scope of the initial quantifier does not extend past the main implication sign. But if we turn now to another proposition which bears a superficial resemblance to the first:

> If something is wrong, then it should be rectified.

it would be *incorrect* to symbolize it as

$$(\exists x)(x \text{ is wrong}) \supset (x \text{ should be rectified})$$

For since the scope of the initial quantifier ends at the implication sign, the occurrence of 'x' in the consequent *cannot* refer back to the initial quantifier because it does not lie within the scope of the quantifier. We have here a free occurrence of a variable, which means that the proposed symbolization is not a proposition and, therefore, not an adequate translation of the given statement. The error is not to be corrected by *simply* extending the scope of the initial quantifier through rebracketing, for the symbolic expression,

$$(\exists x)[(x \text{ is wrong}) \supset (x \text{ should be rectified})]$$

although a proposition, does not have the same meaning as the original proposition in English. Instead, it says merely that there is at least one thing that should be rectified if it is wrong. The sense of the English sentence, however, is clearly that *anything* which is wrong should be rectified. Hence a correct symbolization is neither of the preceding, but rather

$$(x)[(x \text{ is wrong}) \supset (x \text{ should be rectified})]$$

The situation is more complicated, but no different in principle, when one quantifier occurs *within the scope of another quantifier*. Here the same warning against dangling or unquantified variables must be sounded. The proposition

> If something is missing, then if nobody calls the police, someone will be unhappy.

is properly symbolized as

$$(\exists x)(x \text{ is missing}) \supset \{(y)[(y \text{ is a person}) \supset \sim(y \text{ calls the police})] \supset (\exists z)[(z \text{ is a person}) \cdot (z \text{ will be unhappy})]\}$$

But the following proposition, which is superficially analogous to the preceding:

> If something is missing, then if nobody calls the police, it will not be recovered.

is *not* to be symbolized as

$$(\exists x)(x \text{ is missing}) \supset \{(y)[(y \text{ is a person}) \supset \sim(y \text{ calls the police})] \supset \sim(x \text{ will be recovered})\}$$

for the last occurrence of the variable 'x' is outside the scope of the initial quantifier and is left dangling. It cannot be corrected simply by rebracketing, as

$$(\exists x)\{(x \text{ is missing}) \supset \{(y)[(y \text{ is a person}) \supset \sim(y \text{ calls the police})] \supset \sim(x \text{ will be recovered})\}\}$$

for this expression fails equally to preserve the sense of the English sentence, in the same way as in the previous example. That sense is expressed by the formula

$$(x)\{(x \text{ is missing}) \supset \{(y)[(y \text{ is a person}) \supset \sim(y \text{ calls the police})] \supset \sim(x \text{ will be recovered})\}\}$$

which is, therefore, a correct symbolization of the given proposition.

EXERCISES

Symbolize each of the following propositions. In each case, use the suggested notation and make the symbolic formula parallel the English as closely as possible:

*1. If anything is damaged, someone will be blamed. (*Dx*: *x* is damaged. *Px*: *x* is a person. *Bx*: *x* will be blamed.)
2. If anything is damaged, the tenant will be charged for it. (*Dx*: *x* is damaged. *Cx*: *x* will be charged to the tenant.)
3. If nothing is damaged, nobody will be blamed. (*Dx*: *x* is damaged. *Px*: *x* is a person. *Bx*: *x* will be blamed.)
4. If something is damaged, but nobody is blamed, the tenant will not be charged for it. (*Dx*: *x* is damaged. *Px*: *x* is a person. *Bx*: *x* is blamed. *Cx*: *x* will be charged to the tenant.)
*5. If any bananas are yellow, they are ripe. (*Bx*: *x* is a banana. *Yx*: *x* is yellow. Rx: *x* is ripe.)
6. If any bananas are yellow, then some bananas are ripe. (*Bx*: *x* is a banana. *Yx*: *x* is yellow. *Rx*: *x* is ripe.)
7. If any bananas are yellow, then if all yellow bananas are ripe, they are ripe. (*Bx*: *x* is a banana. *Yx*: *x* is yellow. *Rx*: *x* is ripe.)
8. If all ripe bananas are yellow, some yellow things are ripe. (*Rx*: *x* is ripe. *Bx*: *x* is a banana. *Yx*: *x* is yellow.)
9. If all officers present are either captains or majors, then either some captains are present or some majors are present. (*Ox*: *x* is an officer. *Px*: *x* is present. *Cx*: *x* is a captain. *Mx*: *x* is a major.)
*10. If any officer is present, then either no majors are present or he is a major. (*Ox*: *x* is an officer. *Px*: *x* is present. *Mx*: *x* is a major.)
11. If some officers are present, then if all officers present are captains, then some captains are present. (*Ox*: *x* is an officer. *Px*: *x* is present. *Cx*: *x* is a captain.)
12. If some officers are present, then if all officers present are captains, then they are captains. (*Ox*: *x* is an officer. *Px*: *x* is present. *Cx*: *x* is a captain.)
13. If all survivors are fortunate and only women were survivors, then if there are any survivors, then some women are fortunate. (*Sx*: *x* is a survivor. *Fx*: *x* is fortunate. *Wx*: *x* is a woman.)
14. If any survivors are women, then if all women are fortunate, they are fortunate. (*Sx*: *x* is a survivor. *Wx*: *x* is a woman. *Fx*: *x* is fortunate.)
*15. If there are any survivors and only women are survivors, then they are women. (*Sx*: *x* is a survivor. *Wx*: *x* is a woman.)
16. If every position has a future and no employees are lazy, then some employees will be successful. (*Px*: *x* is a position. *Fx*: *x* has a future. *Ex*: *x* is an employee. *Lx*: *x* is lazy. *Sx*: *x* will be successful.)

17. If any employees are lazy, then if some positions have no future, then they will not be successful. (*Ex*: *x* is an employee. *Lx*: *x* is lazy. *Px*: *x* is a position. *Fx*: *x* has a future. *Sx*: *x* will be successful.)
18. If any employees are lazy and some positions have no future, then some employees will not be successful. (*Ex*: *x* is an employee. *Lx*: *x* is lazy. *Px*: *x* is a position. *Fx*: *x* has a future. *Sx*: *x* will be successful.)
19. If any husband is unsuccessful, then if all wives are ambitious, then some wives will be disappointed. (*Hx*: *x* is a husband. *Sx*: *x* is successful. *Wx*: *x* is a wife. *Ax*: *x* is ambitious. *Dx*: *x* will be disappointed.)
20. If any husband is unsuccessful, then if some wives are ambitious he will be unhappy. (*Hx*: *x* is a husband. *Sx*: *x* is successful. *Wx*: *x* is a wife. *Ax*: *x* is ambitious. *Ux*: *x* will be unhappy.)

4.5 Quantification Rules

1. **Inferences Involving Propositional Functions.** In constructing a formal proof of validity for a given argument, the premisses with which we begin and the conclusion with which we end are all propositions. But wherever the rules of Existential Instantiation or Universal Generalization are used, at least some of the intermediate lines must contain free variables and will, therefore, be propositional functions rather than propositions. Each line of a formal proof of validity must be either a premiss, or an assumption of limited scope, or follow validly from preceding lines by an elementary valid argument form accepted as a Rule of Inference, or follow from an immediately preceding sequence of lines by the principle of Conditional Proof. Three questions naturally arise at this point: In what sense can a propositional function be said to follow *validly* from other propositional functions? In what sense can a propositional function be said to follow *validly* from propositions? And in what sense can a proposition be said to follow *validly* from propositional functions?

To answer these questions, we shall introduce a revised, more general sense of the word 'valid'. Propositional functions contain free variables and are, therefore, neither true nor false. But a propositional function is made into a proposition by substituting constants for all of its free variables, and the resulting substitution instance is either true or false. One propositional function can be said to follow *validly*, as conclusion, from one or more other propositional functions, as premisses, when every substitution of constants for free occurrences of variables in it (the same constants being substituted for the same variables in both premisses and conclusion, of course) results in a valid argument. For example, the propositional function '*Gx*' follows validly from the propositional functions '$Fx \supset Gx$' and '*Fx*', because every substitution of a constant for '*x*' results in an argument of the form *Modus Ponens*. We may say of such an inference that it is valid by the principle of *Modus Ponens* despite the fact that propositional functions, rather than propositions, are involved. It should be clear that any inference is valid that proceeds by way of any of the nineteen Rules of Inference on our original list, regardless of whether the premisses and conclusion are propositions or propositional functions. In passing, we may note that this is so, even when the conclusion contains more free variables than are present in the premisses. For example, by the principle of Addition, we validly infer the propositional function of two variables '*Fx* v *Gy*' from the propositional function of one variable '*Fx*'.

The original list of nineteen Rules of Inference also permits the inference of propositional functions from propositions. By the principle of Addition, for example, we can infer the propositional function '*Fa* v *Gx*' from the proposition '*Fa*'. That such inferences as these are valid, in the new sense explained, is obvious. Moreover, propositions can validly be inferred from propositional functions by our Rules of Inference. By the principle of Simplification, for example, we can infer the proposition '*Fa*' from the propositional function '*Fa*·*Gx*'. Hence the letters '*p*', '*q*', '*r*', '*s*' in our nineteen Rules of Inference now range over, or represent, *either* propositions *or* propositional functions.

We can now adopt a more general definition of formal proof of validity, which parallels our earlier definition exactly except that lines of a proof can be either propositions or propositional functions. If each line after the initial premisses follows validly from preceding lines, in the generalized sense of 'valid' already explained, then the last line validly follows from the initial premisses. If our initial premisses and conclusion are propositions rather than propositional functions, then the conclusion validly follows from the initial premisses, in the original sense of 'valid' which applies to arguments whose premisses and conclusions are all statements or propositions. The following considerations demonstrate this conclusion. As we go from our original premisses to propositional functions—if we go validly—then, if the premisses are true, all substitution instances of the inferred propositional functions must be true also. And as we proceed from previously inferred propositional functions to other propositional functions, if we proceed validly, then all substitution instances of these other propositional functions must be true also. Finally, when we go from validly inferred propositional functions to the final conclusion, which is a proposition, then if we proceed validly, since all substitution instances of the former are true, the final conclusion must be true also.

The preceding remarks require some modification to take account of assumptions of limited scope that contain free variables, but the modifications are best introduced as alterations in the quantification rules themselves. The preliminary versions of our quantification rules must be replaced in any case, for as stated, they apply only to propositions and not to propositional functions. The two rules for generalizing, **UG** and **EG**, must now permit the quantification (or binding) of free variables. The instantiation rules **UI** and **EI**, on the other hand, must now permit the freeing of bound variables to permit the introduction of propositional functions themselves, rather than (pretended) substitution instances of them.

In our earlier discussion of propositional functions of the single variable '*x*', we introduced the Greek letters *phi* and *psi*, and let 'Φx' and 'Ψx' denote any propositional function of '*x*', such as '*Fx*', '*Gx*', '*Fx*·*Hx*', '(*Fx* v *Gx*) ⊃ *Hx*', . . . no matter how complicated such functions might be. It will be useful to continue to use these Greek letters: typically 'Φx' will denote any propositional function that contains at least one free occurrence of the variable 'x', even including those propositional functions that contain free occurrences of other variables. Thus 'Φx' may denote any of the following:

$$Fx, \qquad Fx \vee Gx, \qquad Ga \supset Hx, \qquad Fw \cdot Fx, \qquad (\exists z)(Gz \equiv Hx), \ldots$$

Similarly, 'Φy' may denote any of the propositional functions

$$Fy, \qquad Fy \vee Gy, \qquad Ga \supset Hy, \qquad Fw \cdot Fy, \qquad (\exists z)(Gz \equiv Hy), \ldots$$

To be able to refer to *any* propositional function in *either* of the preceding groups, we shall introduce the Greek letters *mu* and *nu* ('μ' and 'ν') to denote individual symbols. Thus '$\Phi\mu$' may denote any of the preceding propositional functions of either 'x' or 'y', according as 'μ' is taken to denote 'x' or 'y'. Similarly, according as 'μ' is taken to denote 'x' or 'y', '$(\mu)(\Phi\mu)$' will denote the universal quantification with respect to 'x' or 'y' of any of the preceding propositional functions of 'x' or 'y'. The existential quantification will be denoted by '$(\exists\mu)(\Phi\mu)$'.

For convenience, the term '$\Phi\mu$' will also denote either a proposition or a propositional function that contains *no* free occurrence of the variable represented by 'μ'. In such a case, $(\mu)(\Phi\mu)$ and $(\exists\mu)(\Phi\mu)$ will be called 'vacuous' quantifications, and will be equivalent to each other and to $\Phi\mu$ itself. This somewhat unnatural notation is included only for the sake of the completeness described and proved for the axiomatic development of Quantification Theory that will be presented in Chapter 10. We shall make no use of it until that time.

The Greek letters *phi* and *psi* may also be used with an individual constant to denote either propositions or propositional functions containing that constant. Thus 'Φa' may symbolize any of the expressions

$$Fa, \qquad Fa \vee Ga, \qquad Gc \supset Ha, \qquad Fw \cdot Fa, \qquad (\exists z)(Gz \equiv Ha), \ldots$$

and 'Φb' may denote any of the expressions

$$Fb, \qquad Fb \vee Gb, \qquad Gc \supset Hb, \qquad Fw \cdot Fb, \qquad (\exists z)(Gz \equiv Hb), \ldots$$

By the convention to be introduced immediately below, '$\Phi\nu$' may denote any expression of the two preceding groups according as 'ν' is taken to denote 'a' or 'b'. This notation will be useful in reformulating our quantification rules.

It will be convenient to establish two definite conventions governing the expressions '$\Phi\mu$' and '$\Phi\nu$', so that each expression may be used in the same sense in the statements of all four quantification rules. The first convention is that *mu* ('μ') denotes individual variables exclusively, whereas *nu* ('ν') can denote either an individual variable or an individual constant. The second convention is this:

The expression '$\Phi\mu$' denotes any proposition or propositional function. The expression '$\Phi\nu$' denotes the result of replacing every free occurrence of μ in $\Phi\mu$ by ν, provided that if ν is a variable, it must occur free in $\Phi\nu$ at all places that μ occurs free in $\Phi\mu$. (If $\Phi\mu$ contains no free occurrence of μ, then $\Phi\nu$ and $\Phi\mu$ are identical. The variables ν and μ may, of course, be the same. If they are, in this case too, $\Phi\nu$ and $\Phi\mu$ are identical.)

This general convention helps to prevent unwanted (that is, invalid) inferences from being allowed by our four quantification rules. How it contributes to this end will be explained after formulating each of the four rules.

EXERCISES

I. For each formula $\Phi\mu$ (where μ is 'x') in the left-hand column, indicate which formulas in the right-hand column are legitimate instances of a $\Phi\nu$ for the given $\Phi\mu$:

*1. $Fx \lor Gx$
2. $Fx \lor Gy$
3. $Fx \supset (\exists y)Gy$
4. $(y)(Fx \supset Gy)$
*5. $Fx \equiv (\exists y)(Gx \lor Hy)$
6. $(x)Fx \supset (\exists y)(Gx \cdot Hy)$

a. $Fa \lor Gy$
b. $(y)(Fy \supset Gy)$
c. $Fa \cdot Gx$
d. $Fb \lor Gb$
e. $Fx \supset (\exists y)Gy$
f. $Fy \lor Gy$
g. $(y)(Fa \supset Gy)$
h. $Fy \supset (\exists y)Gy$
i. $(y)(Fz \supset Gy)$
j. $Fz \equiv (\exists y)(Gx \cdot Hy)$
k. $(x)Fx \supset (\exists y)(Gz \cdot Hy)$
l. $Fa \equiv (\exists y)(Gc \lor Hy)$
m. $(x)(Fx \supset Gy)$

II. For each formula $\Phi\nu$ (where ν is 'y') in the left-hand column, indicate which formulas in the right-hand column are legitimate instances of a $\Phi\mu$ for the given $\Phi\nu$:

*1. Fy
2. $Fx \cdot Gy$
3. $Fy \cdot Gy$
4. $Fy \lor (\exists x)(Gx \supset Hy)$
*5. $Fy \lor (\exists y)(Gc \lor Hy)$
6. $Fy \supset (\exists y)(Gy \cdot Hy)$

a. Fa
b. Fx
c. Fy
d. $Fx \cdot Gx$
e. $Fx \cdot Gz$
f. $Fx \lor (\exists x)(Gx \supset Hx)$
g. $Fx \lor (\exists y)(Gc \lor Hy)$
h. $Fx \supset (\exists y)(Gy \cdot Hx)$
i. $Fz \lor (\exists x)(Gx \supset Hz)$
j. $Fx \supset (\exists y)(Gy \cdot Hy)$
k. $Fz \lor (\exists x)(Gx \supset Hy)$
l. $Fx \supset (\exists y)(Gy \cdot Hz)$
m. $Fy \lor (\exists x)(Gx \supset Hy)$

2. Universal Instantiation. The presentation of our quantification rules will be accompanied by examples of valid arguments which they must permit, and also by examples of invalid arguments to be avoided by restrictions placed on those rules. The following inferences are clearly valid:

$$\frac{(x)Fx}{\therefore Fa}, \quad \frac{(y)(Fy \vee Gb)}{\therefore Fa \vee Gb}, \quad \frac{(z)(Fz \supset Gb)}{\therefore Fb \supset Gb}, \quad \frac{(x)(Fx \equiv Gy)}{\therefore Fc \equiv Gy}, \quad \frac{(x)[Fx \cdot (\exists x)(Gx \cdot Hy)]}{\therefore Fb \cdot (\exists x)(Gx \cdot Hy)}$$

They may generally be described as possessing the form

$$\frac{(\mu)(\Phi\mu)}{\therefore \Phi\nu}$$

where μ is an individual variable, ν is an individual constant, and $\Phi\nu$ results from $\Phi\mu$ by replacing all free occurrences of μ in $\Phi\mu$ by ν. Of course, there can be no free occurrence of μ in $(\mu)(\Phi\mu)$, but there may be any number of free occurrences of μ in $\Phi\mu$. On the other hand, not every occurrence of μ in $\Phi\mu$ need be free: for example, where 'μ' denotes 'x' and '$\Phi\mu$' denotes '$Fx \supset (\exists x)(Gx \vee Hy)$', only the first occurrence of μ in $\Phi\mu$ is free. This is the case because the second occurrence is part of the existential quantifier $(\exists\mu)$ and the third is within the scope of that quantifier.

Also valid are such inferences as

$$\frac{(x)Fx}{\therefore Fy}, \quad \frac{(x)Fx}{\therefore Fx}, \quad \frac{(y)(Fy \vee Gb)}{\therefore Fx \vee Gb}, \quad \frac{(z)(Fz \supset Gx)}{\therefore Fx \supset Gx}, \quad \frac{(x)[Fx \cdot (\exists x)(Gx \cdot Hy)]}{\therefore Fz \cdot (\exists x)(Gx \cdot Hy)}, \ldots$$

which are also of the form

$$\frac{(\mu)(\Phi\mu)}{\therefore \Phi\nu}$$

except that in these instances *both* μ and ν are individual variables. Here, the premiss $(\mu)(\Phi\mu)$ may be a proposition, but the conclusion $\Phi\nu$ must be a propositional function.

Just as we count valid the inference

$$\frac{(z)(Fz \vee Gb)}{\therefore Fb \vee Gb}$$

where the instantiating constant 'b' occurs in the premiss as well as in the conclusion, so we wish to count as valid such inferences as

$$\frac{(x)(Fx \vee Gy)}{\therefore Fy \vee Gy}, \quad \frac{(y)(Fx \supset Gy)}{\therefore Fx \supset Gx}, \quad \frac{(z)[Fx \supset (Gy \cdot Hz)]}{\therefore Fx \supset (Gy \cdot Hy)}, \ldots$$

in which the instantiating *variable* occurs free in the premiss as well as in the conclusion. In general, when $\Phi\nu$ is inferred from $(\mu)(\Phi\mu)$ validly, ν must occur free in $\Phi\nu$ at every place where μ occurs free in $\Phi\mu$. There *may* be more free occurrences of ν in $\Phi\nu$ than there are free occurrences of μ in $\Phi\mu$, however. There will be more occurrences whenever ν occurs free in $\Phi\mu$. All the preceding inferences must be legitimized by our principle of Universal Instantiation.

Our first Quantification Rule, Universal Instantiation, is stated as

$$\textbf{UI: } \frac{(\mu)(\Phi\mu)}{\therefore \Phi\nu}$$

The general convention governing $\Phi\mu$ and $\Phi\nu$ serves to prevent such an erroneous inference as

$$\frac{(x)[(\exists y)(Fx \equiv \sim Fy)]}{\therefore (\exists y)(Fy \equiv \sim Fy)}$$

from being allowed by **UI**, because ν ('y') does not occur free in $\Phi\nu$ ('$(\exists y)(Fy \equiv \sim Fy)$') at all places that μ ('x') occurs free in $\Phi\mu$ ('$(\exists y)(Fx \equiv \sim Fy)$'). Hence '$(\exists y)(Fy \equiv \sim Fy)$' is not a legitimate $\Phi\nu$ for use in applying **UI** where $(\mu)(\Phi\mu)$ is '$(x)[(\exists y)(Fx \equiv \sim Fy)]$'. It should be obvious that the inference is invalid because it fails for a model or possible universe containing some things that are *F* and some things that are not *F*, which would make the premiss true, whereas the conclusion, being self-contradictory, is false for any model or possible universe.

3. Existential Generalization. Turning now to Existential Generalization, we observe that all the following are valid inferences:

$$\frac{Fa}{\therefore (\exists x)Fx}, \frac{Fa}{\therefore (\exists y)Fy}, \frac{Fa \vee Gb}{\therefore (\exists x)(Fx \vee Gb)}, \frac{Fa \supset Gb}{\therefore (\exists y)(Fa \supset Gy)}, \ldots$$

They may generally be described as possessing the form

$$\frac{\Phi\nu}{\therefore (\exists\mu)(\Phi\mu)}$$

Here both premisses and conclusions are propositions. Also valid are such inferences involving propositional functions as

$$\frac{Fx}{\therefore (\exists y)Fy}, \frac{Fa \vee Gy}{\therefore (\exists x)(Fa \vee Gx)}, \frac{Fx \supset Gy}{\therefore (\exists y)(Fx \supset Gy)}, \frac{Fx \cdot Gx}{\therefore (\exists y)(Fy \cdot Gx)}, \ldots$$

which are of the same pattern as the preceding, except that here ν is a variable instead of a constant.

Our second Quantification Rule, Existential Generalization, is stated as

$$\textbf{EG: } \frac{\Phi\nu}{\therefore (\exists\mu)(\Phi\mu)}$$

The general convention governing $\Phi\mu$ and $\Phi\nu$ serves to prevent such an erroneous inference as

$$\frac{Fx \equiv {\sim}Fy}{\therefore (\exists x)(Fx \equiv {\sim}Fx)}$$

from being allowed by **EG**, because ν ('y') does not occur free in $\Phi\nu$ ('$Fx \equiv {\sim}Fy$') at all places that μ ('x') occurs free in $\Phi\mu$ ('$Fx \equiv {\sim}Fx$'). Hence '$Fx \equiv {\sim}Fy$' is not a legitimate $\Phi\nu$ from which '$(\exists x)(Fx \equiv {\sim}Fx)$' can be obtained as $(\exists\mu)(\Phi\mu)$ by **EG**. It should be obvious that the inference is invalid. Even though the premiss has substitution instances that are true, the conclusion is self-contradictory, hence false.

With both **UI** and **EG** available, we may illustrate their use in proving the validity of the argument

All humans are mortal.

Therefore, if Socrates is human, then some humans are mortal.

by the following conditional proof:

1.	$(x)(Hx \supset Mx)$	$/\therefore Hs \supset (\exists x)(Hx \cdot Mx)$
→ 2.	Hs	
3.	$Hs \supset Ms$	1, **UI**
4.	Ms	3, 2, M.P.
5.	$Hs \cdot Ms$	2, 4, Conj.
6.	$(\exists x)(Hx \cdot Mx)$	5, **EG**
7.	$Hs \supset (\exists x)(Hx \cdot Mx)$	2–6, C.P.

A simple pattern of deduction suffices to establish the validity of any argument of the form

$$\frac{(\mu)(\Phi\mu)}{\therefore (\exists\mu)(\Phi\mu)}$$

Here the pattern of proof is

1.	$(\mu)(\Phi\mu)$	$/\therefore (\exists\mu)(\Phi\mu)$
2.	$\Phi\nu$	1, **UI**
3.	$(\exists\mu)(\Phi\mu)$	2, **EG**

4. Existential Instantiation. Before we discuss our new formulation of Existential Instantiation, we shall establish the logical truth of equivalences of the form

$$\text{(E)} \qquad (\nu)(\Phi\nu \supset p) \equiv [(\exists\mu)(\Phi\mu) \supset p]$$

where ν occurs free in $\Phi\nu$ at all *and only* those places that μ occurs free in $\Phi\mu$, and where p contains no free occurrence of the variable ν. In case p is a true proposition, or for any true substitution instance of p, if p is a propositional function, both sides of the equivalence must be true. This is so because if p is true, then $\Phi\nu \supset p$ is true for every value of ν, whence $(\nu)(\Phi\nu \supset p)$ is true. The truth of $(\exists\mu)(\Phi\mu) \supset p$ also follows immediately from the truth of p. In case p is a false proposition, or for any false substitution instance of p if p is a propositional function, and $(\nu)(\Phi\nu \supset p)$ is true, every substitution instance of $\Phi\nu \supset p$ is true and so every substitution instance of $\Phi\nu$ must be false, whence $(\exists\mu)(\Phi\mu)$ must be false and so $(\exists\mu)(\Phi\mu) \supset p$ is true. In case p is a false proposition, or for any false substitution instance of p if p is a propositional function, and $(\exists\mu)(\Phi\mu) \supset p$ is true, $(\exists\mu)(\Phi\mu)$ must be false. Thus every substitution instance of $\Phi\nu$ must be false, whence every substitution instance of $\Phi\nu \supset p$ must be true. Therefore, $(\nu)(\Phi\nu \supset p)$ is true also. This argument establishes the truth of every equivalence of the form (E) because if p is true, both sides are true. If p is false, each side implies the other.

Turning now to Existential Instantiation, we want to permit going from $(\exists x)Fx$ to Fx or Fy only under very stringent restrictions. Not only must the instantiating variable not have any prior free occurrence (as discussed on pages 74–75), but we must not permit $(x)Fx$ to be inferred from $(\exists x)Fx$ by way of Existential Instantiation and Universal Generalization. There are many ways of imposing such restrictions. One method is to formulate the rule of Existential Instantiation in such a way that the formula or line *finally* inferred by its means contains no free variables that are introduced by it. The feasibility of this procedure can be seen by the following considerations.

Here, as in earlier sections, we are concerned with constructing proofs of validity only for arguments whose premisses and conclusions are propositions. We are not concerned with propositional functions that contain free variables. Hence we never end a proof with a propositional function that contains a free variable. Therefore, in any proof of validity in which the rule of Existential Instantiation involves going from $(\exists\mu)(\Phi\mu)$ to $\Phi\nu$, the propositional function $\Phi\nu$ serves only to permit the subsequent inference of a formula that does not contain any free occurrence of the variable ν.

Suppose we already have $(\exists\mu)(\Phi\mu)$ as a line in a proof. Suppose further that in the presence of other lines already obtained, if we also had $\Phi\nu$, we would be able to derive a desired formula p that contains no free occurrence of the variable ν. We can proceed by writing down $\Phi\nu$ as an assumption of limited scope. Then, after p has been derived, we can close the scope of the assumption and infer the formula $\Phi\nu \supset p$ by the strengthened Rule of Conditional Proof.

From this line (subject to reasonable restrictions), the formula $(\nu)(\Phi\nu \supset p)$ can be inferred by Universal Generalization. And from the latter formula, by equivalence (E), we can infer $(\exists\mu)(\Phi\mu) \supset p$. Now, from this formula and the earlier line $(\exists\mu)(\Phi\mu)$, we can obtain p by *Modus Ponens*. This whole process can be represented schematically as

i.	$(\exists\mu)(\Phi\mu)$	
→ j.	$\Phi\nu$	(where ν is a variable and has no free occurrence in any preceding line)
	.	
	.	
	.	
k.	p	(where p contains no free occurrence of ν)
k+1.	$\Phi\nu \supset p$	j–k, C.P.
k+2.	$(\nu)(\Phi\nu \supset p)$	k+1, **UG**
k+3.	$(\exists\mu)(\Phi\mu) \supset p$	k+2, Equivalence (E)
k+4.	p	k+3, i, M.P.

The preceding discussion can be regarded as providing an informal justification for the rule of Existential Instantiation, which we now state as

EI: $(\exists\mu)(\Phi\mu)$
→ $\Phi\nu$
.
.
.
p
$\therefore p$

provided that ν is a variable and does not occur free either in p or in any line preceding $\Phi\nu$.

Before discussing the restrictions in the statement of **EI**, it may be helpful to present a proof of validity that makes use of the new rule:

1.	$(x)(Fx \supset Gx)$	
2.	$(\exists y)Fy$	$/\therefore (\exists z)Gz$
→ 3.	Fu	
4.	$Fu \supset Gu$	1, **UI**
5.	Gu	4, 3, M.P.
6.	$(\exists z)Gz$	5, **EG**
7.	$(\exists z)Gz$	2, 3–6, **EI**

There can be, though there need not be, other lines intervening between the premiss $(\exists\mu)(\Phi\mu)$ (line 2) and the propositional function $\Phi\nu$ (line 3) which is marked as an assumption of limited scope. When the desired formula p (line 6) is reached, the next line is simply p again. The justification written next to this line is the number of the line consisting of $(\exists\mu)(\Phi\mu)$, the hyphenated numbers of the lines consisting of $\Phi\nu$ and the first line consisting of p, and, finally, the notation **EI**.

The general convention that governs $\Phi\mu$ and $\Phi\nu$ serves to prevent such an erroneous "proof" as

1.	$(\exists x)(Fx \cdot Gx)$	$/\therefore (x)Fx$
→ 2.	$Fx \cdot Gy$	(*wrong* as part of **EI**)
3.	Fx	2, Simp.
4.	Fx	1, 2–3, **EI** (*wrong*)
5.	$(x)Fx$	4, **UG**

from being allowed by **EI**. This is so because ν ('y') does not occur free in $\Phi\nu$ ('$Fx \cdot Gy$') at all places that μ ('x') occurs free in $\Phi\mu$ ('$Fx \cdot Gx$'). Hence '$Fx \cdot Gy$' is not a legitimate $\Phi\nu$ for use in applying **EI**, where $(\exists\mu)(\Phi\mu)$ is '$(\exists x)(Fx \cdot Gx)$'. It should be obvious that the argument is invalid. That something is both an F and a G clearly does not entail that everything is an F.

The restriction that ν does not occur free in any line preceding $\Phi\nu$, ensures that ν does not occur free in $(\exists\mu)(\Phi\mu)$. Therefore, if ν is different from μ, ν does not occur free in $\Phi\mu$ either. The general convention already assures us that there are no free occurrences of μ in $\Phi\nu$. Thus the present restriction on **EI**, together with the general convention, entails the restriction associated with the logical equivalence (E), namely, that ν occurs free in $\Phi\nu$ at all *and only* those places that μ occurs free in $\Phi\mu$.

The restriction that ν does not occur free in p serves to prevent any subsequent and erroneous use of Universal Generalization to derive $(\mu)(\Phi\mu)$ as conclusion, from $(\exists\mu)(\Phi\mu)$, as premiss. Without the restriction, such an error would be possible because if p were allowed to contain a free occurrence of ν, it could be $\Phi\nu$ itself, from which **UG** could be used to derive the conclusion $(\mu)(\Phi\mu)$.

The restriction that ν does not occur free in any line preceding $\Phi\nu$ serves to prevent such an erroneous "proof" as

1.	$(x)(\exists y)(Fx \equiv {\sim}Fy)$	$/\therefore (\exists x)(Fx \equiv {\sim}Fx)$
2.	$(\exists y)(Fx \equiv {\sim}Fy)$	1, **UI**
→ 3.	$Fx \equiv {\sim}Fx$	(*wrong* as part of **EI**)
4.	$(\exists x)(Fx \equiv {\sim}Fx)$	3, **EG**
5.	$(\exists x)(Fx \equiv {\sim}Fx)$	2, 3–4, **EI** (*wrong*)

from being allowed by **EI**. This is so because ν ('x') occurs free in line 2, which precedes $\Phi\nu$ ('$Fx \equiv {\sim}Fx$') in line 3. Hence '$Fx \equiv {\sim}Fx$' is not a legitimate $\Phi\nu$ for the given $\Phi\mu$ for use in applying **EI**, where $(\exists\mu)(\Phi\mu)$ is '$(\exists y)(Fx \equiv {\sim}Fy)$'. It should be obvious that the argument is invalid, since it fails for a model containing some things that are F and some things that are not F, which would make the premiss true, whereas the conclusion, being self-contradictory, is false for every model.

5. Universal Generalization. The complicated and highly restricted formulation of **EI** just given permits a somewhat less restricted formulation of the rule of Universal Generalization, which we now state as

$$\mathbf{UG}: \frac{\Phi\nu}{\therefore (\mu)(\Phi\mu)}$$

provided that ν is a variable and does not occur free either in $(\mu)(\Phi\mu)$ or in any assumption within whose scope $\Phi\nu$ lies.

The general convention governing $\Phi\mu$ and $\Phi\nu$ serves to prevent such an erroneous "proof" as

1.	$(\exists x)(y)(Fx \supset {\sim}Fy)$	$/ \therefore (x)(Fx \supset {\sim}Fx)$
→ 2.	$(y)(Fx \supset {\sim}Fy)$	
3.	$Fx \supset {\sim}Fy$	2, **UI**
4.	$(x)(Fx \supset {\sim}Fx)$	3, **UG** (*wrong*)
5.	$(x)(Fx \supset {\sim}Fx)$	1, 2–4, **EI**

from being allowed by **UG**, because ν ('y') does not occur free in $\Phi\nu$ ('$Fx \supset {\sim}Fy$') at all places that μ ('x') occurs free in $\Phi\mu$ ('$Fx \supset {\sim}Fx$'). Hence line 3, '$Fx \supset {\sim}Fy$', is not a legitimate $\Phi\nu$, for the given $\Phi\mu$, from which line 4, '$(x)(Fx \supset {\sim}Fx)$', can be obtained as $(\mu)(\Phi\mu)$ by **UG**. It should be obvious that the argument is invalid, for its premiss is true if there is at least one thing that is not an F, whereas its conclusion states that there are no F's at all.

The restriction that ν does not occur free in $(\mu)(\Phi\mu)$ serves to prevent such an erroneous "proof" as

1.	$(x)(Fx \equiv Fx)$	$/ \therefore (x)(y)(Fx \equiv Fy)$
2.	$Fx \equiv Fx$	1, **UI**
3.	$(y)(Fx \equiv Fy)$	2, **UG** (*wrong*)
4.	$(x)(y)(Fx \equiv Fy)$	3, **UG**

from being allowed by **UG**, because ν ('x') does occur free in $(\mu)(\Phi\mu)$ at line 3, ('$(y)(Fx \equiv Fy)$'). Hence line 2, '$Fx \equiv Fx$', is not a legitimate $\Phi\nu$ for the given $\Phi\mu$, from which '$(y)(Fx \equiv Fy)$' can be derived as $(\mu)(\Phi\mu)$ by **UG**. It should be obvious that the argument is invalid, for its premiss states only that any thing is an F if and only if it is an F, whereas its conclusion states that of any things x and y, x is an F if and only if y is an F also.

The restriction that ν does not occur free in any assumption within whose scope $\Phi\nu$ lies, serves to prevent such an erroneous "proof" as

	1. $(\exists x)Fx$	$/ \therefore (x)Fx$
→	2. Fy	
	3. $(x)Fx$	2, **UG** (*wrong*)
	4. $(x)Fx$	1, 2–3, **EI**

from being allowed by **UG**, because ν ('y') occurs free in the assumption 'Fy' within whose scope the premiss $\Phi\nu$ ('Fy') lies. Hence, in this case, 'Fy' is not a legitimate $\Phi\nu$ for the given $\Phi\mu$, from which '$(x)Fx$' can be derived as $(\mu)(\Phi\mu)$ by **UG**. The argument is, of course, obviously invalid.

Like their preliminary versions, these final versions of our four quantification rules are similar to the first nine Rules of Inference in that they can be applied only to whole lines in a proof.

EXERCISES

Identify and explain all of the mistakes in the following erroneous "proofs":

1.
→ 1. Fx
2. $(y)Fy$ — 1, **UG**
3. $Fx \supset (y)Fy$ — 1–2, C.P.
4. $(x)[Fx \supset (y)Fy]$ — 3, **UG**

2.
1. $(\exists x)(Fx \cdot Gx)$ $/ \therefore (\exists x)Fx$
→ 2. $Fx \cdot Gx$
3. Fx — 2, Simp.
4. Fx — 1, 2–3, **EI**
5. $(\exists x)Fx$ — 4, **EG**

***3.**
1. $(x)(\exists y)(Fx \equiv Gy)$ $/ \therefore (\exists y)(x)(Fx \equiv Gy)$
2. $(\exists y)(Fx \equiv Gy)$ — 1, **UI**
→ 3. $Fx \equiv Gy$
4. $(x)(Fx \equiv Gy)$ — 3, **UG**
5. $(\exists y)(x)(Fx \equiv Gy)$ — 4, **EG**
6. $(\exists y)(x)(Fx \equiv Gy)$ — 2, 3–5, **EI**

4.
1. $(x)(\exists y)(Fx \supset Gy)$ $/ \therefore (\exists y)(x)(Fx \supset Gy)$
2. $(\exists y)(Fx \supset Gy)$ — 1, **UI**
→ 3. $Fx \supset Gx$
4. $(x)(Fx \supset Gx)$ — 3, **UG**
5. $(\exists y)(x)(Fx \supset Gy)$ — 4, **EG**
6. $(\exists y)(x)(Fx \supset Gy)$ — 2, 3–5, **EI**

5.
1. $(y)(\exists x)(Fx \vee Gy)$ $/ \therefore (\exists x)(y)(Fx \vee Gy)$
2. $(\exists x)(Fx \vee Gy)$ — 1, **UI**
→ 3. $Fx \vee Gx$
4. $(y)(Fx \vee Gy)$ — 3, **UG**
5. $(\exists x)(y)(Fx \vee Gy)$ — 4, **EG**
6. $(\exists x)(y)(Fx \vee Gy)$ — 2, 3–5, **EI**

***6.**

1. $(\exists x)(y)[(Fx \cdot Gx) \supset Hy]$ $/\therefore (\exists x)[(Fx \cdot Gx) \supset Hx]$
2. $(y)[(Fz \cdot Gz) \supset Hy]$
3. $(Fz \cdot Gz) \supset Hy$ — 2, **UI**
4. $(\exists x)[(Fx \cdot Gz) \supset Hy]$ — 3, **EG**
5. $(y)(\exists x)[(Fx \cdot Gy) \supset Hy]$ — 4, **UG**
6. $(y)(\exists x)[(Fx \cdot Gy) \supset Hy]$ — 1, 2–5, **EI**
7. $(\exists x)[(Fx \cdot Gx) \supset Hx]$ — 6, **UI**

7.

1. $(\exists x)Fx$
2. $(\exists x)Gx$ $/\therefore (\exists x)(Fx \cdot Gx)$
3. Fy
4. Gy
5. $Fy \cdot Gy$ — 3, 4. Conj.
6. $(\exists x)(Fx \cdot Gx)$ — 5, **EG**
7. $(\exists x)(Fx \cdot Gx)$ — 2, 4–6, **EI**
8. $(\exists x)(Fx \cdot Gx)$ — 1, 3–7, **EI**

8.

1. $(\exists x)(\exists y)[(Fx \vee Gy) \cdot Hy]$ $/\therefore (x)(y)(Fy \vee Gx)$
2. $(\exists y)[(Fx \vee Gy) \cdot Hy]$
3. $(Fx \vee Gx) \cdot Hx$
4. $Fx \vee Gx$ — 3, Simp.
5. $Fx \vee Gx$ — 2, 3–4, **EI**
6. $Fx \vee Gx$ — 1, 2–5, **EI**
7. $(y)(Fy \vee Gx)$ — 6, **UG**
8. $(x)(y)(Fy \vee Gx)$ — 7, **UG**

***9.**

1. $(\exists x)(Fx \cdot Gx)$
2. $(\exists x)(\sim Fx \cdot Gx)$ $/\therefore (\exists x)(Fx \cdot \sim Fx)$
3. $Fx \cdot Gy$
4. Fx — 3, Simp.
5. Fx — 1, 3–4, **EI**
6. $\sim Fx \cdot Gx$
7. $\sim Fx$ — 6, Simp.
8. $\sim Fx$ — 2, 6–7, **EI**
9. $Fx \cdot \sim Fx$ — 5, 8, Conj.
10. $(\exists x)(Fx \cdot \sim Fx)$ — 9, **EG**

10.

1. $(x)[(Fx \supset Gx) \cdot \sim Ga]$ $/\therefore (x)\sim Fx$
2. $(x)[(Fx \supset Gx) \cdot \sim Gy]$
3. $(Fz \supset Gz) \cdot \sim Gy$ — 2, **UI**
4. $(y)[(Fy \supset Gy) \cdot \sim Gy]$ — 3, **UG**
5. $(Fu \supset Gu) \cdot \sim Gu$ — 4, **UI**
6. $Fu \supset Gu$ — 5, Simp.
7. $\sim Gu \cdot (Fu \supset Gu)$ — 5, Com.
8. $\sim Gu$ — 7, Simp.
9. $\sim Fu$ — 6, 8, M.T.
10. $(x)\sim Fx$ — 9, **UG**
11. $(x)[(Fx \supset Gx) \cdot \sim Gy] \supset (x)\sim Fx$ — 2–10, C.P.
12. $(w)\{(x)[(Fx \supset Gx) \cdot \sim Gw] \supset (x)\sim Fx\}$ — 11, **UG**
13. $(x)[(Fx \supset Gx) \cdot \sim Ga] \supset (x)\sim Fx$ — 12, **UI**
14. $(x)\sim Fx$ — 13, 1, M.P.

6. Shorter Proofs of Validity. At this stage of our work, it is desirable to cut down the length of our formal proofs of validity by permitting shortcuts in the application of the original list of Rules of Inference. We can combine any use of the principle of Double Negation with any other step, which will permit us to go directly from '$\sim A \supset B$' to '$A \vee B$' or vice versa, without having to write down the intermediate step '$\sim\sim A \vee B$'. We can short-cut tedious uses of the principle of Commutation by permitting not only '$A \therefore A \vee B$' by the principle of Addition, but also such inferences as '$A \therefore B \vee A$'. Also, we permit such inferences as '$A \vee B, \sim B \therefore A$' by the principle of the Disjunctive Syllogism as well as inferences such as '$A \vee B, \sim A \therefore B$'. Since the definition of Material Implication and the principle of Distribution can always be used to obtain '$A \supset (B \cdot C)$' from '$(A \supset B) \cdot (A \supset C)$', and vice versa, our demonstrations can be shortened by permitting Distribution to be applied directly to conditionals that have conjunctions as their consequents. This is tantamount to adding the form '$[p \supset (q \cdot r)] \equiv [(p \supset q) \cdot (p \supset r)]$' to our list as an alternative version of the principle of Distribution. By repeated application of the principles of Association and Commutation, we can rearrange the terms of any conjunction or disjunction in any way we please. Hence we can shorten our demonstrations by omitting parentheses, brackets, and so forth, from conjunctions of any three or more propositions. Thus such propositions as '$A \cdot \{B \cdot [C \cdot (D \cdot E)]\}$', '$(A \cdot B) \cdot [C \cdot (D \cdot E)]$', '$(A \cdot B) \cdot [(C \cdot D) \cdot E]$', '$[(A \cdot B) \cdot C] \cdot (D \cdot E)$', '$[A \cdot (B \cdot C)] \cdot (D \cdot E)$', '$\{[(A \cdot B) \cdot C] \cdot D\} \cdot E$', '$A \cdot [(B \cdot C) \cdot (D \cdot E)]$', . . . will all be written indifferently as '$A \cdot B \cdot C \cdot D \cdot E$', and *any* permutation will be justified simply by the principle of Commutation. Moreover, if we wish to infer the conjunction of *some* of the indicated conjuncts, in any order, we can do so in one step and justify it by the principle of Simplification. Thus from '$A \cdot B \cdot C \cdot D \cdot E$', we can infer '$E \cdot B \cdot D$' in a single step. Also, the principle of Conjunction can be applied to any number of lines to produce a new line that is the conjunction of all of them. Finally, we shall permit the telescoping of the rules of Material Implication, De Morgan, and Double Negation to obtain '$\sim(A \supset B)$' from '$A \cdot \sim B$', and vice versa, which amounts to adding the form '$\sim(p \supset q) \equiv (p \cdot \sim q)$' to our list as an alternative version of Material Implication.

In proving the validity of some arguments, all four of our quantification rules must be used. Consider the following moderately complex argument:

> If all drugs are contaminated, then all negligent technicians are scoundrels. If there are any drugs that are contaminated, then all drugs are contaminated and unsafe. All germicides are drugs. Only the negligent are absent-minded. Therefore, if any technician is absent-minded, then if some germicides are contaminated, then he is a scoundrel.

By the use of fairly obvious abbreviations, it may be symbolized and proved valid as follows:

1.	$(x)(Dx \supset Cx) \supset (y)[(Ny \cdot Ty) \supset Sy]$	
2.	$(\exists x)(Dx \cdot Cx) \supset (y)[Dy \supset (Cy \cdot Uy)]$	
3.	$(x)(Gx \supset Dx)$	
4.	$(x)(Ax \supset Nx) \quad /\therefore (x)\{(Tx \cdot Ax) \supset [(\exists y)(Gy \cdot Cy) \supset Sx]\}$	
5.	$Tu \cdot Au$	
6.	$(\exists y)(Gy \cdot Cy)$	
7.	$Gw \cdot Cw$	
8.	$Gw \supset Dw$	3, **UI**
9.	Gw	7, Simp.
10.	Dw	8, 9, M.P.
11.	Cw	7, Simp.
12.	$Dw \cdot Cw$	10, 11, Conj.
13.	$(\exists x)(Dx \cdot Cx)$	12, **EG**
14.	$(y)[Dy \supset (Cy \cdot Uy)]$	2, 13, M.P.
15.	$Dz \supset (Cz \cdot Uz)$	14, **UI**
16.	$(Dz \supset Cz) \cdot (Dz \supset Uz)$	15, Dist.
17.	$Dz \supset Cz$	16, Simp.
18.	$(x)(Dx \supset Cx)$	17, **UG**
19.	$(y)[(Ny \cdot Ty) \supset Sy]$	1, 18, M.P.
20.	$(Nu \cdot Tu) \supset Su$	19, **UI**
21.	Au	5, Simp.
22.	$Au \supset Nu$	4, **UI**
23.	Nu	22, 21, M.P.
24.	Tu	5, Simp.
25.	$Nu \cdot Tu$	23, 24, Conj.
26.	Su	20, 25, M.P.
27.	Su	6, 7–26, **EI**
28.	$(\exists y)(Gy \cdot Cy) \supset Su$	6–27, C.P.
29.	$(Tu \cdot Au) \supset [(\exists y)(Gy \cdot Cy) \supset Su]$	5–28, C.P.
30.	$(x)\{(Tx \cdot Ax) \supset [(\exists y)(Gy \cdot Cy) \supset Sx]\}$	29, **UG**

EXERCISES

I. Construct a formal proof of validity for each of the following arguments:

***1.** $(x)(Ax \supset Bx)$
$\therefore (x)(Bx \supset Cx) \supset (Ak \supset Ck)$

2. $(x)(Dx \supset Ex)$
$\therefore Da \supset [(y)(Ey \supset Fy) \supset Fa]$

3. $(x)[Gx \supset (y)(Hy \supset Iy)]$
$\therefore (x)Gx \supset (y)(Hy \supset Iy)$

4. $(\exists x)Jx \supset (\exists y)Ky$
$\therefore (\exists x)[Jx \supset (\exists y)Ky]$

***5.** $(\exists x)Lx \supset (y)My$
$\therefore (x)[Lx \supset (y)My]$

6. $(x)(Nx \supset Ox)$
$\therefore (x)\{Px \supset [(y)(Py \supset Ny) \supset Ox]\}$

7. $(x)(Qx \supset Rx)$
$(x)(Sx \supset Tx)$
$\therefore (x)(Rx \supset Sx) \supset (y)(Qy \supset Ty)$

8. $(\exists x)Ux \supset (y)[(Uy \vee Vy) \supset Wy]$
$(\exists x)Ux \cdot (\exists x)Wx$
$\therefore (\exists x)(Ux \cdot Wx)$

9. $(\exists x)Xx \supset (y)(Yy \supset Zy)$
   $\therefore (\exists x)(Xx \cdot Yx) \supset (\exists y)(Xy \cdot Zy)$

*10. $(\exists x)Ax \supset (y)(By \supset Cy)$
   $(\exists x)Dx \supset (\exists y)By$
   $\therefore (\exists x)(Ax \cdot Dx) \supset (\exists y)Cy$

11. $(x)(\exists y)(Ex \vee Fy)$
   $\therefore (x)Ex \vee (\exists y)Fy$

12. $(\exists x)Gx \vee (y)(Gy \supset Hy)$
   $(x)(Ix \supset \sim Gx)$
   $\therefore (x)(Gx \supset Ix) \supset (y)(Gy \supset Hy)$

13. $(\exists x)Jx \vee (\exists y)Ky$
   $(x)(Jx \supset Kx)$
   $\therefore (\exists y)Ky$

14. $(x)(Lx \supset Mx)$
   $(x)(Mx \supset Nx)$
   $\therefore (\exists x)Lx \supset (\exists y)Ny$

*15. $(x)\{Ox \supset [(y)(Py \supset Qy) \supset Rx]\}$
   $(x)\{Rx \supset [(y)(Py \supset Sy) \supset Tx]\}$
   $\therefore (y)[Py \supset (Qy \cdot Sy)] \supset (x)(Ox \supset Tx)$

16. $(\exists x)[Ux \cdot (y)(Vy \supset Wy)]$
   $(x)\{Ux \supset [(\exists y)(Xy \cdot Wy) \supset Yx]\}$
   $\therefore (\exists y)(Xy \cdot Vy) \supset (\exists x)Yx$

17. $(x)\{Ax \supset [(\exists y)By \supset Cx]\}$
   $(x)\{Cx \supset [(\exists y)Dy \supset Ex]\}$
   $\therefore (\exists x)(Bx \cdot Fx) \supset [(y)(Fy \supset Dy) \supset (z)(Az \supset Ez)]$

18. $(x)(\exists y)(Gx \cdot Hy)$
   $\therefore (x)Gx \cdot (\exists y)Hy$

19. $(\exists x)(y)(Ix \equiv Jy)$
   $\therefore (y)(\exists x)(Ix \equiv Jy)$

20. $(x)(\exists y)(Kx \cdot Ly)$
   $\therefore (\exists y)(x)(Kx \cdot Ly)$

II. Construct a formal proof of validity for each of the following arguments. In each case use the suggested notation and make the symbolic formulas parallel the English as closely as possible:

1. No acrobats are clumsy. Therefore, if Al is a waiter, then if all waiters are clumsy, Al is not an acrobat. (*Ax*, *Cx*, *Wx*, *a*.)
2. All pets are gentle. Therefore, if any dogs are excitable and no excitable dogs are gentle, they are not pets. (*Px*, *Gx*, *Dx*, *Ex*.)
3. All the accused are guilty. All who are convicted will hang. Therefore, if all who are guilty are convicted, then all the accused will hang. (*Ax*, *Gx*, *Cx*, *Hx*.)
*4. If there are any geniuses, then all great composers are geniuses. If anyone is temperamental, all geniuses are temperamental. Therefore, if anyone is a temperamental genius, then all great composers are temperamental. (*Gx*: *x* is a genius. *Cx*: *x* is a great composer. *Px*: *x* is a person. *Tx*: *x* is temperamental.)
5. Any car with good brakes is safe to drive and safe to ride in. So, if a car is new, then if all new cars have good brakes, it is safe to drive. (*Cx*: *x* is a car. *Bx*: *x* has good brakes. *Dx*: *x* is safe to drive. *Rx*: *x* is safe to ride in. *Nx*: *x* is new.)
6. Either all the guests enjoyed themselves or some of the guests concealed their real feelings. No honest person would conceal his real feelings. Therefore, if the guests were all honest persons, then all the guests enjoyed themselves. (*Gx*: *x* is a guest. *Ex*: *x* enjoyed himself. *Cx*: *x* conceals his real feelings. *Hx*: *x* is honest. *Px*: *x* is a person.)
7. Any businessman who is a poet must be a wealthy man. Wealthy men are all

conservatives. If some conservative does not like poetry, then no poets are conservatives. Therefore, if there is a wealthy man who does not like poetry, then no businessmen are poets. (*Bx*: *x* is a businessman. *Px*: *x* is a poet. *Wx*: *x* is a wealthy man. *Cx*: *x* is a conservative. *Lx*: *x* likes poetry.)

*8. All radioactive substances either have a very short life or have medical value. No uranium isotope that is radioactive has a very short life. Therefore, if all uranium isotopes are radioactive, then all uranium isotopes have medical value. (*Rx*: *x* is radioactive. *Sx*: *x* has very short life. *Mx*: *x* has medical value. *Ux*: *x* is a uranium isotope.)

9. No sane witness would lie if his lying would implicate him in a crime. Therefore, if any witness implicated himself in a crime, then if all witnesses were sane, that witness did not lie. (*Sx*: *x* is sane. *Wx*: *x* is a witness. *Lx*: *x* lies. *Ix*: *x* implicates himself in a crime.)

10. If any jewelry is missing, then if all the servants are honest, it will be returned. If any servant is honest, they all are. So if any jewelry is missing, then if at least one servant is honest, it will be returned. (*Jx*: *x* is jewelry. *Mx*: *x* is missing. *Sx*: *x* is a servant. *Hx*: *x* is honest. *Rx*: *x* will be returned.)

11. If there are any liberals, then all philosophers are liberals. If there are any humanitarians, then all liberals are humanitarians. So if there are any humanitarians who are liberals, then all philosophers are humanitarians. (*Lx*: *x* is a liberal. *Px*: *x* is a philosopher. *Hx*: *x* is a humanitarian.)

*12. If something is lost, then if everyone values his possessions, it will be missed. If anyone values his possessions, so does everyone. Therefore, if something is lost, then if someone values his possessions, then something will be missed. (*Lx*: *x* is lost. *Px*: *x* is a person. *Vx*: *x* values his possessions. *Mx*: *x* is missed.)

4.6 More on Proving Invalidity

In Section 4.3 a method was presented for proving the invalidity of invalid arguments that involve quantifiers. There, we were dealing only with singly general propositions. That method, however, is easily adapted to deal with arguments that contain multiply general propositions.

Here, as before, we assume that there is at least one individual. That assumption is satisfied by models containing one, two, or any larger number of objects. For each such model, any given proposition containing quantifiers is logically equivalent to a truth-functional compound of singular propositions. Hence for each such model, any given argument whose propositions contain quantifiers is logically equivalent to some argument that contains only singular propositions and truth-functional compounds of singular propositions. An argument involving quantifiers is valid if and only if, for every model, it is logically equivalent to a valid truth-functional argument. Hence a given argument that involves quantifiers is proved invalid by describing a model for which the given argument is logically equivalent to an *invalid* truth-functional argument. We do this by translating the given argument involving quantifiers into a logically equivalent argument that contains only singular propositions and their truth-functional compounds. Then, we use the method of assigning truth values to prove that the truth-functional argument is invalid. The assignment of truth values to the singular propositions that are involved adequately *describes* the model that is being used.

We already know how to translate singly-general propositions into logically equivalent truth-functional propositions for any given model. As explained on page 78 for a model containing exactly k objects, $a, b, c, \ldots, k$, we use the logical equivalences

$$(x)(\Phi x) \equiv (\Phi a \cdot \Phi b \cdot \Phi c \cdot \ldots \cdot \Phi k)$$

and

$$(\exists x)(\Phi x) \equiv (\Phi a \vee \Phi b \vee \Phi c \vee \ldots \vee \Phi k)$$

There is nothing fundamentally new in translating multiply-general propositions into logically equivalent truth-functional propositions. The translation process is only more complicated. For example, the multiply-general proposition

$$(x)Fx \supset (\exists y)Gy$$

is logically equivalent for a one-element model that contains just a to

$$Fa \supset Ga$$

For a two-element model containing just a and b, it is logically equivalent to

$$(Fa \cdot Fb) \supset (Ga \vee Gb)$$

And for a three-element model that contains a, b, and c, it is logically equivalent to

$$(Fa \cdot Fb \cdot Fc) \supset (Ga \vee Gb \vee Gc)$$

For a multiply-general proposition with *nested* quantifiers, that is, some of whose quantifiers lie within the scope of other quantifiers, a stepwise process will be found convenient, beginning with the outer or leftmost quantifier. For example, although the multiply-general proposition

$$(x)(\exists y)(Fx \supset Gy)$$

is logically equivalent for the one-element model that contains just a to

$$Fa \supset Ga,$$

for the two-element model containing just a and b, it is logically equivalent first to

$$[(\exists y)(Fa \supset Gy)] \cdot [(\exists y)(Fb \supset Gy)]$$

and then to

$$[(Fa \supset Ga) \vee (Fa \supset Gb)] \cdot [(Fb \supset Ga) \vee (Fb \supset Gb)]$$

Some invalid arguments that contain multiply-general propositions can be proved invalid using a one-element model. For example, the argument

$$(x)(\exists y)(Px \supset Qy)$$
$$(y)(\exists z)(Ry \supset Qz)$$
$$\therefore (\exists x)(z)(Px \supset Rz)$$

is logically equivalent in the one-element model that contains just a to the truth-functional argument

$$Pa \supset Qa$$
$$Ra \supset Qa$$
$$\therefore Pa \supset Ra$$

which is proved invalid by assigning the truth value **T** to '*Pa*' and '*Qa*' and the truth value **F** to '*Ra*'.

As in the case of arguments containing singly-general propositions, a given argument may be valid for a one-element model but invalid for a two-element model. For example,

$$(\exists x)(y)(Fx \supset Gy)$$
$$(y)(\exists z)(Gy \supset Hz)$$
$$\therefore (x)(\exists z)(Fx \supset Hz)$$

is logically equivalent for the one-element model that contains just a to the valid truth-functional argument

$$Fa \supset Ga$$
$$Ga \supset Ha$$
$$\therefore Fa \supset Ha$$

but, for the two-element model that contains just a and b, it is logically equivalent to the truth-functional argument

$$[(Fa \supset Ga) \cdot (Fa \supset Gb)] \vee [(Fb \supset Ga) \cdot (Fb \supset Gb)]$$
$$[(Ga \supset Ha) \vee (Ga \supset Hb)] \cdot [(Gb \supset Ha) \vee (Gb \supset Hb)]$$
$$\therefore [(Fa \supset Ha) \vee (Fa \supset Hb)] \cdot [(Fb \supset Ha) \vee (Fb \supset Hb)]$$

which is proved invalid by assigning the truth value **T** to '*Fa*' and the truth value **F** to all the others, that is, to '*Fb*', '*Ga*', '*Gb*', '*Ha*', and '*Hb*'.

EXERCISES

Prove that each of the following arguments is invalid:

*1. $(x)(y)[Ax \supset (By \vee Cy)]$
$(z)\{[(y)By \vee (y)Cy] \supset Dz\}$
$\therefore (\exists x)(\exists z)(Ax \supset Dz)$

2. $(x)(\exists y)(Ex \supset Fy)$
$(\exists y)(z)(Fy \supset \sim Gz)$
$\therefore (x)(z)(\sim Ex \supset Gz)$

3. $(x)(\exists y)(Hx \supset Iy)$
$(\exists y)(z)(Iy \supset Jz)$
$\therefore (x)Hx \supset (z)Jz$

4. $(x)(Kx \supset Lx)$
$(\exists x)(\exists y)(Lx \cdot My)$
$\therefore (y)(Kx \supset My)$

*5. $(x)(\exists y)(Fx \equiv Gy)$
$\therefore (\exists y)(x)(Fx \equiv Gy)$

6. $(x)Nx \supset (\exists y)Oy$
$(y)Oy \supset (\exists z)Pz$
$\therefore (\exists x)Nx \supset (z)Pz$

7. $(x)Qx \supset [(\exists y)Ry \cdot (\exists y)Sy]$
$(\exists y)(Ry \cdot Sy) \supset (z)Tz$
$\therefore (x)Qx \supset (\exists z)Tz$

8. $(\exists x)(\exists y)\{Ux \supset [Vy \cdot (\exists y)Wy]\}$
$(y)(z)[(Vy \cdot Wy) \supset Xz]$
$\therefore (x)(\exists z)(Ux \supset Xz)$

9. $(\exists x)(\exists y)(Yx \supset Zy)$
$(\exists y)(z)(Zy \supset Az)$
$\therefore (\exists x)Yx \supset (z)Az$

10. $(x)(y)(Bx \supset Cy)$
$(x)Cx \supset [(\exists y)(Dy \cdot Ey) \cdot (\exists z)(Dz \cdot \sim Ez)]$
$\therefore (x)(Bx \supset Dx)$

4.7 Logical Truths Involving Quantifiers

In Chapter 2 truth tables were used, not only to establish the *validity* of arguments, but also to certify the *logical truth* of propositions (tautologies such as '$A \vee \sim A$'). The notion of a logically true proposition is, therefore, familiar. As we have seen, not every valid argument can be established by the method of truth tables. Some of them are proved valid by using quantification rules. Similarly, not every logically true proposition can be certified by the method of truth tables. Some of them are *demonstrated* by using quantification rules.

The method used in demonstrating the logical truth of tautologies was set forth in Chapter 3. A demonstration of the logical truth of the tautology '$A \supset (A \vee B)$' can be set down as

→ 1.	A	
2.	$A \vee B$	1, Add.
3.	$A \supset (A \vee B)$	1–2, C.P.

In demonstrating the logical truth of propositions involving quantifiers, we appeal not only to the original list of elementary valid argument forms and the strengthened principle of Conditional Proof, but to our quantification rules as well. Thus a demonstration of the logical truth of the proposition '$(x)Fx \supset (\exists x)Fx$' can be set down as

→ 1.	$(x)Fx$	
2.	Fy	1, **UI**
3.	$(\exists x)Fx$	2, **EG**
4.	$(x)Fx \supset (\exists x)Fx$	1–3, C.P.

(Just as in discussing the validity of arguments, so in discussing the logical truth of propositions, we explicitly limit our consideration to possible nonempty universes or models.)

Other logically true propositions that involve quantifiers require more complicated demonstrations. For example, the logically true proposition '$(x)Fx \supset {\sim}(\exists x){\sim}Fx$' has the following demonstration:

→ 1.	$(\exists x){\sim}Fx$	
→ 2.	${\sim}Fy$	
→ 3.	$(x)Fx$	
4.	Fy	3, **UI**
5.	$(x)Fx \supset Fy$	3–4, C.P.
6.	${\sim}(x)Fx$	5, 2, M.T.
7.	${\sim}(x)Fx$	1, 2–6, **EI**
8.	$(\exists x){\sim}Fx \supset {\sim}(x)Fx$	1–7, C.P.
9.	$(x)Fx \supset {\sim}(\exists x){\sim}Fx$	8, Trans., D.N.

Similarly, the truth of '${\sim}(\exists x){\sim}Fx \supset (x)Fx$' is demonstrated by the following:

→ 1.	${\sim}(\exists x){\sim}Fx$	
→ 2.	${\sim}Fy$	
3.	$(\exists x){\sim}Fx$	2, **EG**
4.	${\sim}Fy \supset (\exists x){\sim}Fx$	2–3, C.P.
5.	Fy	4, 1, M.T., D.N.
6.	$(x)Fx$	5, **UG**
7.	${\sim}(\exists x){\sim}Fx \supset (x)Fx$	1–6, C.P.

Given the logical truths established by the two preceding demonstrations, we conjoin them to obtain the equivalence '$(x)Fx \equiv {\sim}(\exists x){\sim}Fx$', a logical truth already noted in Section 4.1. Since our proof of this equivalence does not depend upon any peculiarities of the propositional function 'Fx', the equivalence holds for *any* propositional function. Also, since our proof does not depend upon any peculiarities of the variable 'x', the equivalence holds not only for any propositional function, but also for any individual variable. The equivalence *form* $(\nu)(\Phi\nu) \equiv {\sim}(\exists\nu){\sim}(\Phi\nu)$ is thus seen to be *logically true*, and can be added to the other logical equivalences in our list of Rules of Inference. It permits us (under the Rule of Replacement) validly to interchange $(\nu)(\Phi\nu)$ and ${\sim}(\exists\nu){\sim}(\Phi\nu)$ wherever they may occur. This connection between the two quantifiers, by way of negation, will now be adopted as an additional rule of inference and may be used in constructing formal proofs of validity and demonstrations of logical truth. When it is so used, the letters '**QN**' (for *quantifier negation*) should be written to indicate the principle that is appealed to. It should be obvious that the forms

$$\sim(\nu)(\Phi\nu) \equiv (\exists\nu)\sim(\Phi\nu)$$
$$(\nu)\sim(\Phi\nu) \equiv \sim(\exists\nu)(\Phi\nu)$$
$$\sim(\nu)\sim(\Phi\nu) \equiv (\exists\nu)(\Phi\nu)$$

are all logically equivalent to each other and to the form **QN**, and that they are, therefore, logically true.

Some fairly obvious logical truths are simply stated and are easily demonstrated with our present symbolic apparatus. A logically true biconditional for any propositional functions '*Fx*' and '*Gx*' is

$$[(x)Fx \cdot (x)Gx] \equiv (x)(Fx \cdot Gx)$$

which asserts that everything has the attribute *F* and everything has the attribute *G* if, and only if, everything has both the attributes *F* and *G*. The demonstrations of the two implications involved may be written side by side:

$$\begin{array}{lll}
\rightarrow 1. & (x)Fx \cdot (x)Gx & \\
2. & (x)Fx & 1, \text{Simp.} \\
3. & (x)Gx & 1, \text{Simp.} \\
4. & Fy & 2, \textbf{UI} \\
5. & Gy & 3, \textbf{UI} \\
6. & Fy \cdot Gy & 4, 5, \text{Conj.} \\
7. & (x)(Fx \cdot Gx) & 6, \textbf{UG} \\
8. & [(x)Fx \cdot (x)Gx] \supset (x)(Fx \cdot Gx) & \\
 & & 1\text{–}7, \text{C.P.}
\end{array}$$

$$\begin{array}{lll}
\rightarrow 1. & (x)(Fx \cdot Gx) & \\
2. & Fy \cdot Gy & 1, \textbf{UI} \\
3. & Fy & 2, \text{Simp.} \\
4. & Gy & 2, \text{Simp.} \\
5. & (x)Fx & 3, \textbf{UG} \\
6. & (x)Gx & 4, \textbf{UG} \\
7. & (x)Fx \cdot (x)Gx & 5, 6, \text{Conj.} \\
8. & (x)(Fx \cdot Gx) \supset [(x)Fx \cdot (x)Gx] & \\
 & & 1\text{–}7, \text{C.P.}
\end{array}$$

Another logical truth is in the form of a conditional rather than a biconditional. Written as

$$[(x)Fx \vee (x)Gx] \supset (x)(Fx \vee Gx)$$

it asserts that if either everything is an *F* or everything is a *G*, then everything is either an *F* or a *G*. Its demonstration involves making several assumptions of limited scope, and can be written as follows:

$$\begin{array}{lll}
\rightarrow 1. & (x)Fx \vee (x)Gx & \\
\rightarrow 2. & (x)Fx & \\
3. & Fy & 2, \textbf{UI} \\
4. & Fy \vee Gy & 3, \text{Add.} \\
5. & (x)(Fx \vee Gx) & 4, \textbf{UG} \\
6. & (x)Fx \supset (x)(Fx \vee Gx) & 2\text{–}5, \text{C.P.} \\
\rightarrow 7. & (x)Gx & \\
8. & Gy & 7, \textbf{UI} \\
9. & Fy \vee Gy & 8, \text{Add.}
\end{array}$$

10.	$(x)(Fx \vee Gx)$	9, UG
11.	$(x)Gx \supset (x)(Fx \vee Gx)$	7–10, C.P.
12.	$[(x)Fx \supset (x)(Fx \vee Gx)] \cdot [(x)Gx \supset (x)(Fx \vee Gx)]$	6, 11, Conj.
13.	$(x)(Fx \vee Gx) \vee (x)(Fx \vee Gx)$	12, 1, C.D.
14.	$(x)(Fx \vee Gx)$	13, Taut.
15.	$[(x)Fx \vee (x)Gx] \supset (x)(Fx \vee Gx)$	1–14, C.P.

The converse of this conditional is *not* logically true, however. The converse states that if everything is either *F* or *G*, then either everything is *F* or everything is *G*. That this converse is not always true can be seen by replacing '*G*' by '$\sim F$', for '$(x)(Fx \vee \sim Fx)$' is true for any predicate '*F*', whereas there are few predicates for which '$(x)Fx \vee (x)\sim Fx$' holds. Another logical truth that is conditional in form is

$$(\exists x)(Fx \cdot Gx) \supset [(\exists x)Fx \cdot (\exists x)Gx]$$

Its demonstration is perfectly straightforward and can be left as an exercise for the reader. That its converse is not true in general can be seen by again replacing '*G*' by '$\sim F$'. For most predicates '*F*', the proposition '$(\exists x)Fx \cdot (\exists x)\sim Fx$' is true (e.g. 'something is round and something is not round'). However, for any '*F*', the proposition '$(\exists x)(Fx \cdot \sim Fx)$' is logically false.

It has already been observed that propositional functions can contain propositions and/or other propositional functions as components. Examples of such propositional functions are

$$Fx \cdot Ga, \qquad Fx \vee Fy, \qquad Gy \vee (z)Hz, \qquad Gw \supset Fz, \ldots$$

When such propositional functions as these are quantified, to obtain

$$(x)[Fx \cdot Ga], \qquad (x)[Fx \vee Fy], \qquad (\exists y)[Gy \vee (z)Hz], \qquad (z)[Gw \supset Fz], \ldots$$

we have propositions and/or propositional functions lying within the scopes of quantifiers, although the quantifiers have no real affect on those expressions. When a quantifier with respect to a given variable is prefixed to an expression, its only effect is to bind previously free occurrences of that variable. In the expressions written above, the propositions '*Ga*' and '$(z)Hz$', and the propositional functions '*Fy*' and '*Gw*'—although lying within the scopes of the quantifiers '(x)', '$(\exists y)$', '(x)' and '(z)', respectively—are not really affected by them. Wherever we have an expression that contains a quantifier on the variable μ within whose scope lies either a proposition or a propositional function not containing any free occurrence of μ, the entire expression is logically equivalent to another expression in which the scope of the quantifier on μ does *not* extend over that proposition or propositional function. An example or two will make this clear. In the following, let '*Q*' be either a proposition or a propositional function containing no free occurrence of the variable 'x', and let 'Fx' be any propositional function containing at least one free occurrence of the variable 'x'. Our first logical equivalence here is between the universal quantification of '$Fx \cdot Q$' and the conjunction of the universal quantification of 'Fx' with 'Q', which is more briefly expressed as

$$(x)(Fx \cdot Q) \equiv [(x)Fx \cdot Q]$$

The demonstration of this equivalence can be written as

→ 1.	$(x)(Fx \cdot Q)$	
2.	$Fx \cdot Q$	1, **UI**
3.	Fx	2, Simp.
4.	$(x)Fx$	3, **UG**
5.	Q	2, Simp.
6.	$(x)Fx \cdot Q$	4, 5, Conj.
7.	$(x)(Fx \cdot Q) \supset [(x)Fx \cdot Q]$	1–6, C.P.

→ 1.	$(x)Fx \cdot Q$	
2.	$(x)Fx$	1, Simp.
3.	Fx	2, **UI**
4.	Q	1, Simp.
5.	$Fx \cdot Q$	3, 4, Conj.
6.	$(x)(Fx \cdot Q)$	5, **UG**
7.	$[(x)Fx \cdot Q] \supset (x)(Fx \cdot Q)$	1–6, C.P.

Another logical equivalence holds between the universal quantification of '$Q \supset Fx$' and the conditional statement whose antecedent is 'Q' and whose consequent is the universal quantification of 'Fx'. The first asserts that *given any individual* x, *Q implies that* x *has F,* and is equivalent to *Q implies that given any individual* x, x *has F.* Our symbolic expression of this equivalence is

$$(x)(Q \supset Fx) \equiv [Q \supset (x)Fx]$$

Its demonstration is easily constructed:

→ 1.	$(x)(Q \supset Fx)$	
2.	$Q \supset Fx$	1, **UI**
→ 3.	Q	
4.	Fx	2, 3, M.P.
5.	$(x)Fx$	4, **UG**
6.	$Q \supset (x)Fx$	3–5, C.P.
7.	$(x)(Q \supset Fx) \supset [Q \supset (x)Fx]$	1–6, C.P.

→ 1.	$Q \supset (x)Fx$	
→ 2.	Q	
3.	$(x)Fx$	1, 2, M.P.
4.	Fx	3, **UI**
5.	$Q \supset Fx$	2–4, C.P.
6.	$(x)(Q \supset Fx)$	5, **UG**
7.	$[Q \supset (x)Fx] \supset (x)(Q \supset Fx)$	1–6, C.P.

The same pattern of equivalence holds for the existential quantification of '$Q \supset Fx$' and the conditional statement '$Q \supset (\exists x)Fx$'. The first asserts that *there is at least one individual* x *such that Q implies that* x *has F,* and is equivalent to *Q implies that there is at least one individual* x *such that* x *has F,* which is asserted by the second. Its demonstration is very easily constructed and will be left as an exercise.

However, the pattern of equivalence is different when 'Q' occurs as consequent, rather than antecedent. Although the universal quantification of '$Fx \supset Q$' implies '$(x)Fx \supset Q$', it is not implied by the latter. There is an equivalence, however, between *given any x, if x has F, then Q* and *if there is at least one x such that x has F, then Q*, which was established informally as (E) on page 96 and is expressed symbolically as

$$(x)(Fx \supset Q) \equiv [(\exists x)Fx \supset Q]$$

Although the existential quantification of '$Fx \supset Q$' is implied by '$(\exists x)Fx \supset Q$', it does not imply the latter. There is an equivalence, however, between *there is at least one x such that if x has F, then Q* and *if given any x, x has F, then Q*, which is expressed symbolically as

$$(\exists x)(Fx \supset Q) \equiv [(x)Fx \supset Q]$$

The first of these two logical equivalences supplies an alternative method of symbolizing one of the propositions discussed in Section 4.4:

> If something is wrong with the house, then everyone in the house complains.

The translation given there abbreviates to

$$(\exists x)Wx \supset (y)(Py \supset Cy)$$

which, as we have just remarked, is logically equivalent to

$$(x)[Wx \supset (y)(Py \supset Cy)]$$

We shall conclude our discussion of logically true propositions involving quantifiers by turning our attention to four logically true propositions that are neither equivalences nor conditionals. They correspond, in a sense, to our quantification rules:

1. $(y)[(x)Fx \supset Fy]$
2. $(y)[Fy \supset (\exists x)Fx]$
3. $(\exists y)[Fy \supset (x)Fx]$
4. $(\exists y)[(\exists x)Fx \supset Fy]$

The first of these corresponds to **UI**, and says that given any individual y, if everything is an F, then y is an F. Its demonstration is almost trivially obvious, proceeding:

→ 1.	$(x)Fx$	
2.	Fz	1, **UI**
3.	$(x)Fx \supset Fz$	1–2, C.P.
4.	$(y)[(x)Fx \supset Fy]$	3, **UG**

The second corresponds to **EG**, and says that given any thing y, if y is an F, then something is an F. The third and fourth correspond to **UG** and **EI**, respectively. They are not so immediately obvious, but are logically true and quite easily demonstrated. An intuitive explanation can be given by reference to the ancient Athenian general and statesman Aristides, who was often called 'the just'. So outstanding was Aristides for his rectitude that the Athenians had a saying:

> If anyone is just, Aristides is just.

With respect to *any* attribute, there is always some individual y, such that if anything has that attribute, y has it. That is what is asserted by the fourth proposition listed above, which corresponds to **EI**. The matter can be put another way. If we turn our attention not to the attribute of being just, but to its reverse, the attribute of being corruptible, then the sense of the first Athenian saying is also expressible as

> If Aristides is corruptible, then everyone is corruptible.

Again generalizing, we may observe that, with respect to *any* attribute, there is always some individual y, such that if y has that attribute, everything has it. That is what is asserted by the third proposition listed above, which corresponds to **UG**. Its demonstration proceeds:

→ 1.	$\sim(\exists y)[Fy \supset (x)Fx]$	
2.	$(y)\sim[Fy \supset (x)Fx]$	1, **QN**
3.	$\sim[Fy \supset (x)Fx]$	2, **UI**
4.	$Fy \cdot \sim(x)Fx$	3, Impl.
5.	Fy	4, Simp.
6.	$(x)Fx$	5, **UG**
7.	$(x)Fx \vee (\exists y)[Fy \supset (x)Fx]$	6, Add.
8.	$\sim(x)Fx$	4, Simp.
9.	$(\exists y)[Fy \supset (x)Fx]$	7, 8, D.S.
10.	$\sim(\exists y)[Fy \supset (x)Fx] \supset (\exists y)[Fy \supset (x)Fx]$	1–9, C.P.
11.	$(\exists y)[Fy \supset (x)Fx] \vee (\exists y)[Fy \supset (x)Fx]$	10, Impl., D.N.
12.	$(\exists y)[Fy \supset (x)Fx]$	11, Taut.

Although we shall not prove it until Chapter 10, the methods of proof so far assembled (techniques for 'Natural Deduction', as they are sometimes called) permit the demonstration of all logically true propositions that are constructed out of truth-functional connectives and the quantification of individual variables. It will also be proved that *only* propositions that are logically true can be demonstrated by these techniques.

EXERCISES

Construct demonstrations for each of the following. (Each occurrence of 'Q' represents either a proposition or a propositional function containing no free occurrence of the variable 'x'.)

*1. $(y)[Fy \supset (\exists x)Fx]$
2. $(x)(Fx \supset Gx) \supset [(x)Fx \supset (x)Gx]$
3. $(x)(Fx \supset Q) \equiv [(\exists x)Fx \supset Q]$
4. $(\exists x)(Fx \cdot Q) \equiv [(\exists x)Fx \cdot Q]$
*5. $(\exists x)(Fx \cdot Gx) \supset [(\exists x)Fx \cdot (\exists x)Gx]$
6. $(x)(\exists y)(Fx \supset Gy) \supset [(x)Fx \supset (\exists y)Gy]$
7. $(\exists x)(Fx \supset Q) \equiv [(x)Fx \supset Q]$
8. $(\exists x)(Fx \vee Q) \equiv [(\exists x)Fx \vee Q]$
9. $[(\exists x)Fx \supset (\exists x)Gx] \supset (\exists x)(Fx \supset Gx)$
*10. $(\exists y)[(\exists x)Fx \supset Gy]$
11. $[(\exists x)Fx \vee (\exists x)Gx] \equiv (\exists x)(Fx \vee Gx)$
12. $(x)(Fx \vee Q) \equiv [(x)Fx \vee Q]$
13. $(x)(\exists y)(Fx \cdot Gy) \equiv (\exists y)(x)(Fx \cdot Gy)$
14. $(\exists x)(Q \supset Fx) \equiv [Q \supset (\exists x)Fx]$
*15. $[(\exists x)Fx \supset (y)Gy] \equiv (x)(y)(Fx \supset Gy)$
16. $[(x)Fx \supset (y)Gy] \equiv (y)(\exists x)(Fx \supset Gy)$
17. $[(\exists x)Fx \supset (\exists y)Gy] \equiv (x)(\exists y)(Fx \supset Gy)$
18. $[(x)Fx \supset (\exists y)Gy] \equiv (\exists x)(\exists y)(Fx \supset Gy)$
19. $(x)[Fx \supset (\exists y)Gy] \equiv (\exists y)[(\exists x)Fx \supset Gy]$
20. $(x)(\exists y)(Fx \vee Gy) \equiv (\exists y)(x)(Fx \vee Gy)$

5

The Logic of Relations

5.1 Symbolizing Relations

Some propositions containing two or more proper names (of individuals) are rightly understood to be truth-functional compounds of singular propositions that have different subject terms. For example, the proposition

Lincoln and Grant were presidents.

is rightly understood to be the conjunction of the two singular propositions

Lincoln was a president and Grant was a president.

But for some other propositions that have the same verbal pattern that analysis is wholly unsatisfactory. Thus the proposition

Lincoln and Grant were acquainted.

is definitely *not* a conjunction or any other truth function of

Lincoln was acquainted and Grant was acquainted.

On the contrary, dividing the proposition in this way destroys its significance because its meaning is not that both Lincoln and Grant were (or had) acquaintances, but rather that they were *acquainted with each other.* The given proposition does not assert that Lincoln and Grant both had a certain *attribute,* but that they stood in a certain *relation.* Lincoln is not said simply to be acquainted (whatever that might mean), but is said to be *acquainted with Grant.* Other propositions that express relations between two individuals are

John loves Mary.
Plato was a student of Socrates.
Isaac was a son of Abraham.
New York is east of Chicago.
Chicago is smaller than New York.

Relations such as these, that can hold between two individuals, are called 'binary' or 'dyadic' relations. Other relations may relate three or more individuals. For example, the propositions

Detroit is between New York and Chicago.
Helen introduced John to Mary.
America won the Philippines from Spain.

express *ternary* or *triadic* relations, while *quaternary* or *tetradic* relations are expressed by the propositions

America bought Alaska from Russia for seven million dollars.
Jack traded his cow to the peddler for a handful of beans.
Al, Bill, Charlie, and Doug played bridge together.

Relations enter into arguments in various ways. One example of a relational argument is

Al is older than Bill.
Bill is older than Charlie.

Therefore, Al is older than Charlie.

A slightly more complex example, which involves quantification, is

Helen likes David.
Whoever likes David likes Tom.
Helen likes only good-looking men.

Therefore, Tom is a good-looking man.

A still more complex relational inference, which involves multiple quantification, is

All horses are animals.

Therefore, the head of a horse is the head of an animal.

The latter is a valid inference that, as De Morgan observed, all the logic of Aristotle will not permit one to draw. Its validation by our apparatus of quantifiers and propositional functions, will be set forth in the next section.

Before discussing proofs of validity for relational arguments (and they will require no methods of proof beyond those developed in the preceding chapter), the problem of *symbolizing* relational propositions must be dealt with. Just as a single predicate symbol can occur in different propositions, so a single relation symbol can occur in different propositions. Just as we have the predicate 'human' common to the propositions:

Aristotle is human.
Plato is human.
Socrates is human.

so we have the relation word, 'taught', common to the propositions:

Socrates taught Plato.
Plato taught Aristotle.

Also, just as we regard the three subject-predicate propositions as different substitution instances of the propositional function 'x is human', so we can regard the two relational propositions as different substitution instances of the propositional function 'x taught y'. Substituting the constant 'Socrates' for the variable x, and the constant 'Plato' for the variable 'y', gives us the first proposition. Substituting 'Plato' for 'x' and 'Aristotle' for 'y' gives the second. The *order* of replacement is of great importance here. If 'x' has 'Aristotle' substituted for it, and 'y' has 'Plato' substituted for it, the result is the *false* proposition

Aristotle taught Plato.

Just as a propositional function of one variable like 'x is human' was abbreviated as 'Hx', so a propositional function of two variables like 'x taught y' is abbreviated as 'Txy'. Similarly, the propositional function 'x is between y and z' will be abbreviated as '$Bxyz$', and the propositional function 'x traded y to z for w' will be abbreviated as '$Txyzw$'. Our first example of a relational argument involves no quantifiers and is very easily symbolized. Using the individual constants 'a', 'b', and 'c' to denote Al, Bill, and Charlie, and the expression 'Oxy' to abbreviate 'x is older than y', we have

$$\begin{array}{l} Oab \\ Obc \\ \hline \therefore Oac \end{array}$$

Our second argument is not much more difficult, since none of its propositions contains more than a single quantifier. Using the individual constants 'h', 'd', and 't' to denote Helen, David, and Tom, respectively, 'Gx' to abbreviate 'x is a good-looking man', and the symbol 'Lxy' to abbreviate 'x likes y', the argument can be symbolized as

$$\begin{array}{ll} 1. & Lhd \\ 2. & (x)(Lxd \supset Lxt) \\ 3. & (x)(Lhx \supset Gx) \\ \cline{2-2} & \therefore Gt \end{array}$$

A proof of its validity is so easily constructed that it may well be set down now. Referring back to the numbered premisses, the proof proceeds:

4. $Lhd \supset Lht$ 2, **UI**
5. Lht 4, 1, M.P.
6. $Lht \supset Gt$ 3, **UI**
7. Gt 6, 5, M.P.

Symbolizing relational propositions becomes more complicated when several quantifiers occur in a single proposition. Our discussion of the problem will be simplified by confining attention at first to the two individual constants, '*a*' and '*b*', and the propositional function '*x* attracts *y*', abbreviated as '*Axy*'. The two statements '*a* attracts *b*' and '*b* is attracted by *a*' obviously have the same meaning. The first statement expresses that meaning by use of the *active voice*, while the second statement expresses that meaning by use of the *passive voice*. Both statements translate directly into the single formula '*Aab*'. Similarly, the two statements '*b* attracts *a*' and '*a* is attracted by *b*' are both symbolized by the formula '*Aba*'. These two different substitution instances of '*Axy*' are logically independent of each other. Either can be true, without entailing the truth of the other.

We are still on elementary and familiar ground when we come to symbolize

'*a* attracts everything'
'everything is attracted by *a*' } as '$(x)Aax$',

'*a* attracts something'
'something is attracted by *a*' } as '$(\exists x)Aax$',

'everything attracts *a*'
'*a* is attracted by everything' } as '$(x)Axa$',

'something attracts *a*'
'*a* is attracted by something' } as '$(\exists x)Axa$'.

But the problem of symbolizing becomes more complex when we dispense entirely with individual constants and consider relational propositions that are completely general. The simplest propositions of this kind are

1. Everything attracts everything.
2. Everything is attracted by everything.
3. Something attracts something.
4. Something is attracted by something.
5. Nothing attracts anything.
6. Nothing is attracted by anything.

which are symbolized by the following formulas:

1. $(x)(y)Axy$
2. $(y)(x)Axy$
3. $(\exists x)(\exists y)Axy$

4. $(\exists y)(\exists x)Axy$
5. $(x)(y){\sim}Axy$
6. $(y)(x){\sim}Axy$

In their English formulations, propositions 1 and 2 are clearly equivalent to each other, as are 3 and 4, and 5 and 6. The first two equivalences are easily established for the corresponding symbolic translations:

→1.	$(x)(y)Axy$	
2.	$(y)Awy$	1, **UI**
3.	Awv	2, **UI**
4.	$(x)Axv$	3, **UG**
5.	$(y)(x)Axy$	4, **UG**
6.	$(x)(y)Axy \supset (y)(x)Axy$	1–5, C.P.

→1.	$(\exists x)(\exists y)Axy$	
→2.	$(\exists y)Awy$	
→3.	Awv	
4.	$(\exists x)Axv$	3, **EG**
5.	$(\exists y)(\exists x)Axy$	4, **EG**
6.	$(\exists y)(\exists x)Axy$	2, 3–5, **EI**
7.	$(\exists y)(\exists x)Axy$	1, 2–6, **EI**
8.	$(\exists x)(\exists y)Axy \supset (\exists y)(\exists x)Axy$	1–7, C.P.

These demonstrate the logical truth of conditionals rather than of equivalences, but their converses can be established by the same patterns of argument. (The equivalence between formulas 5 and 6 is clearly established by the same pattern of argument that proves 1 equivalent to 2.)

When we turn to the next pair of statements,

7. Everything attracts something.
8. Something is attracted by everything.

there is no longer any logical equivalence or sameness of meaning. Sentence 7 is not entirely unambiguous, and some exceptional contexts might shift its meaning. Its most natural interpretation, however, is *not* that there is some one thing that is attracted by everything, but rather that everything attracts *something or other*. We symbolize it by way of successive paraphrasings, writing first

(x)(x attracts something)

and then symbolizing the expression 'x attracts something' the same way in which we symbolized 'a attracts something'. This gives us the formula

7. $(x)(\exists y)Axy$

Sentence 8 is also susceptible to alternative interpretations, one of which would make it synonymous with sentence 7, meaning that something or other is attracted by any (given) thing. But a perfectly straightforward way of understanding sentence 8 is to take it as asserting that there is some *one thing* that is attracted by all things. It, too, is symbolized in a stepwise fashion, writing first

$$(\exists y)(y \text{ is attracted by everything})$$

and then symbolizing the expression 'y is attracted by everything' the same way that we symbolized 'a is attracted by everything'. This gives us the formula

8. $(\exists y)(x)Axy$

There is a certain *misleading* similarity between formulas 7 and 8. They both consist of the propositional function 'Axy', to which are applied a universal quantifier with respect to 'x' and an existential quantifier with respect to 'y'. But the *order* in which the quantifiers are written is different in each case, and that makes a world of difference in their meanings. Formula 7, in which the universal quantifier comes first, asserts that given anything whatever, there is something or other that it attracts. But formula 8, in which the existential quantifier comes first, asserts that there is some one thing such that everything attracts *it*. Where two quantifiers are applied successively to one propositional function, if they are both universal or both existential, their order does not matter, as is shown by the equivalence of formulas 1 and 2, 3 and 4, and 5 and 6. But where one is universal and the other existential, the order of generalization or quantification is very important.

Although formulas 7 and 8 are not equivalent, they are not independent. The former is validly deducible from the latter. The demonstration is easily constructed as follows:

	1. $(\exists y)(x)Axy$	
→	2. $(x)Axv$	
	3. Auv	2, **UI**
	4. $(\exists y)Auy$	3, **EG**
	5. $(\exists y)Auy$	1, 2–4, **EI**
	6. $(x)(\exists y)Axy$	5, **UG**

The inference is valid only one way, however. Any attempt to derive formula 8 from 7 must inevitably run afoul of one of the restrictions on **UG**. The argument

$$(x)(\exists y)Axy$$
$$\therefore (\exists y)(x)Axy$$

is easily proved invalid by the method of Section 4.6. For a model containing just the two individuals a and b, the given argument is logically equivalent to the truth-functional argument

$$(Aaa \lor Aab) \cdot (Aba \lor Abb)$$
$$\therefore (Aaa \cdot Aba) \lor (Aab \cdot Abb)$$

which is proved invalid by assigning the truth value **T** to '*Aaa*' and '*Abb*', and the truth value **F** to '*Aab*' and '*Aba*'.

A similar pair of inequivalent propositions may be written as

9. Everything is attracted by something.
10. Something attracts everything.

These are clearly inequivalent when the 'something' in 9, coming at the end, is understood as 'something or other', and the 'something' in 10, coming at the beginning, is understood as 'there is at least one thing'. They are symbolized as

9. $(y)(\exists x)Axy$
10. $(\exists x)(y)Axy$

Relational propositions are sometimes formulated as though they were simple subject-predicate assertions. For example, '*a* was struck' is most plausibly interpreted to assert that *something struck a*. Such implicit occurrences of relations are often marked by the passive voice of a transitive verb. Our symbolizing of propositions containing implicit relations should be guided by consideration of the use to which they are to be put. Our purpose in symbolizing arguments is to put them into a form convenient for testing their validity. Our goal, therefore, with respect to a given argument, is not to provide a theoretically complete analysis, but rather to provide one complete enough for the purpose at hand—the testing of validity. Consequently, some implicit relations may be left implicit, while others require a more thorough analysis, as may be made clear by an example. Consider the argument

> Whoever visited the building was observed. Anyone who had observed Andrews would have remembered him. Nobody remembered Andrews. Therefore, Andrews didn't visit the building.

The first proposition of this argument contains two relations, one explicit, the other implicit. Explicitly, we have the relation of *someone visiting the building*. It is explicit because mention is made of both the visitor and what was visited by him. Implicitly, we have the relation of *someone observing someone*, which is implicit because no mention is made of the someone who does the observing—the omission being marked by the use of the passive voice. However, because the only other occurrence of '*x* visited the building' is also as a *unit*, in the conclusion, it need not be treated as a relation at all, but may be symbolized as a simple predicate. On the other hand, '*x* observed *y*', despite its merely implicit occurrence in the first premiss, must be explicitly symbolized as a relation if the argument is to be proved valid. For its second occurrence is not a simple repetition of the original unit. It appears instead as an explicit relation, with the first variable quantified and the second replaced by the proper name 'Andrews'. Using 'a' to denote Andrews, 'Vx' to abbreviate 'x visited the building', 'Oxy' to abbreviate 'x observed y', and 'Rxy' to abbreviate 'x remembers y', a symbolic translation and proof of validity for the given argument may be written as

1.	$(x)[Vx \supset (\exists y)Oyx]$	
2.	$(x)[Oxa \supset Rxa]$	
3.	$(x){\sim}Rxa$	$/ \therefore {\sim}Va$
4.	$Oza \supset Rza$	2, **UI**
5.	${\sim}Rza$	3, **UI**
6.	${\sim}Oza$	4, 5, M.T.
7.	$(y){\sim}Oya$	6, **UG**
8.	${\sim}(\exists y)Oya$	7, **QN**
9.	$Va \supset (\exists y)Oya$	1, **UI**
10.	${\sim}Va$	9, 8, M.T.

Our proof of validity for this argument would not have been helped at all by symbolizing 'Andrews visited the building' as a substitution instance of the relational 'x visited y' rather than of the simpler 'Vx'. But our proof of validity depended upon our symbolizing 'was observed' explicitly as a relation.

Most of our previous examples were illustrations of *unlimited* generality, in which it was asserted that *everything* stood in such-and-such a relation, or something did, or nothing did. A great many relational propositions are not so sweeping. Most assertions are more modest, claiming not that *everything* stands in such-and-such a relation, but that everything does *if* it satisfies certain conditions or restrictions. Thus we may say either that

Everything is attracted by all magnets.

or that

Everything made of iron is attracted by all magnets.

The second, of course, is the more modest assertion, being less *general* than the first. While the first is adequately symbolized, where 'Mx' abbreviates 'x is a magnet', as

$$(x)(y)[My \supset Ayx]$$

the second is symbolized, where 'Ix' abbreviates 'x is made of iron', as

$$(x)[Ix \supset (y)(My \supset Ayx)]$$

which is equivalent to $(x)(y)[Ix \supset (My \supset Ayx)]$ as was proved on page 112. That this symbolic translation is correct can be seen by paraphrasing the second proposition in English as

> Given anything whatever, *if* it is made of iron, then it is attracted by all magnets.

Perhaps the best way to translate relational propositions into our logical symbolism is by the kind of stepwise process that has already been illustrated. Let us use it again, this time for propositions of limited generality. First let us consider the proposition

> Any good amateur can beat some professional.

As a first step we may write

$$(x)\{(x \text{ is a good amateur}) \supset (x \text{ can beat some professional})\}$$

Next, the consequent of the conditional between the braces

$$x \text{ can beat some professional}$$

is symbolized as a generalization or quantified expression:

$$(\exists y)[(y \text{ is a professional}) \cdot (x \text{ can beat } y)]$$

Now, using the obvious abbreviations, 'Gx', 'Px', and 'Bxy' for 'x is a good amateur', 'x is a professional', and 'x can beat y', the given proposition is symbolized by the formula

$$(x)[Gx \supset (\exists y)(Py \cdot Bxy)]$$

Using the same method of paraphrasing by successive steps, we may symbolize

> Some professionals can beat all amateurs.

first as

$$(\exists x)[(x \text{ is a professional}) \cdot (x \text{ can beat all amateurs})]$$

then as

$$(\exists x)\{(x \text{ is a professional}) \cdot (y)[(y \text{ is an amateur}) \supset (x \text{ can beat } y)]\}$$

and finally (using abbreviations) as

$$(\exists x)[Px \cdot (y)(Ay \supset Bxy)]$$

The same method is applicable in more complex cases, where more than one relation is involved. We symbolize the proposition

> Anyone who promises everything to everyone is certain to disappoint somebody.

first by paraphrasing it as

$$(x)\{[(x \text{ is a person}) \cdot (x \text{ promises everything to everyone})] \supset [x \text{ disappoints somebody}]\}$$

The second conjunct of the antecedent

$$x \text{ promises everything to everyone}$$

may be further paraphrased, first as

$$(y)[(y \text{ is a person}) \supset (x \text{ promises everything to } y)]$$

and then as

$$(y)[(y \text{ is a person}) \supset (z)(x \text{ promises } z \text{ to } y)]$$

The consequent in our first paraphrase

$$x \text{ disappoints somebody}$$

has its structure made more explicit by being rewritten as

$$(\exists u)[(u \text{ is a person}) \cdot (x \text{ disappoints } u)]$$

The original proposition can now be rewritten as

$$(x)\{\{(x \text{ is a person}) \cdot (y)[(y \text{ is a person}) \supset (z)(x \text{ promises } z \text{ to } y)]\} \supset (\exists u)[(u \text{ is a person}) \cdot (x \text{ disappints } u)]\}$$

Using the obvious abbreviations, 'Px', '$Pxyz$', 'Dxy' for 'x is a person', 'x promises y to z', and 'x disappoints y', we can express the proposition more compactly in the formula

$$(x)\{\{Px \cdot (y)[Py \supset (z)Pxzy]\} \supset (\exists u)(Pu \cdot Dxu)\}$$

With practice, of course, not all such intermediate steps need be written out explicitly.

Quantification words such as 'everyone', 'anyone', 'everybody', 'anybody', and 'whoever', refer to *all persons* rather than to *all things*. Such quantification words as 'someone' and 'somebody' refer to *some persons* rather than to *some things*. It is frequently desirable to represent this reference in our symbolism. But doing so is not always necessary for the purpose of evaluating arguments that contain these words, however, and the choice of symbols is determined on the same grounds on which one decides if a relational clause or phrase is to be symbolized explicitly as a relation or as a mere attribute.

The words 'always', 'never', and 'sometimes' frequently have nontemporal significance, as in the propositions

> Good men always have friends.
> Bad men never have friends.
> Men who have no wives sometimes have friends.

which may be symbolized, using obvious abbreviations, as

$$(x)[(Gx \cdot Mx) \supset (\exists y)Fxy]$$
$$(x)[(Bx \cdot Mx) \supset \sim(\exists y)Fxy]$$
$$(\exists x)\{[Mx \cdot \sim(\exists y)(Wy \cdot Hxy)] \cdot (\exists z)Fxz\}$$

However, some uses of these words are definitely temporal, and when they are, they can be symbolized using the logical machinery already available, as can other temporal words like 'while', 'when', 'whenever', and the like. An example or two should serve to make this clear. Thus the proposition

> Dick always writes Joan when they are separated.

asserts that all times when Dick and Joan are separated are times when Dick writes Joan. This can be symbolized using 'Tx' for 'x is a time', '$Wxyz$' for 'x writes y at (time) z', and '$Sxyz$' for 'x and y are separated at (time) z', as

$$(x)\{Tx \supset [Sdjx \supset Wdjx]\}$$

Perhaps the most vivid illustration of the adaptability of the present notation is in symbolizing the following remark, usually attributed to Lincoln:

> You can fool some of the people all of the time, and all of the people some of the time, but you cannot fool all of the people all of the time.

The first conjunct: 'You can fool some of the people all of the time' is ambiguous. It may be taken to mean either that *there is at least one person who can always be fooled* or that, *for any time, there is at least one person* (*or other*) *who can be fooled at that time.* Adopting the first interpretation and using 'Px' for 'x

is a person', '*Tx*' for '*x* is a time', and '*Fxy*' for 'you can fool *x* at (or during) *y*', the above may be symbolized as

$$\{(\exists x)[Px \cdot (y)(Ty \supset Fxy)] \cdot (\exists y)[Ty \cdot (x)(Px \supset Fxy)]\} \cdot (\exists y)(\exists x)(Ty \cdot Px \cdot {\sim}Fxy)$$

The actual testing of relational arguments presents no new problems once the translations into logical symbolism are effected. The latter is the more troublesome part, and so a number of exercises are provided for the student to do before going on.

EXERCISES

I. Using the following 'vocabulary', translate the given formulas into idiomatic English sentences:

Ax-*x* is silver	*Axy*-*x* helps *y*
Bx-*x* is blissful	*Bxy*-*x* belongs to *y*
Cx-*x* is a cloud	*Bxyz*-*x* borrows *y* from *z*
Dx-*x* is a dog	*Cxy*-*x* can command *y*
Ex-*x* is smoke	*Dxy*-*x* is done at (or by) *y*
Fx-*x* is fire	*Exy*-*x* shears *y*
Gx-*x* is glass	*Fxy*-*x* is fair for *y*
Hx-*x* is home	*Gxy*-*x* gathers *y*
Ix-*x* is ill	*Hxy*-*x* hears *y*
Jx-*x* is work	*Ixy*-*x* lives in *y*
Kx-*x* is a lining	*Jxy*-*x* is jack of *y*
Lx-*x* is a lamb	*Kxy*-*x* knows *y*
Mx-*x* is moss	*Lxy*-*x* likes *y*
Nx-*x* is good	*Mxy*-*x* is master of *y*
Ox-*x* is a fool	*Nxy*-*x* loses *y*
Px-*x* is a person	*Oxy*-*x* is judged by *y*
Qx-*x* is a place	*Pxyz*-*x* blows *y* to *z*
Rx-*x* rolls	*Qxy*-*x* keeps company with *y*
Sx-*x* is a stone	*Rxy*-*x* is like *y*
Tx-*x* is a trade	*Sxy*-*x* says *y*
Ux-*x* is a house	*Txy*-*x* should throw *y*
Vx-*x* is a woman	*Txyz*-*x* tempers *y* to *z*
Wx-*x* is wind	*Uxy*-*x* comes to *y*
Xx-*x* is a time	*Vxy*-*x* ventures *y*
Yx-*x* is a day	*Wxy*-*x* is at (or in) *y*
Zx-*x* waits	*Xxy*-*x* is parent of *y*

g-*God*

Formulas

***1.** $(x)[Dx \supset (\exists y)(Yy \cdot Byx)]$
2. $(x)[(\exists y)(Py \cdot Fxy) \supset (z)(Pz \supset Fxz)]$
3. $(x)[(Rx \cdot Sx) \supset (y)(My \supset {\sim}Gxy)]$
4. $(x)[(Px \cdot Axx) \supset (Agx)]$

*5. $(x)[(Px \cdot Zx) \supset (y)(Uyx)]$
6. $(x)[Hx \supset (y)(Qy \supset \sim Ryx)]$
7. $(x)[(Px \cdot \sim Nxg) \supset (y)(\sim Nxy)]$
8. $(x)[(Px \cdot \sim Cxx) \supset (y)(Py \supset \sim Cxy)]$
9. $(x)\{Cx \supset (\exists y)[(Ay \cdot Ky) \cdot Byx]\}$
*10. $(x)[Px \supset (y)(Qxy \supset Oxy)]$
11. $(x)\{Qx \supset [(\exists y)(Ey \cdot Wyx) \supset (\exists z)(Fz \cdot Wzx)]\}$
12. $(x)\{[Px \cdot (y)(Ty \supset Jxy)] \supset (z)(Tz \supset \sim Mxz)\}$
13. $(x)\{[Px \cdot (\exists y)[(Gy \cdot Uy) \cdot Ixy]] \supset (z)(Sz \supset \sim Txz)\}$
14. $(x)\{[Px \cdot (y)(Lxy \supset Sxy)] \supset (\exists z)(Hxz \cdot \sim Lxz)\}$
*15. $(x)\{[Wx \cdot (y)[Py \supset \sim(\exists z)(Nz \cdot Pxzy)]] \supset Ix\}$
16. $(x)\{[Px \cdot (y)(\sim Vxy)] \supset (z)(\sim Gxz)\}$
17. $(x)\{Vx \supset (y)[Xy \supset (\exists z)[(Jz \cdot Bzx) \cdot \sim Dzy]]\}$
18. $(x)\{[Lx \cdot (\exists y)(Py \cdot Eyx)] \supset (z)(Wz \supset Tgzx)\}$
19. $(x)\{Px \supset (\exists y)[Py \cdot (\exists z)(Bxzy)]\}$
*20. $(x)\{Px \supset (\exists y)[Py \cdot (\exists z)(\sim Bxzy)]\}$
21. $(x)\{Px \supset (y)[Py \supset (z)(\sim Bxzy)]\}$
22. $(x)\{Px \supset (y)[Py \supset (\exists z)(\sim Bxzy)]\}$
23. $(x)[(Nx \cdot Dx) \supset (y)(Lxy \supset Myx)]$
24. $(x)[Px \supset (\exists y)(Py \cdot Xyx)] \cdot (\exists u)[Pu \cdot (v)(Pv \supset \sim Xuv)]$
25. $(x)\{[Qx \cdot (y)\{[(Py \cdot Wyx) \cdot (z)(\sim Kyz)] \supset By\}] \supset$
$\cdot (u)\{[(Pu \cdot Wux) \cdot (v)(Kuv)] \supset Ou\}\}$

II. Symbolize the following sentences, in each case using the indicated symbols:

*1. Dead men tell no tales. (*Dx*: *x* is dead. *Mx*: *x* is a man. *Tx*: *x* is a tale. *Txy*: *x* tells *y*.)
2. A lawyer who pleads his own case has a fool for a client. (*Lx*: *x* is a lawyer. *Fx*: *x* is a fool. *Pxy*: *x* pleads *y*'s case. *Cxy*: *x* is a client of *y*.)
3. A dead lion is more dangerous than a live dog. (*Lx*: *x* is a lion. *Ax*: *x* is alive. *Dx*: *x* is a dog. *Dxy*: *x* is more dangerous than *y*.)
4. Uneasy lies the head that wears a crown. (*Ux*: *x* lies uneasy. *Hx*: *x* is a head. *Cx*: *x* is a crown. *Wxy*: *x* wears *y*.)
*5. If a boy tells only lies, none of them will be believed. (*Bx*: *x* is a boy. *Lx*: *x* is a lie. *Txy*: *x* tells *y*. *Bxy*: *x* believes *y*.)
6. Anyone who consults a psychiatrist ought to have his head examined. (*Px*: *x* is a person. *Sx*: *x* is a psychiatrist. *Ox*: *x* ought to have his head examined. *Cxy*: *x* consults *y*.)
7. No one ever learns anything unless he teaches it to himself. (*Px*: *x* is a person. *Lxy*: *x* learns *y*. *Txyz*: *x* teaches *y* to *z*.)
8. Delilah wore a ring on every finger and had a finger in every pie. (*d*: Delilah. *Rx*: *x* is a ring. *Fxy*: *x* is a finger of *y*. *Oxy*: *x* is on *y*. *Px*: *x* is a pie. *Ixy*: *x* is in *y*.)
9. Any man who hates children and dogs cannot be all bad. (*Mx*: *x* is a man. *Cx*: *x* is a child. *Dx*: *x* is a dog. *Bx*: *x* is all bad. *Hxy*: *x* hates *y*.)
*10. Anyone who accomplishes anything will be envied by everyone. (*Px*: *x* is a person. *Axy*: *x* accomplishes *y*. *Exy*: *x* envies *y*.)
11. To catch a fish, one must have some bait. (*Px*: *x* is a person. *Fx*: *x* is a fish. *Bx*: *x* is bait. *Cxy*: *x* catches *y*. *Hxy*: *x* has *y*.)

12. Every student does some problems, but no student does all of them. (*Sx*: *x* is a student. *Px*: *x* is a problem. *Dxy*: *x* does *y*.)
13. Any contestant who answers all the questions put to him will win any prize he chooses. (*Cx*: *x* is a contestant. *Qx*: *x* is a question. *Px*: *x* is a prize. *Axy*: *x* answers *y*. *Pxy*: *x* is put to *y*. *Wxy*: *x* wins *y*. *Cxy*: *x* chooses *y*.)
14. Every son has a father, but not every father has a son. (*Px*: *x* is a person. *Mx*: *x* is male. *Pxy*: *x* is a parent of *y*.)
*15. A person is maintaining a nuisance if he has a dog that barks at everyone who visits its owner. (*Px*: *x* is a person. *Nx*: *x* is a nuisance. *Mxy*: *x* maintains *y*. *Dx*: *x* is a dog. *Bxy*: *x* barks at *y*. *Vxy*: *x* visits *y*. *Hxy*: *x* has *y*.)
16. A doctor has no scruples who treats a patient who has no ailment. (*Dx*: *x* is a doctor. *Sx*: *x* is a scruple. *Hxy*: *x* has *y*. *Px*: *x* is a patient. *Ax*: *x* is an ailment. *Txy*: *x* treats *y*.)
17. A doctor who treats a person who has every ailment has a job for which no one would envy him. (*Dx*: *x* is a doctor. *Px*: *x* is a person. *Txy*: *x* treats *y*. *Ax*: *x* is an ailment. *Hxy*: *x* has *y*. *Jx*: *x* is a job. *Exyz*: *x* envies *y* his *z*.)
18. If a farmer keeps only hens, none of them will lay eggs that are worth setting. (*Fx*: *x* is a farmer. *Kxy*: *x* keeps *y*. *Hx*: *x* is a hen. *Ex*: *x* is an egg. *Lxy*: *x* lays *y*. *Wx*: *x* is worth setting.)

In symbolizing the following, use only the abbreviations: *Px*: *x* is a person. *Sx*: *x* is a store, *Bxyz*: *x* buys *y* from *z*.

19. Everyone buys something from some store (or other).
*20. There is a store from which everyone buys something (or other).
21. Some people make all their purchases from a single store.
22. No one buys everything that it sells from any store.
23. No one buys things from every store.
24. No store has everyone for a customer.
*25. No store makes all its sales to a single person.

In symbolizing the following, use only the abbreviations: *Cx*: *x* is a charity; *Mx*: *x* is money; *Px*: *x* is a person; *Bxy*: *x* belongs to *y*; *Dxyz*: *x* donates *y* to *z*.

26. Nobody donates to every charity.
27. Nobody donates money to every charity.
28. Nobody donates all of his money to charity.
29. Nobody donates all of his money to any single charity.
*30. Nobody donates all of his belongings to any single charity.
31. Nobody gives all of his donations to any single charity.
32. No charity receives all of its money from any single person.
33. No charity receives all of his money from any single person.
34. No charity receives all of his donations from any single person.
*35. No charity receives all of his donations from any one donor.
36. No charity receives only money as donations.
37. Somebody gives money to charity.
38. Somebody donates all of his money to charity.
39. At least one person donates all of his belongings to a single charity.
*40. At least one person gives all of his donations to a single charity.
41. Some charities receive donations from everybody.

42. Some charities receive donations from every donor.
43. Some donations are not given to a charity.
44. Some donors donate to every charity.
45. Every charity receives donations from at least one donor.

III. Symbolize each of the following:

*1. Whoso sheddeth man's blood, by man shall his blood be shed. (Genesis 9:6)
2. His hand will be against every man, and every man's hand against him. (Genesis 16:12)
3. The sun shall not smite thee by day, nor the moon by night. (Psalm 121:6)
4. A wise son maketh a glad father. (Proverbs 10:1)
*5. He that spareth his rod hateth his son. (Proverbs 13:24)
6. The borrower is servant to the lender. (Proverbs 22:7)
7. Whoso diggeth a pit shall fall therein: and he that rolleth a stone, it will return upon him. (Proverbs 26:27)
8. The fathers have eaten sour grapes, and the children's teeth are set on edge. (Ezekiel 18:2)
9. The foxes have holes, and the birds of the air have nests; but the Son of man hath not where to lay his head. (Matthew 8:20)
10. . . . the good that I would, I do not; but the evil which I would not, that I do. (Romans 7:19)

5.2 Arguments Involving Relations

No new principles need be introduced to deal with relational arguments. The original list of nineteen Rules of Inference, together with the strengthened method of Conditional Proof, and our quantification rules plus Quantifier Negation, enable us (if we have sufficient ingenuity) to construct a formal proof of validity for every valid argument in which only individual variables are quantified and only truth-functional connectives occur.

However, a certain change of technique is advisable in working with arguments involving relations. In most of our previous sample proofs and demonstrations, **UI** and **EI** were used to instantiate with respect to a variable different from any quantified in the premiss, and **UG** and **EG** were used to quantify with respect to a variable different from any that occurred free in the premiss. Our inferences were for the most part of the following forms:

$$\frac{(x)Fx}{\therefore Fy}, \qquad \begin{array}{l} (\exists x)Fx \\ \left|\begin{array}{l} \rightarrow Fy \\ \quad\vdots \\ \quad p \end{array}\right. \\ \hline \therefore p \end{array}, \qquad \frac{Fx}{\therefore (y)Fy}, \qquad \frac{Fy}{\therefore (\exists w)Fw}$$

But our statement of the quantification rules does not require that μ and ν be different variables; they may well be the same. On the whole, it is simpler (wherever it is legitimate) to instantiate with respect to the same variable that had been quantified, and to quantify with respect to the same variable that had been free in the premiss. Thus the above inferences may also take the following forms:

$$\frac{(x)Fx}{\therefore Fx}, \quad \begin{array}{l}(\exists x)Fx \\ \rightarrow Fx \\ \quad \vdots \\ \quad p \\ \hline \therefore p\end{array}, \quad \frac{Fx}{\therefore (x)Fx}, \quad \frac{Fy}{\therefore (\exists y)Fy}$$

In this way instantiation is accomplished by simply dropping a quantifier, and generalization is accomplished by simply adding a quantifier. Of course, our restrictions on the quantification rules must still be observed. For example, where we have two premisses, '$(\exists x)Fx$' and '$(\exists x){\sim}Fx$', we can begin our use of **EI** with respect to one by simply dropping the quantifier. When that has been done, however, if within the scope of that assumption 'Fx', **EI** is subsequently used on the other, a new variable must be used instead of 'x', for the latter will already have a free occurrence in the proof under construction. Of course, we remain perfectly free to use **UI** to instantiate with respect to any variable or constant we choose. The preceding remarks can be illustrated by constructing a proof of validity for the argument

There is a man whom all men despise.
____
Therefore, at least one man despises himself.

Its symbolic translation and proof, using 'Mx' and 'Dxy' to abbreviate 'x is a man' and 'x despises y' may be written as follows:

1.	$(\exists x)[Mx \cdot (y)(My \supset Dyx)]$	$/ \therefore (\exists x)(Mx \cdot Dxx)$
→ 2.	$Mx \cdot (y)(My \supset Dyx)$	
3.	$(y)(My \supset Dyx)$	2, Simp.
4.	$Mx \supset Dxx$	3, **UI**
5.	Mx	2, Simp.
6.	Dxx	4, 5, M.P.
7.	$Mx \cdot Dxx$	5, 6, Conj.
8.	$(\exists x)(Mx \cdot Dxx)$	7, **EG**
9.	$(\exists x)(Mx \cdot Dxx)$	1, 2–8, **EI**

In the foregoing proof, the only use of a quantification rule accompanied by a change of variable was the use of **UI** in going from line 3 to line 4. This was done because we needed the expression 'Dxx' thus obtained.

Another formal proof serves to establish the validity of the third specimen argument stated at the beginning of the present chapter. Its premiss, 'All horses are animals' will be symbolized as '$(x)(Ex \supset Ax)$', where 'Ex' and 'Ax' abbreviate 'x is a horse' and 'x is an animal', respectively. In its conclusion

The head of a horse is the head of an animal.

the word 'the' has the same sense that it does in such propositions as 'The whale is a mammal' or 'The burnt child dreads the fire'. We may paraphrase it, therefore, first as

All heads of horses are heads of animals.

then as

$$(x)[(x \text{ is a head of a horse}) \supset (x \text{ is a head of an animal})]$$

and finally, writing 'Hxy' for 'x is a head of y', we may express the conclusion by the formula

$$(x)[(\exists y)(Ey \cdot Hxy) \supset (\exists y)(Ay \cdot Hxy)]$$

Once it is symbolized, the argument is easily proved valid by the techniques already available:

1.	$(x)(Ex \supset Ax)$ $\quad /\therefore (x)[(\exists y)(Ey \cdot Hxy) \supset (\exists y)(Ay \cdot Hxy)]$	
→ 2.	$(\exists y)(Ey \cdot Hxy)$	
→ 3.	$Ey \cdot Hxy$	
4.	Ey	3, Simp.
5.	$Ey \supset Ay$	1, **UI**
6.	Ay	5, 4, M.P.
7.	Hxy	3, Simp.
8.	$Ay \cdot Hxy$	6, 7, Conj.
9.	$(\exists y)(Ay \cdot Hxy)$	8, **EG**
10.	$(\exists y)(Ay \cdot Hxy)$	2, 3–9, **EI**
11.	$(\exists y)(Ey \cdot Hxy) \supset (\exists y)(Ay \cdot Hxy)$	2–10, C.P.
12.	$(x)[(\exists y)(Ey \cdot Hxy) \supset (\exists y)(Ay \cdot Hxy)]$	11, **UG**

The first specimen argument presented in this chapter, which dealt with the relation of *being older than*, raises a new problem, which will be discussed in the following section.

EXERCISES

I. Construct a formal proof of validity for each of the following arguments:

***1.** $(\exists x)(y)[(\exists z)Ayz \supset Ayx]$
$(y)(\exists z)Ayz$
$\therefore (\exists x)(y)Ayx$

2. $(x)[(\exists y)Byx \supset (z)Bxz]$
$\therefore (y)(z)(Byz \supset Bzy)$

3. $(x)(Cax \supset Dxb)$
$(\exists x)Dxb \supset (\exists y)Dby$
$\therefore (\exists x)Cax \supset (\exists y)Dby$

4. $(x)[Ex \supset (y)(Fy \supset Gxy)]$
$(\exists x)[Ex \cdot (\exists y) \sim Gxy]$
$\therefore (\exists x) \sim Fx$

*5. $(\exists x)[Hx \cdot (y)(Iy \supset Jxy)]$
$\therefore (x)(Hx \supset Ix) \supset (\exists y)(Iy \cdot Jyy)$

6. $(x)\{Kx \supset [(\exists y)Lxy \supset (\exists z)Lzx]\}$
$(x)[(\exists z)Lzx \supset Lxx]$
$\sim(\exists x)Lxx$
$\therefore (x)(Kx \supset (y) \sim Lxy)$

7. $(x)[Mx \supset (y)(Ny \supset Oxy)]$
$(x)[Px \supset (y)(Oxy \supset Qy)]$
$\therefore (\exists x)(Mx \cdot Px) \supset (y)(Ny \supset Qy)$

8. $(x)[(Rx \cdot \sim Sx) \supset (\exists y)(Txy \cdot Uy)]$
$(\exists x)[Vx \cdot Rx \cdot (y)(Txy \supset Vy)]$
$(x)(Vx \supset \sim Sx)$
$\therefore (\exists x)(Vx \cdot Ux)$

9. $(x)(Wx \supset Xx)$
$(x)[(Yx \cdot Xx) \supset Zx]$
$(x)(\exists y)(Yy \cdot Ayx)$
$(x)(y)[(Ayx \cdot Zy) \supset Zx]$
$\therefore (x)[(y)(Ayx \supset Wy) \supset Zx]$

10. $(x)\{[Bx \cdot (\exists y)[Cy \cdot Dyx \cdot (\exists z)(Ex \cdot Fxz)]] \supset (\exists w)Gxwx\}$
$(x)(y)(Hxy \supset Dyx)$
$(x)(y)(Fxy \supset Fyx)$
$(x)(Ix \supset Ex)$
$\therefore (x)\{Bx \supset [[(\exists y)(Cy \cdot Hxy) \cdot (\exists z)(Iz \cdot Fxz)] \supset (\exists u)(\exists w)Gxwu]\}$

II. Construct a formal proof of validity for each of the following arguments:

1. Whoever supports Ickes will vote for Jones. Anderson will vote for no one but a friend of Harris. No friend of Kelly has Jones for a friend. Therefore, if Harris is a friend of Kelly, Anderson will not support Ickes. (*Sxy*: *x* supports *y*. *Vxy*: *x* votes for *y*. *Fxy*: *x* is a friend of *y*. *a*: Anderson *i*: Ickes. *j*: Jones. *h*: Harris, *k*: Kelly.)
2. All circles are figures. Therefore, all who draw circles draw figures. (*Cx*: *x* is a circle. *Fx*: *x* is a figure. *Dxy*: *x* draws *y*.)
3. Any friend of Al is a friend of Bill. Therefore, anyone who knows a friend of Al knows a friend of Bill. (*Px*: *x* is a person. *Fxy*: *x* is a friend of *y*. *Kxy*: *x* knows *y*. *a*: Al. *b*: Bill.)

*4. Only a fool would lie about one of Bill's fraternity brothers to him. A classmate of Bill's lied about Al to him. Therefore, if none of Bill's classmates are fools, then Al is not a fraternity brother of Bill. (*Fx*: *x* is a fool. *Lxyz*: *x* lies about *y* to *z*. *Cxy*: *x* is a classmate of *y*. *Bxy*: *x* is a fraternity brother of *y*. *a*: Al. *b*: Bill.)

5. Whoever belongs to the Country Club is wealthier than any member of the Elks Lodge. Not everyone who belongs to the Country Club is wealthier than anyone who does not belong. Therefore, not everyone belongs either to the Country Club or the Elks Lodge. (*Cx*: *x* belongs to the Country Club. *Ex*: *x* belongs to the Elks Lodge. *Px*: *x* is a person. *Wxy*: *x* is wealthier than *y*.)

*6. It is a crime to sell an unregistered gun. All the weapons that Red owns were purchased by him from either Lefty or Moe. So if one of Red's weapons is an unregistered gun, then if Red never bought anything from Moe, Lefty is a criminal. (*Rx*: *x* is registered. *Gx*: *x* is a gun. *Cx*: *x* is a criminal. *Wx*: *x* is a weapon. *Oxy*: *x* owns *y*. *Sxyz*: *x* sells *y* to *z*. *r*: Red. *l*: Lefty. *m*: Moe.)

7. Everything on my desk is a masterpiece. Anyone who writes a masterpiece is a genius. Someone very obscure wrote some of the novels on my desk.

Therefore, some very obscure person is a genius. (*Dx*: x is on my desk. *Mx*: x is a masterpiece. *Px*: x is a person. *Gx*: x is a genius. *Ox*: x is very obscure. *Nx*: x is a novel. *Wxy*: x wrote y.)

8. There is a professor who is liked by every student who likes at least one professor. Every student likes some professor or other. Therefore, there is a professor who is liked by all students. (*Px*: x is a professor. *Sx*: x is a student. *Lxy*: x likes y.)

*9. No one respects a person who does not respect himself. No one will hire a person he does not respect. Therefore, a person who respects no one will never be hired by anybody. (*Px*: x is a person. *Rxy*: x respects y. *Hxy*: x hires y.)

10. Whoever donates to the United Fund gives all of his donations to the United Fund. Therefore, if Jones is a person who does not give all of his donations to the United Fund, then he does not donate anything at all to the United Fund. (*Px*: x is a person. *Dxyz*: x donates y to z. *f*: The United Fund. *j*: Jones.)

11. Any book that is approved by all critics is read by every literary person. Anyone who reads anything will talk about it. A critic will approve any book written by any person who flatters him. Therefore, if someone flatters every critic, then any book he writes will be talked about by all literary persons. (*Bx*: x is a book. *Cx*: x is a critic. *Lx*: x is literary. *Px*: x is a person. *Axy*: x approves y. *Rxy*: x reads y. *Txy*: x talks about y. *Fxy*: x flatters y. *Wxy*: x writes y.)

12. A work of art that tells a story can be understood by everyone. Some religious works of art have been created by great artists. Every religious work of art tells an inspirational story. Therefore, if some people admire only what they cannot understand, then some great artists' creations will not be admired by everyone. (*Ax*: x is a great artist. *Px*: x is a person. *Sx*: x is a story. *Ix*: x is inspirational. *Rx*: x is religious. *Wx*: x is a work of art. *Cxy*: x creates y. *Axy*: x admires y. *Txy*: x tells y. *Uxy*: x can understand y.)

5.3 Some Attributes of Relations

There are many interesting attributes that relations themselves may possess. We shall consider only a few of the more familiar ones, and our discussion will be confined to attributes of *binary* or *dyadic* relations.

Binary relations may be characterized as *symmetrical*, *asymmetrical*, or *nonsymmetrical*. Various symmetrical relations are designated by the phrases: 'is next to', 'is married to', and 'has the same weight as'. A *symmetrical* relation is a relation such that if one thing has that relation to a second, then the second *must* have that relation to the first. A propositional function '*Rxy*' designates a symmetrical relation if and only if

$$(x)(y)(Rxy \supset Ryx)$$

On the other hand, an *asymmetrical* relation is a relation such that if one thing has that relation to a second, then the second *cannot* have that relation to the first. Various asymmetrical relations are designated by the phrases 'is north of', 'is older than', and 'weighs more than'. A propositional function '*Rxy*' designates an asymmetrical relation if and only if

$$(x)(y)(Rxy \supset \sim Ryx)$$

Not all relations are either symmetrical or asymmetrical, however. If one individual loves a second, or is a brother of a second, or weighs no more than a second, it does not follow that the second loves the first, or is a brother to the first (possibly being a sister instead), or weighs no more than the first. Nor does it follow that the second does *not* love the first, or is *not* a brother to him, or *does* weigh more than the first. Such relations as these are *nonsymmetrical*, and are defined as those which are neither symmetrical nor asymmetrical.

Binary relations may also be characterized as *transitive*, *intransitive*, or *nontransitive*. Various transitive relations are designated by the phrases: 'is north of', 'is an ancestor of', and 'weighs the same as'. A *transitive* relation is a relation such that if one thing has it to a second, and the second to a third, then the first must have it to the third. A propositional function 'Rxy' designates a transitive relation if and only if

$$(x)(y)(z)[(Rxy \cdot Ryz) \supset Rxz]$$

An *intransitive* relation, on the other hand, is a relation such that if one thing has it to a second, and the second to a third, then the first *cannot* have it to the third. Some intransitive relations are designated by the phrases 'is mother of', 'is father of', and 'weighs exactly two pounds more than'. A propositional function 'Rxy' designates an intransitive relation if and only if

$$(x)(y)(z)[(Rxy \cdot Ryz) \supset \sim Rxz]$$

Not all relations are either transitive or intransitive. We define a *nontransitive* relation as one that is neither transitive nor intransitive. Examples of nontransitive relations are designated by 'loves', 'is discriminably different from', and 'has a different weight than'.

Finally, relations may be *reflexive*, *irreflexive*, or *nonreflexive*. Various definitions of these properties have been proposed by different authors, and there seems to be no standard terminology established. It is convenient to distinguish between reflexivity and total reflexivity. A relation is *totally reflexive* if everything has that relation to itself. For example, the phrase 'is identical with' designates the totally reflexive relation of identity. A propositional function 'Rxy' designates a totally reflexive relation if and only if

$$(x)Rxx$$

On the other hand, a relation is said to be *reflexive* if any thing a has that relation to itself if there is something b such that either Rab or Rba. Obvious examples of reflexive relations are designated by the phrases 'has the same color hair as', 'is the same age as', and 'is a contemporary of'. A propositional function 'Rxy' designates a reflexive relation if and only if

$$(x)[(\exists y)(Rxy \vee Ryx) \supset Rxx]$$

It is obvious that all totally reflexive relations are reflexive.

An irreflexive relation is a relation that nothing has to itself. A propositional function 'Rxy' designates an irreflexive relation if and only if

$$(x){\sim}Rxx$$

Examples of irreflexive relations are common indeed; the phrases: 'is north of', 'is married to', and 'is parent of' all designate irreflexive relations. Relations that are neither reflexive nor irreflexive are said to be *nonreflexive*. The phrases 'loves', 'hates', and 'criticizes' designate nonreflexive relations.

Relations may have various combinations of the attributes described. The relation of *weighing more than* is asymmetrical, transitive, and irreflexive, while the relation of *having the same weight as* is symmetrical, transitive, and reflexive. However, some attributes entail the presence of others. For example, all asymmetrical relations must be irreflexive, as can easily be demonstrated. Let 'Rxy' designate any asymmetrical relation. Then, by definition:

1. $(x)(y)(Rxy \supset {\sim}Ryx)$

From this premiss we can deduce that R is irreflexive, that is, that $(x){\sim}Rxx$:

2.	$(y)(Rxy \supset {\sim}Ryx)$	1, **UI**
3.	$Rxx \supset {\sim}Rxx$	2, **UI**
4.	${\sim}Rxx \vee {\sim}Rxx$	3, Impl.
5.	${\sim}Rxx$	4, Taut.
6.	$(x){\sim}Rxx$	5, **UG**

Other logical connections among these attributes of relations are easily stated and proved, but our present interest lies in another direction.

At this point, it should be remarked that the method of proving invalidity that was presented in Chapter 4 (pages 78–81 and pages 105–107), which is *always* applicable to nonrelational arguments, is *not always* applicable to relational arguments. Some sets of relational premisses, although perfectly consistent, cannot all be true for any finite model. Thus premisses asserting that the relation R is transitive, asymmetrical, and irreflexive, and further that $(x)(\exists y)Rxy$, cannot all be true for any finite model. Consider any model that is a nonempty set. Since it is nonempty, it contains at least one object, which we shall call 'a'. From $(x)(\exists y)Rxy$ we know that $(\exists y)Ray$. By irreflexiveness of R, ${\sim}Raa$, so y cannot be a. Hence the model must contain another object, b, such that Rab. Now again, from $(x)(\exists y)Rxy$ we know that $(\exists y)Rby$. By irreflexiveness, y cannot be b. Since Rab, by asymmetry $\sim Rba$, so y cannot be a either. Hence the model must contain still another object c such that Rbc. By transitivity, Rac also. Again from $(x)(\exists y)Rxy$, we know that $(\exists y)Rcy$. But by asymmetry and irreflexiveness, y cannot be a or b or c, so the model must contain *still* another object, d, such that Rcd. It is clear that this pattern of argument can be repeated indefinitely to show that however large a finite model we consider, the premisses require a still larger model. Yet the premisses are perfectly consistent, being *true* of the natural number 0, 1, 2, 3, . . . where 'Rxy' is interpreted to mean *x is smaller than y*. Hence the method of proving invalidity discussed in Chapter 4 is not always applicable to arguments involving relations.

The relevance of these various attributes of relations to relational arguments is easily seen. An argument to which one of them is relevant might be stated thus:

Tom has the same weight as Dick.
Dick has the same weight as Harry.
The relation of *having the same weight as* is transitive.
———
Therefore, Tom has the same weight as Harry.

When it is translated into our symbolism as

$$
\begin{array}{l}
Wtd \\
Wdh \\
(x)(y)(z)[(Wxy \cdot Wyz) \supset Wxz] \\
\hline
\therefore Wth
\end{array}
$$

the method of proving its validity is immediately obvious. We said that the argument 'might' be stated in the way indicated. But such a way of stating the argument would be the exception rather than the rule. The ordinary way of propounding such an argument would be to state only the first two premisses and the conclusion, on the grounds that *everyone knows* that *having the same weight as* is a transitive relation. Relational arguments are often used, and many of them depend essentially on the transitivity, or symmetry, or one of the other attributes of the relations involved. But *that* the relation in question *has* the relevant attribute is seldom—if ever—stated explicitly as a premiss. The reason is easy to see. In most discussions, a large body of propositions can be presumed to be common knowlege. The majority of speakers and writers save themselves trouble by not repeating well-known and perhaps trivially true propositions that their hearers or readers can perfectly well be expected to supply for themselves. An argument that is incompletely expressed, part of it being 'understood', is an *enthymeme*.

Because it is incomplete, an enthymeme must have its suppressed premiss or premisses taken into account when the question of its validity arises. Where a necessary premiss is missing, the inference is technically invalid. But where the unexpressed premiss is easily supplied and obviously true, in all fairness it ought to be included as part of the argument in any appraisal of it. In such a case, one assumes that the maker of the argument did have more 'in mind' than he stated explicitly. In most cases there is no difficulty in supplying the tacit premiss that the speaker intended but did not express. Thus the first specimen argument stated at the beginning of this chapter:

> Al is older than Bill.
> Bill is older than Charlie.
> ______________________________
> Therefore, Al is older than Charlie.

ought to be counted as valid, since it becomes so when the trivially true proposition that *being older than* is a transitive relation, is added as an auxiliary premiss. When the indicated missing premiss is supplied, a formal proof of the argument's validity is very easily set down.

Of course, premisses other than relational ones are often left unexpressed. For example, in the argument

> Any horse can outrun any dog. Some greyhounds can outrun any rabbit. Therefore, any horse can outrun any rabbit.

not only is the needed premiss about the transitivity of *being able to outrun* left unexpressed, but also left unexpressed is the nonrelational premiss that all greyhounds are dogs. When these are added—and they are certainly not debatable issues—the validity of the argument can be demonstrated as follows:

1.	$(x)[Hx \supset (y)(Dy \supset Oxy)]$	premisses
2.	$(\exists y)[Gy \cdot (z)(Rz \supset Oyz)]$	$/\therefore (x)[Hx \supset (z)(Rz \supset Oxz)]$
3.	$(x)(y)(z)[(Oxy \cdot Oyz) \supset Oxz]$	additional premisses
4.	$(y)(Gy \supset Dy)$	
→ 5.	Hx	
→ 6.	Rz	
→ 7.	$Gy \cdot (z)(Rz \supset Oyz)$	
8.	Gy	7, Simp.
9.	$Gy \supset Dy$	4, **UI**
10.	Dy	9, 8, M.P.
11.	$Hx \supset (y)(Dy \supset Oxy)$	1, **UI**
12.	$(y)(Dy \supset Oxy)$	11, 5, M.P.
13.	$Dy \supset Oxy$	12, **UI**
14.	Oxy	13, 10, M.P.
15.	$(z)(Rz \supset Oyz)$	7, Simp.
16.	$Rz \supset Oyz$	15, **UI**
17.	Oyz	16, 6, M.P.
18.	$Oxy \cdot Oyz$	14, 17, Conj.

19.	$(y)(z)[(Oxy \cdot Oyz) \supset Oxz]$	3, **UI**
20.	$(z)[(Oxy \cdot Oyz) \supset Oxz]$	19, **UI**
21.	$(Oxy \cdot Oyz) \supset Oxz$	20, **UI**
22.	Oxz	21, 18, M.P.
23.	Oxz	2, 7–22, **EI**
24.	$Rz \supset Oxz$	6–23, C.P.
25.	$(z)(Rz \supset Oxz)$	24, **UG**
26.	$Hx \supset (z)(Rz \supset Oxz)$	5–25, C.P.
27.	$(x)[Hx \supset (z)(Rz \supset Oxz)]$	26, **UG**

Missing premisses are not always so easily noticed and supplied as in the present example. When it is not so obvious which needed premisses are missing from an enthymematically expressed argument, it is a good policy in beginning a proof of the argument's validity to leave a little space just below the given premisses. Additional premisses can be written in that space when need arises. The only point to be stressed is that no statement that is as doubtful or debatable as the argument's own conclusion is to be admitted as a supplementary premiss, for in a valid argument that is enthymematically stated, only the sheerest platitudes should be left unexpressed for the hearer or reader to fill in for himself.

EXERCISES

Prove the validity of the following enthymemes—adding obviously true premisses where necessary:

1. A Cadillac is more expensive than any low-priced car. Therefore, no Cadillac is a low-priced car. (*Cx*: *x* is a Cadillac. *Lx*: *x* is a low-priced car. *Mxy*: *x* is more expensive than *y*.)
2. Alice is Betty's mother. Betty is Charlene's mother. Therefore, if Charlene loves only her mother, then she does not love Alice. (*a*: Alice. *b*: Betty. *c*: Charlene. *Mxy*: *x* is mother of *y*. *Lxy*: *x* loves *y*.)

*3. Any man on the first team can outrun every man on the second team. Therefore, no man on the second team can outrun any man on the first team. (*Fx*: *x* is a man on the first team. *Sx*: *x* is a man on the second team. *Oxy*: *x* can outrun *y*.)

4. Every boy at the party danced with every girl who was there. Therefore, every girl at the party danced with every boy who was there. (*Bx*: *x* is a boy. *Gx*: *x* is a girl. *Px*: *x* was at the party. *Dxy*: *x* danced with *y*.)

*5. Anyone is unfortunate who bears the same name as a person who commits a crime. Therefore, anyone who commits a burglary is unfortunate. (*Px*: *x* is a person. *Ux*: *x* is unfortunate. *Cx*: *x* is a crime. *Bx*: *x* is a burglary. *Cxy*: *x* commits *y*. *Nxy*: *x* bears the same name as *y*.)

6. All the watches sold by Kubitz are made in Switzerland. Anything made in a foreign country has a tariff paid on it. Anything on which a tariff was paid costs its purchaser extra. Therefore, it costs extra to buy a watch from Kubitz. (*Wx*: *x* is a watch. *Tx*: *x* has a tariff paid on it. *Fx*: *x* is a foreign country. *Cxy*: *x* costs *y* extra. *Mxy*: *x* is made in *y*. *Bxyz*: *x* buys *y* from *z*. *s*: Switzerland. *k*: Kubitz.)

*7. Vacant lots provide no income to their owners. Any owner of real estate must pay taxes on it. Therefore, any owner of a vacant lot must pay taxes on something that provides no income to him. (Vx: x is a vacant lot. Rx: x is real estate. Ixy: x provides income to y. Txy: x pays taxes on y. Oxy: x owns y.)

8. All admirals wear uniforms having gold buttons. Therefore, some naval officers wear clothes that have metal buttons. (Ax: x is an admiral. Ux: x is a uniform. Gx: x is gold. Bx: x is a button. Nx: x is a naval officer. Cx: x is clothing. Mx: x is metal. Wxy: x wears y. Hxy: x has y.)

9. Whenever Charlie moved to Boston, it was after he had met Al. Whenever Charlie got married, it was before he ever saw Dave. Therefore, if Charlie moved to Boston and subsequently got married, then he met Al before he ever saw Dave. (Tx: x is a time. Ax: Charlie met Al at (time) x. Bx: Charlie moved to Boston at (time) x. Mx: Charlie got married at (time) x. Dx: Charlie saw Dave at (time) x. Pxy: x precedes y.)

10. A fish that chases every shiner will be hooked by an angler who uses a shiner for bait. A greedy fish will chase every shiner. So if all anglers are sportsmen, then no pike that is not hooked by a sportsman who uses minnows for bait is greedy. (Fx: x is a fish. Sx: x is a shiner. Cxy: x chases y. Hxy: x hooks y. Ax: x is an angler. Bxy: x uses y for bait. Gx: x is greedy. Px: x is a pike. Rx: x is a sportsman. Mx: x is a minnow.)

5.4 Identity and the Definite Description

The notion of *identity* is a familiar one. Perhaps the most natural occasion for its use is in the process of *identification,* as in a police line-up when a witness identifies a suspect and asserts that

The man on the right *is* the man who snatched my purse.

Other identifications are common, as in a geography class when it is asserted that

Mt. Everest *is* the tallest mountain in the world.

or when it is asserted in a literature class that

Scott *is* the author of *Waverly.*

Each of these propositions asserts that a relation obtains between the things referred to by its two terms. The relation asserted to hold is that of *identity.* In each of the preceding propositions, at least one term was a *definite description,* which is a phrase of the form 'the so-and-so'. In identifications, however, both terms may be proper names. Just as the two propositions

Brutus killed Caesar.

and

Booth killed Lincoln.

assert the relation of *killing* to hold between the individuals referred to by the proper names that appear in the propositions, so the propositions

Lewis Carroll was Charles Lutwidge Dodgson.

and

Mark Twain was Samuel Clemens.

assert the relation of *identity* to hold between the individuals referred to by the proper names that appear in them.

The usual notation for the relation of identity is the ordinary equals-sign '='. It is intuitively obvious that the relation of identity is transitive, symmetrical, and totally reflexive. In our symbolic notation, we write

$$(x)(y)(z)\{[(x = y)\cdot(y = z)] \supset (x = z)\}$$
$$(x)(y)[(x = y) \supset (y = x)]$$
$$(x)(x = x)$$

All of these are immediate consequences of the definition of identity that is contained in Leibniz's principle of the Identity of Indiscernibles:

$x = y$ if and only if every attribute of x is an attribute of y, and conversely.

This principle permits us to infer, from the premisses $\nu = \mu$ and any formula containing an occurrence of the symbol ν, as conclusion any formula that results from replacing any occurrences of ν in the second premiss by the symbol μ.[1] Any inference of this pattern is valid, and in a proof, it should have the letters 'Id.' written beside it. A specimen deduction or two will make this clear. The argument

O. Henry was William Sidney Porter.
O. Henry was a writer.
Therefore, William Sidney Porter was a writer.

may be symbolized and proved valid by the following, in which we use the letters '*h*' and '*p*' to abbreviate the proper names 'O. Henry' and 'William

[1] Here we are using the letters *mu* and *nu* to represent any individual symbols, constants or variables.

Sidney Porter', and the symbol 'Wx' for 'x was a writer':

1. $h = p$
2. $Wh \quad /\therefore Wp$
3. $Wp \quad$ 1, 2, Id.

Another illustration is provided by the argument

> George Eliot wrote *The Mill on the Floss.*
> George Eliot is Mary Ann Evans.
> Mary Ann Evans is a woman.
> ________________________________
> Therefore, a woman wrote *The Mill on the Floss.*

Using the symbols 'g', 'f', 'm', 'Wxy', 'Wx' to abbreviate 'George Eliot', '*The Mill on the Floss*', 'Mary Ann Evans', 'x wrote y', and 'x is a woman', we may formulate the given argument and demonstrate its validity as follows:

1. Wgf
2. $g = m$
3. $Wm \quad /\therefore (\exists x)(Wx \cdot Wxf)$
4. $Wg \quad$ 2, 3, Id.
5. $Wg \cdot Wgf \quad$ 4, 1, Conj.
6. $(\exists x)(Wx \cdot Wxf) \quad$ 5, **EG**

An alternative proof for the second argument would be the following:

4. $Wmf \quad$ 1, 2, Id.
5. $Wm \cdot Wmf \quad$ 3, 4, Conj.
6. $(\exists x)(Wx \cdot Wxf) \quad$ 5, **EG**

A third illustration is provided by the argument:

> Only a bald man wears a wig. Kaplan is a man who wears a wig. This man is not bald. Therefore, this man is not Kaplan.

Using the symbols 't', 'k', 'Mx', 'Bx', 'Wx', to abbreviate 'this man', 'Kaplan', 'x is a man', 'x is bald', and 'x wears a wig', we may symbolize the present argument and prove its validity as follows:

1. $(x)[(Mx \cdot Wx) \supset Bx]$
2. $Mk \cdot Wk$
3. $\sim Bt \quad /\therefore \sim(t = k)$
4. $(Mk \cdot Wk) \supset Bk \quad$ 1, **UI**
5. $Bk \quad$ 4, 2, M.P.
6. $\sim(t = k) \quad$ 3, 5, Id.

This last proof serves to show that we use the principle of Identity not only to infer $\Phi\nu$ from $\Phi\mu$ and $\nu = \mu$, but also to infer $\sim(\nu = \mu)$ from $\Phi\nu$ and $\sim\Phi\mu$. For completeness and convenience, we include in our statement of the principle of Identity its symmetry and its total reflexiveness, even though its symmetry is readily derived from the others. Our formulation is

Id. $\dfrac{\Phi\mu \quad \nu = \mu}{\therefore \Phi\nu}$, $\dfrac{\Phi\mu \quad \sim\Phi\nu}{\therefore \sim(\nu = \mu)}$, $\dfrac{\nu = \mu}{\therefore \mu = \nu}$, and $\dfrac{p}{\therefore \mu = \mu}$

Now that the symbol for identity is available, we can define still another attribute of binary relations that is of importance in characterizing 'partial order' in Set Theory, as we shall see in Chapter 7. A propositional function '*Rxy*' designates an *antisymmetrical* relation if and only if

$$(x)(y)[(Rxy \cdot Ryx) \supset x = y]$$

An important use for the identity symbol is in the formulation of certain common types of exceptive statements. If we wish to symbolize the proposition

Al is on the team and can outrun anyone else on it.

using '*a*' for 'Al', '*Tx*' for '*x* is on the team', and '*Oxy*' for '*x* can outrun *y*', we cannot simply write down

$$Ta \cdot (x)(Tx \supset Oax)$$

for that would entail

$$Oaa$$

which is false because *being able to outrun* is an irreflexive relation. The preceding formula does not translate the given proposition, but rather the necessarily false one:

Al is on the team and can outrun anyone on it.

In this second proposition, the important word 'else' is missing. The first proposition does not assert that Al can outrun *everyone* on the team, but everyone *else* on it, that is, everyone on it who is *other than* or *not identical with* Al. The proper translation of the first proposition is

$$Ta \cdot (x)\{[Tx \cdot \sim(x = a)] \supset Oax\}$$

If we adopt the convention of abbreviating $\sim(\nu = \mu)$ as $\nu \neq \mu$, the preceding formula may be written

$$Ta \cdot (x)[(Tx \cdot x \neq a) \supset Oax]$$

Making similar use of the identity sign, we may symbolize the propositions

Only John loves Mary.

and

John loves only Mary.

as

$$Ljm \cdot (x)(x \neq j \supset \sim Lxm)$$

and

$$Ljm \cdot (x)(x \neq m \supset \sim Ljx)$$

A similar technique can be used in symbolizing the notion expressed by the English phrases 'at most' and 'no more than'. Thus the statement

There is at most one opening.

is interpreted not as asserting that there *is* an opening, but that there is *no more* than one. This statement may be symbolized as

$$(x)(y)[(Ox \cdot Oy) \supset x = y]$$

Similarly, the statement

No more than two visitors are permitted.

is interpreted as leaving open the question of whether there are any at all. This may be symbolized as

$$(x)(y)(z)[(Vx \cdot Vy \cdot Vz) \supset (x = z \vee y = z \vee x = y)]$$

The identity sign is also useful in symbolizing the concept of *at least*. It is not needed for 'at least one', because the existential quantifier, by itself, suffices. The statement

There is at least one applicant.

is symbolized as

$$(\exists x)Ax$$

But to symbolize

There are at least two applicants.

we use the identity sign, writing

$$(\exists x)(\exists y)[Ax \cdot Ay \cdot x \neq y]$$

Putting together the notations for 'at least one' and 'at most one', we have a method for symbolizing definite numerical propositions. Thus the statement

There is one book on my desk.

meaning *exactly* one, is symbolized, using 'Bx' for 'x is a book' and 'Dx' for 'x is on my desk', as

$$(\exists x)\{Bx \cdot Dx \cdot (y)[(By \cdot Dy) \supset y = x]\}$$

And the statement

Every state elects two Senators.

meaning *exactly* two, is symbolized, using 'Sx' for 'x is a state', 'Nx' for 'x is a Senator', and 'Exy' for 'x elects y', as

$$(x)\{Sx \supset (\exists y)(\exists z)[Ny \cdot Nz \cdot Exy \cdot Exz \cdot y \neq z \cdot (w)[(Nw \cdot Exw) \supset (w = y \vee w = z)]]\}$$

Finally, the (presumably false) statement

Cerberus has three heads.

is symbolized, using 'Hx' for 'x is a head of Cerberus', as

$$(\exists x)(\exists y)(\exists z)\{Hx \cdot Hy \cdot Hz \cdot x \neq y \cdot y \neq z \cdot x \neq z \cdot (w)[Hw \supset (w = x \vee w = y \vee w = z)]\}$$

The present notation is adequate for symbolizing arithmetical statements about individuals, but to symbolize propositions of pure arithmetic requires an expanded logical apparatus like that suggested in the following section.

Very rarely does an individual have two different proper names ν and μ, which would make $\nu = \mu$ a significant and informative statement. However, individuals are often referred to by means of descriptive phrases rather than by their proper names. Thus when a crime has been committed, and the police have not yet learned who committed it, the newspapers are not silenced for want of a name—the individual in question is referred to as 'the kidnapper of the Lindbergh baby', or 'the driver of the missing car'.

The word 'the' has a variety of uses. In one sense, it has the force of 'all' or 'any', as in 'The whale is a mammal'. But in another sense, it serves to indicate *existence* and *uniqueness*, as in the phrases

> The author of *Waverley*,
> The man who shot Lincoln,
> The largest city in Illinois,

which refer to Scott, Booth, and Chicago, respectively. One fairly standard notation for *this* sense of the word 'the' involves an upside-down iota. The three phrases above are (partially) symbolized as

> $(\iota x)(x$ wrotc *Waverley*$)$,
> $(\iota x)(x$ is a man$\cdot x$ shot Lincoln$)$,
> $(\iota x)(x$ is a city in Illinois$\cdot x$ is larger than any other city in Illinois$)$.

In general, a formula like '$(\iota x)(x$ wrote *Waverley*$)$' is read as 'the x that wrote *Waverley*', and is treated as a proper name. Thus we can replace the individual variable μ in the propositional function $\Psi\mu$ by $(\iota\nu)(\Phi\nu)$ to obtain $\Psi(\iota\nu)(\Phi\nu)$ as a substitution instance.

Normally, a definite description functions in argument the same way that a proper name does. The principle of Identity permits us to infer the conclusion 'Scott wrote *Marmion*' from the premisses 'Scott is the author of *Waverley*' and 'the author of *Waverley* wrote *Marmion*'. The proposition 'Something is larger than Detroit' follows validly by **EG** from the premiss 'The largest city in Illinois is larger than Detroit'. Of course, **EI** instantiates only with a variable and **UG** generalizes only from a variable. Therefore, there is no difference, with respect to these two principles, between a proper name and a definite description. But with respect to the principle of Universal Instantiation, certain differences and difficulties arise.

Where the letter 'F' designates an attribute, the symbol complex '$(\iota x)(Fx)$' refers to the individual having the attribute F *if* there is one and only one such individual. But what if no individual, or more than one individual, has this attribute? In such a case, since there is no unique individual to be referred to by the expression '$(\iota x)(Fx)$', that expression does not refer. The problem of interpreting phrases that purport to refer but do not really do so can be handled in the following fashion, attributable to Russell.[2] An explicit definition of a symbol is given by presenting another symbol that is equivalent to it in meaning. Thus the symbol 'bachelor' is defined explicitly by equating it to the

[2]See 'On Denoting,' *Mind*, n.s., vol. 14 (1905). Russell's paper, together with alternative treatments attributed to G. Frege, Keith Donnellan, and Hilary Putnam is reprinted in Part 3 of *Contemporary Philosophical Logic*, I. M. Copi and J. A. Gould eds., St. Martin's Press, New York, 1978.

phrase 'unmarried man'. An alternative method of explaining the meaning of a symbol is to give a *contextual* rather than an explicit definition for it. A contextual definition of a symbol does not explain the symbol's meaning in isolation, but rather explains the meaning of any statement or context in which that symbol occurs. We do not give an explanation of the word 'the' (or the iota symbol) in isolation, but instead, we present a method for interpreting any sentence (or formula) in which it appears. A contextual definition is also called a 'definition in use'. Russell's analysis of the definite description consists of a contextual definition, or definition in use, of the word 'the'—in the sense in which it signifies existence and uniqueness. Consider the proposition

The author of *Waverley* was a genius.

It seems to assert three things: first

There is an individual who wrote *Waverley*,

second

only that individual wrote *Waverley*,

and finally

that individual was a genius.

The three parts of its meaning can be symbolized—without using the iota—as follows:

$$(\exists x)(x \text{ wrote } \textit{Waverley})$$
$$(y)(y \text{ wrote } \textit{Waverley} \supset y = x)$$

and

$$x \text{ was a genius.}$$

Putting these three parts together, we obtain

$$(\exists x)\{(x \text{ wrote } \textit{Waverley}) \cdot (y)(y \text{ wrote } \textit{Waverley} \supset y = x) \cdot (x \text{ was a genius})\}$$

Here we have a symbolic translation of the given statement that contains neither the troublesome word 'the' nor any synonym for it. In general, any statement of the form

The so-and-so is such-and-such,

or any formula such as

$$\Psi(\imath x)(\Phi x)$$

is regarded as logically equivalent to

$$(\exists x)\{(x \text{ is a so-and-so})\cdot(y)(y \text{ is a so-and-so} \supset y = x)\cdot(x \text{ is a such-and-such})\}$$

or as

$$(\exists x)\{\Phi x\cdot(y)(\Phi y \supset y = x)\cdot\Psi x\}$$

Incidentally, when a property is expressed in the superlative form as 'best', 'fastest', 'heaviest', or the like, any proposition containing it can be expressed by using only the comparative forms 'better', 'faster', 'heavier', or the like. Thus the statement

The largest ocean is to the west of America.

can be symbolized by using 'Ox' for 'x is an ocean', 'Wx' for 'x is to the west of America', and 'Lxy' for 'x is larger than y', as

$$(\exists x)\{Ox\cdot(y)[(Oy\cdot y \neq x) \supset Lxy]\cdot Wx\}$$

A definite description is ordinarily used only when it is believed to refer. One normally uses the words 'the so-and-so' only when he believes that there is one and only one so-and-so. But beliefs are often mistaken, and one sometimes uses such a phrase even when it does not refer. When it does not, any sentence affirming that *the-so-and-so* has such-and-such an attribute or stands in this or that relation, is false. Thus, even though it might be true that everything has mass, it is *false* that

The immortal man has mass.

for this sentence affirms the existence of exactly one immortal man, when there is none. And unless it occurs in a context that makes clear either that some particular mountain is being referred to or that *mountains in general* are under discussion, the statement

The mountain has mass.

is false. This is so because it affirms that there is only one mountain, whereas in fact there are many. These remarks should serve to make clear that a phrase of the form 'the so-and-so' or a symbol such as '$(\imath x)(Fx)$' cannot be instantiated by the principle of Universal Instantiation alone. In order to deduce the conclusion '$G(\imath x)(Fx)$' from '$(x)Gx$', we require the additional premiss that there is exactly one thing which is an F. Where that premiss is missing, the inference is invalid. But where it is present, the argument is easily proved valid as follows:

1.	$(x)Gx$	
2.	$(\exists x)[Fx \cdot (y)(Fy \supset y = x)]$	$/ \therefore G(\imath x)(Fx)$
→ 3.	$Fx \cdot (y)(Fy \supset y = x)$	
4.	Gx	1, **UI**
5.	$Fx \cdot (y)(Fy \supset y = x) \cdot Gx$	3, 4, Conj.
6.	$(\exists x)\{Fx \cdot (y)(Fy \supset y = x) \cdot Gx\}$	5, **EG**
7.	$(\exists x)\{Fx \cdot (y)(Fy \supset y = x) \cdot Gx\}$	2, 3–6, **EI**
8.	$G(\imath x)(Fx)$	7—definition

Any proposition of the form $\Psi(\imath x)(\Phi x)$ is false either if there is no x that is Φ or if there is more than one of them. In such a proposition, the descriptive phrase $(\imath x)(\Phi x)$ is said to have a *primary* occurrence. But a proposition containing a descriptive phrase may be part of a larger (truth functional) context, in which the descriptive phrase is said to have a *secondary* occurrence. A proposition in which $(\imath x)(\Phi x)$ has a secondary occurrence may be true even if there is no x that is Φ or there is more than one of them. One of the simplest examples, $\sim\Psi(\imath x)(\Phi x)$, might mean either that

$$(\exists x)\{\Phi x \cdot (y)(\Phi y \supset y = x) \cdot \sim\Psi x\} \tag{1}$$

or that

$$\sim(\exists x)\{\Phi x \cdot (y)(\Phi y \supset y = x) \cdot \Psi x\} \tag{2}$$

If either there is no x that is Φ or there is more than one of them, (1) is false and (2) is true. To make otherwise ambiguous propositions definite and unambiguous, Russell proposed using formulas of the form $(\imath x)(\Phi x)$ as a kind of 'scope indicator'. But the rules governing this use are quite complicated, and we shall have no use for them here. Indeed, we shall make no further use of formulas of the form $(\imath x)(\Phi x)$ and shall use instead the explicit symbolic translation that involves quantifiers and the identity symbol.

EXERCISES

Prove the validity of the following arguments—using only the identity symbol in addition to the indicated abbreviations:

1. The architect who designed Tappan Hall designs only office buildings. Therefore, Tappan Hall is an office building. (Ax: x is an architect. t: Tappan Hall. Dxy: x designed y. Ox: x is an office building.)

*2. The Professor of Greek at Siwash is very learned. Therefore, all Professors of Greek at Siwash are very learned. (Px: x is a Professor of Greek. Sx: x is at Siwash. Lx: x is very learned.)

3. The smallest state is in New England. All states in New England are primarily industrial. Therefore, the smallest state is primarily industrial. (*Sx*: *x* is a state. *Nx*: *x* is in New England. *Ix*: *x* is primarily industrial. *Sxy*: *x* is smaller than *y*.)
*4. The fastest running person is a Scandinavian. Therefore, anyone who is not a Scandinavian can be outrun by someone (or other). (*Sx*: *x* is a Scandinavian. *Px*: *x* is a person. *Fxy*: *x* can run faster than *y*.)
5. All entrants will win. There will be, at most, one winner. There is at least one entrant. Therefore, there is exactly one entrant. (*Ex*: *x* is an entrant. *Wx*: *x* will win.)
*6. Any fish can swim faster than any smaller one. Therefore, if there is a largest fish, then there is a fastest fish. (*Fx*: *x* is a fish. *Lxy*: *x* is larger than *y*. *Sxy*: *x* can swim faster than *y*.)
7. Adams and Brown were the only men at the banquet who drank. All the men at the banquet who brought liquor drank. Adams did not bring any liquor. If any man at the banquet drank, then some man at the banquet who drank must have brought liquor. All who drank became ill. Therefore, the man at the banquet who brought liquor became ill. (*a*: Adams. *b*: Brown. *Mx*: *x* was a man at the banquet. *Dx*: *x* drank. *Lx*: *x* brought liquor. *Ix*: *x* became ill.)
8. Anyone who has climbed Mt. Blanc is braver than anyone who has not. Of all those on our team, only the youngest member has climbed Mt. Blanc. Everyone on our team is a veteran. Therefore, the bravest member of our team is a veteran. (*Cx*: *x* has climbed Mt. Blanc. *Tx*: *x* is on our team. *Vx*: *x* is a veteran. *Bxy*: *x* is braver than *y*. *Oxy*: *x* is older than *y*.)
9. There is exactly one penny in my right hand. There is exactly one penny in my left hand. Nothing is in both my hands. Therefore, there are exactly two pennies in my hands. (*Px*: *x* is a penny. *Rx*: *x* is in my right hand. *Lx*: *x* is in my left hand.)
10. All accompanists were bagpipers. All bagpipers were in the cabin. At most two individuals were in the cabin. There were at least two accompanists. Therefore, there were exactly two bagpipers. (*Ax*: *x* was an accompanist. *Bx*: *x* was a bagpiper. *Cx*: *x* was in the cabin.)

5.5 Predicate Variables and Attributes of Attributes

In all of the preceding discussion, quantification has been confined to individual variables. With the exception of the Greek letters *phi* and *psi*, all attribute symbols and relation symbols introduced thus far have been *constants*. The letters '*W*', '*S*', and '*B*' in our usage of '*Wx*', '*Sxy*', and '*Bxyz*' as abbreviations for '*x* is wise', '*x* is a son of *y*', and '*x* is between *y* and *z*', have designated definite attributes of, or relations between, individuals. However, attribute variables and relation variables can be introduced also, and they too may be quantified.

If we set aside the capital letters '*F*', '*G*', '*H*', as attribute or relation variables and refer to them indifferently as 'predicate variables', then the same quantification techniques that are already familiar will permit symbolizing a greater variety of statements. We shall be able to symbolize statements about all or some attributes or relations that things may have or stand in. The expression

$$Fx$$

consisting of a predicate variable and an individual variable juxtaposed in that order may be regarded as a propositional function of *two* variables. By instantiation with respect to both variables, we obtain such singular propositions as 'Socrates is mortal' and 'Plato is wise', expressed symbolically as 'Ms' and 'Wp'. By instantiation with respect to the predicate variable and generalization with respect to the individual variable, we obtain such familiar singly-general propositions as

$(x)Mx$ (everything is mortal)
$(\exists x)Wx$ (something is wise)

We have seen all these before. A new kind of proposition, however, although still singly general, is obtained by instantiating with respect to the individual variable and generalizing with respect to the predicate variable. Here we have

$(F)Fs$ (Socrates has all attributes)
$(\exists F)Fp$ (Plato has some attribute)

Finally, by generalizing with respect to *both* variables, the following doubly-general propositions are obtained:

(1)	$(x)(F)Fx$	(5)	$(F)(x)Fx$
(2)	$(x)(\exists F)Fx$	(6)	$(F)(\exists x)Fx$
(3)	$(\exists x)(F)Fx$	(7)	$(\exists F)(x)Fx$
(4)	$(\exists x)(\exists F)Fx$	(8)	$(\exists F)(\exists x)Fx$

Of these, (1) and (5) are clearly equivalent, (1) stating that

Every thing has every attribute.

and (5) stating that

Every attribute belongs to every thing.

Propositions (4) and (8) are also equivalent to each other, with (4) stating that

Some thing has some attribute.

and (8) stating that

Some attribute belongs to some thing.

The remaining propositions, however, are all distinct. They may be expressed in English as

(2) Every thing has some attribute (or other).
(3) There is a thing that has every attribute.
(6) Every attribute belongs to some thing (or other).
(7) There is an attribute that belongs to every thing.

There are no equivalences here, but proposition (3) logically entails proposition (6), and proposition (7) logically entails proposition (2). These implications can be formally established by means of our familiar quantification rules, allowing the symbols 'μ' and 'ν' of the rules to denote predicate variables as well as individual variables—with the same restrictions on their application, of course.

When predicate variables are introduced and their quantification permitted, a formal, that is, completely symbolic definition, can be given for the identity symbol. The definition is

$$(x = y) = df\,(F)(Fx \equiv Fy)$$

From this definition we have

$$(x)(y)[(x = y) \equiv (F)(Fx \equiv Fy)]$$

following as a logical consequence. And from the latter all the attributes of the identity relation can be deduced.

For the most part, our preceding discussion has concerned only attributes of things. But things are not the only entities that have attributes. In Section 5.3 we discussed various attributes that can be attributed to, or predicated of, *relations*. Attributes, themselves, can have attributes. Thus the attribute of being *honest* has the attribute of being *desirable;* a *virtue* (which is an attribute) may, itself, have the attribute of being *rare;* and the attribute of being *inconsiderate* is *common*.

Once it has been noted that some attributes can be predicated of other attributes, the temptation arises to predicate certain attributes of themselves. For example, the attribute of being abstract seems, itself, to be abstract, whereas the attribute of being concrete seems not to be itself concrete. Any attribute that can be truly predicated of itself will be said to be a *predicable* attribute. Thus being predicable is an attribute that belongs to all those and only those attributes that can be truly predicated of themselves. On the other hand, any attribute that cannot be truly predicated of itself will be said to be an *impredicable* attribute. Thus being impredicable is an attribute that belongs to all those and only those attributes that cannot be truly predicated of themselves.

If we now ask whether the attribute of being impredicable can be truly predicated of itself or not, we are led to the following unhappy conclusion. If the attribute of being impredicable *can* be truly predicated of itself, then it *has* the attribute of being impredicable. It follows then by definition that the attribute *cannot* be truly predicated of itself. On the other hand, if the attribute of being impredicable *cannot* be truly predicated of itself, then like all other attributes that cannot be truly predicated of themselves, it *has* the attribute of being impredicable. This means that it *can* be truly predicated of itself. We have thus been led to a contradiction.

The contradiction can be derived more clearly by symbolizing the attribute of being impredicable as 'I', and defining it formally as

$$IF = df \sim FF$$

This definition has the following general proposition as an immediate logical consequence:

$$(F)(IF \equiv \sim FF)$$

From the latter, by the principle of Universal Instantiation, we can instantiate with respect to 'I' itself to obtain

$$II \equiv \sim II$$

which is an explicit contradiction.[3]

Several methods have been proposed for avoiding such contradictions. One of the best known is Russell's (simple) Theory of Logical Types,[4] which can be given the following rough formulation. According to Russell, entities are divided into a hierarchy of different logical types, the lowest of which consists of all individuals. The next level of logical types consists of all attributes of individuals, the next of all attributes of attributes of individuals, the next of all attributes of attributes of attributes of individuals, and so on. Relations and their attributes give us additional hierarchies, but they will be ignored in the present discussion. The essential point to the theory of types is not merely the

[3] See Russell's *Principles of Mathematics,* Cambridge University Press, Cambridge, England, 1903, pp. 79–80, 97–98, and 102. See also Carnap's *Logical Syntax of Language,* Harcourt Brace Jovanovich, New York, 1937, p. 211.

[4] Russell first formulated his theory of logical types in Appendix B of his *Principles of Mathematics.* A more complex version of the theory, designed to cope with certain other problems as well, will be presented in Appendix C of the present volume. See also Russell's 'Mathematical Logic as Based on the Theory of Types,' *American Journal of Mathematics,* vol. 30 (1908), pp. 222–262, reprinted in *Bertrand Russell: Logic and Knowledge. Essays 1901–1950,* Robert Charles Marsh, ed., George Allen & Unwin Ltd., London, 1956, and Chapter 2 of the Introduction, *Principia Mathematica,* 1st ed., Whitehead and Russell, Cambridge University Press, Cambridge, England, 1910–1913. The interested reader should also consult Irving M. Copi, *The Theory of Logical Types,* Routledge and Kegan Paul, London, 1971. The best-known alternative resolution of the contradiction is the kind of axiomatic set theory first published by E. Zermelo in 'Untersuchungen über die Grundlagen der Mengenlehre I', *Mathematische Annalen,* vol. 65 (1908), pp. 261–281. Zermelo's theory, as modified by Fraenkel and others, is discussed in Chapter 7.

division of all entities into different logical types, but the restriction that any attribute that may significantly be predicated of an entity of one logical type cannot significantly be predicated of any entity of any other logical type. For some attributes, the theory of types seems perfectly obvious. Thus an individual thing may be orange in color, but it clearly does not make sense to say of any attribute that *it* is orange in color. An attribute may have many instances, but it does not make sense either to affirm or to deny of an individual thing that *it* has many instances.

The primary motivation for accepting the theory of logical types, however, is not its naturalness or obviousness, but the fact that it permits the avoidance of such contradictions as that of the alleged attribute 'impredicable'. According to this theory, the type of an attribute must be higher than the type of any entity of which it can significantly be predicated. Consequently, it does not make sense either to affirm or to deny of any attribute that it belongs to itself. Such expressions as '*FF*' and '$\sim$*FF*' must be dismissed as meaningless. Consequently, no such attribute as 'impredicable' can be defined, and the contradiction vanishes. The version of the theory of types sketched above—sometimes called the 'Simple Theory of Types'—suffices to rule out all contradictions of this kind. It also has a certain consonance with common sense. However, there are alternative solutions or ways to avoid the contradictions, so the theory of logical types cannot be regarded as *the* solution. It is widely accepted, however, and we shall follow it in introducing a new type of symbol to represent attributes of attributes of individuals.

Some words have more than a single meaning, of course, and may in one sense designate an attribute of individuals, and in another sense designate an attribute of attributes of individuals. Thus the word 'rare' in one sense designates an attribute of attributes of individuals. An attribute of individuals is rare if it is exemplified by only a few individuals and, in this sense, it cannot be affirmed or denied of individuals, themselves. On the other hand, there is a *different* sense of the word 'rare', in which it designates an attribute of an individual piece of meat that has been cooked for only a short while, and in this sense it cannot significantly be affirmed or denied of any attribute. To avoid ambiguity, we symbolize attributes of attributes of individuals by boldface italic capital letters '***A***', '***B***', '***C***', . . . , to prevent their being confused with attributes of individuals. With this additional symbolic apparatus, we can translate into our notation such propositions as 'Unpunctuality is a fault', and 'Honesty is a virtue'. Here we use '*Ux*', '*Hx*', '***F**F*', and '***V**F*', to abbreviate '*x* is unpunctual', '*x* is honest', '*F* is a fault', and '*F* is a virtue', and symbolize the two stated propositions as '***F**U*', and '***V**H*'. More complex propositions can also be symbolized. The propositions

> All useful attributes are desirable.

and

> Some desirable attributes are not useful.

can be symbolized, using the symbols '$\boldsymbol{U}F$' for 'F is useful' and '$\boldsymbol{D}F$' for 'F is desirable', as

$$(F)(\boldsymbol{U}F \supset \boldsymbol{D}F)$$

and

$$(\exists F)(\boldsymbol{D}F \cdot \sim \boldsymbol{U}F)$$

Finally, the proposition

Tom has all of his mother's good qualities.

can be symbolized, using the additional symbols 't' for 'Tom' and 'Mxy' for 'x is mother of y', either as

$$(F)\{(\exists x)[Mxt \cdot (y)(Myt \supset y = x) \cdot Fx \cdot \mathbf{G}\mathrm{F}] \supset Ft\}$$

or as

$$(\exists x)\{Mxt \cdot (y)(Myt \supset y = x) \cdot (F)[(Fx \cdot \boldsymbol{G}F) \supset Ft]\}$$

EXERCISES

I. Symbolize the following propositions:

1. Nothing has all attributes.
2. Some attributes belong to nothing.
***3.** No two things have all their attributes in common.
4. Any two things have some common attribute.
5. Napoleon had all the attributes of a great general. (n: Napoleon. Gx: x is a great general.)
***6.** David has all of his father's virtues and none of his faults. (d: David. Fxy: x is father of y. $\boldsymbol{F}F$: F is a fault. $\boldsymbol{V}F$: F is a virtue.)
7. Jones and Smith share all their good qualities but have no bad qualities in common. (j: Jones. s: Smith. $\boldsymbol{G}F$: F is a good quality. $\boldsymbol{B}F$: F is a bad quality.)
8. Nothing that possesses every rare attribute has any ordinary attribute. ($\boldsymbol{R}F$: F is rare. $\boldsymbol{O}F$: F is ordinary.)
***9.** A man who possesses all virtues is a virtuous man, but there are virtuous men who do not possess all virtues. (Mx: x is a man. Vx: x is a virtuous individual. $\boldsymbol{V}F$: F is a virtue.)
10. Everyone has some unusual attribute or other. It would be an unusual person who had no unusual attributes. (Px: x is a person. Ux: x is an unusual individual. $\boldsymbol{U}F$: F is an unusual attribute.)

II. Prove the following:

1. $[(x)(F)Fx] \equiv [(F)(x)Fx]$
***2.** $[(\exists x)(\exists F)Fx] \equiv [(\exists F)(\exists x)Fx]$
3. $[(\exists x)(F)Fx] \supset [(F)(\exists x)Fx]$
4. $[(\exists F)(x)Fx] \supset [(x)(\exists F)Fx]$
5. $[(\exists R)(x)(\exists y)Rxy] \supset [(x)(\exists y)(\exists R)Rxy]$

*6. Every (binary) relation that is transitive and irreflexive, is asymmetric.
7. All intransitive binary relations are irreflexive.
8. Any binary relation that every individual has to some individual or other is totally reflexive if it is both symmetric and transitive.
9. From the premiss '$(x)(y)[(x = y) \equiv (F)(Fx \equiv Fy)]$' (which is true by definition), deduce that the identity relation is symmetric, without using the principle Id.
*10. From the same premiss as in 9 and under the same restriction, deduce that the identity relation is totally reflexive.
11. From the same premiss as in 9 and under the same restriction, deduce that the identity relation is transitive.
12. If circles are ellipses, then circles have all properties of ellipses.
13. All (binary) relations that are both symmetric and transitive are reflexive.

6

Deductive Systems

> Science is built up with facts, as a house is with stones. But a collection of facts is no more a science than a heap of stones is a house.
>
> *Henri Poincaré*

6.1 Definition and Deduction

In the preceding chapters, a number of principles of logic have been set forth. These principles embody some knowledge about logic, but they do not constitute a *science* of logic, for *science* is *organized* knowledge. No mere list or catalog of truths is ever said to constitute a system of knowledge or a science. We have scientific knowledge only when the propositions setting forth what we know are organized in a systematic way, to display their interrelations. If a system of logic or a science of logical principles is to be achieved, those principles must be arranged or organized in a systematic fashion. This task will be attempted, on a limited scale, in the following chapters. But first, it will be of interest to consider the general questions of what interrelations are important, and how propositions may be organized into a system or science.

All knowledge that we possess can be formulated in propositions, and these propositions consist of terms. In any science, some propositions can be deduced from or proved on the basis of other propositions. For example, Galileo's laws of falling bodies and Kepler's laws of planetary motion are all derivable from Newton's more general laws of gravitation and motion. The discovery of these deductive interrelationships was an exciting phase in the development of the *science* of physics. Thus one important relationship among the propositions of a science is deducibility. Propositions that embody knowledge about a subject become a *science* of that subject when they are arranged or ordered by displaying some of them as conclusions deduced from others.

In any science, some of the terms involved in its propositions can be defined on the basis of other terms. For example, in physics again, *density* is defined as *mass per unit volume*. *Acceleration* is defined as the *time rate of change in velocity*, and *velocity* is in turn defined as the *time rate of change of position*. This definition of some terms by means of others also serves to reveal interrelations among the propositions. It shows their concern with a common subject matter and integrates the concepts of the science just as deductions integrate its laws or statements. Propositions that embody knowledge are helped to become a *science* when some of the words or symbols they contain are defined in terms of their other symbols.

The recognition that definition and deduction are important to science may suggest a specious ideal for scientific systems. It may be imagined that in an ideal science, *all* propositions should be *proved* by deducing them from others and *all* terms should be *defined*. But this would be 'ideal' only in the pejorative sense that this goal is impossible to realize. Terms can be defined only by means of other terms, whose meanings are presupposed and must be understood beforehand if the definitions are to explain the meanings of the terms being defined. Deductions can establish their conclusions only on the basis of premisses that must already have been verified if the conclusions are really to be established by the proofs. Hence, if all terms or symbols of a system are to be defined *within the system*, there must be either infinite sequences of definitions, or circular definitions, as in a pocket dictionary that defines the word 'big' as meaning *large*, and the word 'large' as meaning *big*. Circular definitions are obviously worthless as explanations, and infinite sequences of definitions are worthless also. No term can really be explained until the end of the sequence is reached and an infinite sequence has no end. Similarly, to prove *all* propositions, there must be either infinite regressions of proofs or circular proofs. And these are equally objectionable.

It must be admitted that *within* a system of propositions that constitutes a science, not all propositions can be proved and not all terms can be defined. It is not that there is some particular proposition that cannot be proved or some particular term that cannot be defined, but rather that they cannot *all* be proved or defined without a vicious regression or circularity. The ideal of science, then, cannot be a system in which *every* proposition is proved and *every* term defined. The ideal is rather one in which a minimum number of propositions suffices for the deduction of all the rest and a minimum number of terms suffices for the definition of all the others. This ideal of knowledge is described as a *deductive system*.

6.2 Euclidean Geometry

Euclidean Geometry is the oldest example of systematized knowledge or science. Of historical interest and importance in its own right, it has the advantage (for our purpose) of being an example with which the reader has already come into contact in high school.

It is generally recognized that geometry, as a science, was originated and developed by the Greeks. Among the most important contributors to its development were the mathematicians Pythagoras and Euclid. And yet, geometrical truths were known to the Egyptians thousands of years earlier, as attested by their pyramids that were already ancient in the time of Pythagoras (6th century B.C.). Records reveal that the Babylonians, even earlier, were familiar with various principles of geometry. If geometrical knowledge already existed before their time, in what sense did the Greeks originate the science of geometry? The answer has already been indicated. Before Pythagoras, man's geometrical knowledge consisted of a collection or catalog of almost wholly isolated facts. Geometrical truths constituted a mere list of useful empirical rules-of-thumb for surveying land or constructing bridges or buildings, and there was no system to their knowledge of geometrical truths. By introducing order into the subject, the Greeks transformed it from a mere body of isolated bits of knowledge into a science.

System was introduced into geometry by the deduction of some of its propositions from others. The propositions of geometry were ordered by listing first those that could be used as premisses in the demonstrations of propositions that were put later. This systematization of geometry was begun by Pythagoras and continued by his successors. It culminated in the *Elements* of Euclid (c.300 B.C.), in which all geometrical propositions were arranged in order, beginning with Axioms, Definitions, and Postulates, and continuing with Theorems deduced from the initial propositions. Geometry was cast by the Greeks into the form of a deductive system. Theirs was the first deductive system ever devised. So great was the achievement that it has served as a model for scientific thought down to the present time. Even today the most advanced sciences are those which most nearly approximate the form of a deductive system. These are the sciences that have achieved a relatively small number of very general principles from which a relatively large number of other laws and special cases have been derived. Parts of physics have actually been formulated as deductive systems and similar attempts have been made, with somewhat less impressive results, in parts of biology and psychology. Perhaps the boldest attempt in this direction was that of Spinoza, whose most important work, the *Ethics*, was written in 'geometrical' form. Starting with axioms and definitions, Spinoza attempted to deduce the rest of his metaphysical and ethical doctrines as theorems that were provable on the basis of those initial assumptions.

Euclid begins his geometry with definitions of some of the terms used in its development. Thus Definition 1 reads: 'A point is that which has no parts', and Definition 2 reads: 'A line is length without breadth'.[1] Euclid does not attempt to define *all* his terms, of course. The first two definitions define the terms 'point' and 'line', respectively. The words *used* in these definitions, such as 'parts', 'length', and 'breadth', are not, themselves, defined but are among the *undefined terms* of the system for Euclid. As more new terms are introduced, their definitions make use of previously defined terms as well as the original undefined ones. Thus Definition 4: 'A straight line is . . . [a line] . . . which lies evenly between its extreme points', makes use not only of such undefined terms as 'evenly' and 'between', but also the previously defined terms 'point' and 'line'.

The use of *defined* terms is, from the point of view of logic, a matter of convenience only. Theoretically, every proposition that contains defined terms

[1] These and the following are quoted from the Todhunter edition of *The Elements of Euclid*, No. 891, Everyman's Library, London and New York, 1933.

can be translated into one that contains only undefined ones. This can be done by replacing each occurrence of a defined term by the sequence of undefined terms that were used to define it. For example, Postulate 1: 'Let it be granted that a straight line may be drawn from any one point to any other point'. This postulate contains the defined terms 'straight', 'line', and 'point', and can be expressed, without using those defined terms, as: 'Let it be granted that a length without breadth which lies evenly between its extreme parts, which (themselves) have no parts, may be drawn from any one thing which has no parts to any other thing which has no parts'. This version of the Postulate is extremely awkward, however. Although they are theoretically eliminable, in actual practice a considerable saving of space, time, and effort is effected by using relatively brief, defined terms to replace lengthy sequences or combinations of undefined ones.

In setting up his deductive system of geometry, Euclid divided his unproved propositions into two groups, one called 'Axioms', the other called 'Postulates'. He gave, however, no reason for making this division, and there seems to be no very clear basis for distinguishing between them. Possibly, he felt that some were more *general* than others or psychologically more *obvious*. The contemporary practice is to draw no such distinction but to regard all the unproved, initial propositions of a deductive system as having the same standing. Now, it is customary to refer to them all, indifferently, as 'axioms' or as 'postulates', without attaching any difference in meaning to those two terms.

Every deductive system, to avoid falling into circularity or into a vicious regression, must contain some axioms (or postulates) that are assumed but are not proved within the system. They need not be *precarious* assumptions or *mere* assumptions. They may be very carefully and convincingly established—but they are *not proved within the system itself*. Any argument intended to establish the truth of the axioms is definitely *outside* the system, or *extra-systematic*.

The older conception of Euclidean geometry held that all of its theorems followed logically from its axioms and were, therefore, just as *true* as the axioms. Euclidean geometry also held that the axioms were *self-evident*. It is in this tradition that any statement was regarded as 'axiomatic' when its truth is beyond all doubt, since it was self-evident and did not require any proof. It should be clear from what has already been said, however, that we are *not* using the word 'axiom' in *that* sense. No claim is made that the axioms of any system are self-evidently true. Any proposition of a deductive system is an axiom of that system, if it is assumed, rather than proved, in that system. This modern point of view has arisen largely as a consequence of the historical development of geometry and physics.

The self-evident truth of the Euclidean axioms (and postulates) was long believed. It was not believed quite whole-heartedly, however. Most of the axioms, such as Axiom 9, 'The whole is greater than its part', were not questioned. Although there was no doubt about the *truth* of Axiom 12 (the famous 'parallel Postulate'), there was considerable scepticism about its 'self-evidence'. Axiom 12 reads: 'If a straight line meet two straight lines, so as to make the two interior angles on the same side of it, taken together, less than two right angles, these straight lines, being continually produced, shall at length meet on that side on which are the angles which are less than two right angles'.[2] Proclus, a fifth century A.D. commentator, wrote of it: 'This ought even to be struck out of the Postulates altogether; for it is a theorem involving many difficulties . . .'.[3] That is, although its *truth* was not questioned, its *self-evidence* was denied, which was deemed sufficient reason to demote it from its exalted position as an axiom to the more humble status of a mere theorem.

The history of mathematics is filled with attempts to prove the proposition in question as a theorem, either by deducing it from the remaining axioms of Euclid or from those axioms supplemented by some more nearly 'self-evident' additional assumption. The latter kind of attempt was pretty uniformly unsuccessful, because every additional or alternative assumption that proved strong enough to permit the deduction of the parallel postulate turned out to be no more self-evident than Euclid's own hypothesis. The first kind of attempt failed also. It was just not possible to deduce the parallel postulate from the others. The most fruitful attempt was that of the Italian mathematician Gerolamo Saccheri (1667–1733), who *replaced* the parallel postulate by alternative, contrary assumptions, and then sought to derive a contradiction from them, together with Euclid's other axioms. Had he succeeded in doing so, he would have obtained a *reductio ad absurdum* proof of the parallel postulate. He derived many theorems that he regarded as *absurd* because they were so different from common sense or ordinary geometrical intuition. He believed that he had succeeded thereby in demonstrating the parallel postulate and in 'vindicating Euclid'. Although 'absurd' in the sense that they violated ordinary geometrical intuitions, his derived theorems were *not* 'absurd' in the logical or mathematical sense of being self-contradictory. Instead of proving the parallel postulate, Saccheri (unknowingly) did something more important: He was the first to set up and develop a system of non-Euclidean geometry.

The parallel postulate is, in fact, *independent* of the other Euclidean postulates—although it was not *proved* to be so until the modern period. It is independent of the other postulates in the sense that neither it, nor its denial, is deducible from them. Alternative systems of 'geometry', non-Euclidean geometries, were subsequently developed, notably, by Lobachevsky and Riemann. These were long regarded as ingenious fictions, mere mathematical playthings, in contrast with the Euclidean geometry, which was thought to be 'true' of the real space about us. But subsequent physical and astronomical research along lines suggested by Einstein's theory of relativity have tended to show that—to the extent that the question is significant—'real' or physical

[2] Listed as Postulate 5 by Sir Thomas L. Heath in *The Thirteen Books of Euclid's Elements*, Cambridge University Press, Cambridge, England, 1926. For an interesting discussion of the history of the parallel postulate, the reader is referred to pp. 202 ff. of Volume 1 of that work.

[3] Ibid., p. 202.

space is more probably non-Euclidean than Euclidean. In any event, the truth or falsehood of its axioms is a purely *external* property of any deductive system. The truth of its propositions is an extrasystematic consideration. It is no doubt important, to the extent that a deductive system is ordered *knowledge*. When we concentrate our attention on the system as such, however, its *order* is its more important characteristic.

From the purely mathematical or logical point of view, a deductive system can be regarded as a vast and complex argument. Its premisses are the axioms, and its conclusion is the conjunction of all the theorems that are deduced. As with any other argument, the logical question does not concern the truth or falsehood of the premisses, but deals with the validity of the inference. Granted the truth of the axioms, does the truth of the theorems necessarily follow? That is the question with which the logician and the mathematician are concerned. The answer is, of course, yes—*if* the demonstrations of the theorems are all valid arguments. Hence the most important aspect of any deductive system is the cogency with which its theorems are proved. During the rigorous development of deductive systems in abstraction from the extrasystematic explanation of their undefined terms, the question of truth or falsehood is obviously irrelevant.

6.3 Formal Deductive Systems

There are serious errors in the system of geometry set forth by Euclid in his *Elements*. Indeed, a mistake occurs in his very first proof. The flaw in his proof, paradoxically enough, was the result of his knowing too much about his subject. He did not appeal to his explicitly stated axioms, alone, as premisses, but depended, also, upon what might be called his geometrical intuition.[4] Where a chain of argument involves familiar notions, there is always the danger of assuming more than the explicitly stated premisses warrant. This is particularly serious in the development of a deductive system, for any attempted systematization that appeals to unacknowledged assumptions in the derivations of its theorems, thereby *fails* to achieve its aim. In a deductive system, the theorems must be deduced *rigorously* from the stated postulates. If they are not, however true they may be, the result falls short of the goal of systematization.

Since lapses from rigor are most often occasioned by too great familiarity with the subject matter, mathematicians have found it helpful to minimize or eliminate such familiarity in the interest of achieving greater rigor. In the case of geometry, that end is accomplished by abstracting from the meanings of such geometrical words as 'point', 'line', and 'plane', and developing the

[4] Euclid's proof and a brief discussion of his mistake can be found on pp. 241–243 of vol. 1 of *The Thirteen Books of Euclid's Elements*, by Heath, op. cit. An example of how the same type of error can lead to conclusions that are false or even self-contradictory can be found on pp. 77–78, W. W. Rouse Ball, *Mathematical Recreations and Essays*, Macmillan Publishing Co., Inc., New York, 1940.

theorems as purely formal consequences of the postulates. The familiar geometrical words, with all their associations and suggestions, are replaced by arbitrary symbols. Instead of deductive systems explicitly and avowedly concerned with geometrical entities, mathematicians today develop *formal* deductive systems whose primitive or undefined terms include arbitrary, uninterpreted symbols—usually letters of the Greek or Latin alphabets. Since the undefined terms of a *formal* deductive system include arbitrary symbols, its postulates are not propositions at all, but mere formulas. The theorems also are formulas. The deductive system for Class Algebra in Section 7.2 is a formal deductive system in the sense just explained.

Deductive relationships can exist, of course, among mere formulas as well as among propositions. Thus the formula 'all *F*'s are *H*'s' is logically deducible from the formulas 'all *F*'s are *G*'s' and 'all *G*'s are *H*'s'. Because the postulates and theorems of a formal deductive system are formulas rather than propositions, the proofs of theorems can proceed unhampered by familiar associations and unconscious assumptions. Moreover, because the formulas are not propositions, the question of their truth is strictly irrelevant and does not arise.

More than rigor is gained by the formal development of deductive systems. Since some of the symbols of a formal deductive system are arbitrary, uninterpreted symbols, it may be possible to give them different, alternative interpretations. And since the theorems are formal consequences of the axioms, any interpretation of the arbitrary symbols that makes the axioms true will necessarily make the theorems true, also. The additional advantage of generality is thus gained. An example may help to make this clear. With some knowledge about astronomy, a deductive system for that subject can be set up. To avoid the errors into which familiarity with the subject matter may lead in deducing theorems from the axioms chosen, the system may be developed *formally*. Instead of taking, say, 'stars' and 'planets' among the undefined terms, one may take '*A*'s and '*B*'s. The axioms and theorems will contain these symbols, and when the system is developed, all its formulas may be interpreted by letting the symbol '*A*' designate stars and the symbol '*B*' designate planets. Now, if the axioms are *true* when so interpreted, the theorems must be true also, and the formal system with this interpretation will constitute a science or deductive system of astronomy. It may be possible on the other hand, to find *different* interpretations of the symbols '*A*' and '*B*' that also make the axioms true (and hence the theorems also). The formulas of the system might be made into different, but equally true, statements by letting the symbol '*A*' designate atomic nuclei and the symbol '*B*' designate electrons. If this could be done (and at one stage in the history of atomic physics it seemed highly plausible), the original formal system with this second interpretation would constitute a science or deductive system of atomic physics. Hence developing a deductive system formally, that is, not interpreting its undefined terms until after its theorems have all been derived, not only helps achieve rigor in its development, but also achieves greater generality because of the possibility of finding alternative interpretations for and applications of the deductive system that is formulated. This kind of advantage is often realized in pure mathematics. For example, different interpretations of its arbitrary primitive symbols will transform the same formal deductive system into the theory of real numbers, on the one hand, or into the theory of points on a straight line, on the other. That fact provides the theoretical foundation for the branch of mathematics called *Analytical Geometry*. In Section 7.2, we note that the formal deductive system for Class Algebra is also a Propositional Calculus.

As the term is being used here, a *formal deductive system* is a deductive system that consists of axioms and theorems, some of whose undefined or primitive terms are arbitrary symbols, the interpretation of which is completely extrasystematic. In addition to those special undefined terms, and others defined by means of them, the formulas (axioms and theorems) of the system contain only such logical terms as 'if . . . then . . .', 'and', 'or', 'not', 'all', 'are', and the like, and possibly (unless the system is intended for arithmetic itself) such arithmetical terms as 'sum' and 'product', and numerical symbols.

6.4 Attributes of Formal Deductive Systems

Usually, though not always, a formal deductive system is set up with some particular interpretation 'in mind'. That is, the investigator has some knowledge about a certain subject, and wishes to set up a system adequate for its expression. When the formal system has been constructed, the question naturally arises whether or not it is adequate for the formulation of all the propositions it is intended to express. If it is, it may be said to be 'expressively complete' *with respect to that subject matter*. We are here discussing what can be *said* in the system, *not* what can be proved. With respect to a given subject matter, a formal deductive system is 'expressively complete' when it is possible to assign meanings to its undefined terms in such a way that every proposition about that subject matter can be *expressed* as a formula of the system. Whether the *true* propositions can be *proved as theorems* or not is another question that will be discussed later in this section.

A system is said to be *inconsistent* if two formulas—one of which is the denial or contradictory of the other—can both be proved as theorems within it. A system is *consistent* if it contains no formula such that both the formula and its negation are provable as theorems within it. As was remarked in Section 3.7, a contradiction logically entails any proposition whatever. Hence a derivative definition or criterion for consistency can be formulated as follows: Any system is consistent if it contains (that is, can express) a formula that is not provable as a theorem within the system. This is known as the 'Post criterion for consistency', having been enunciated by the American mathematician and logician, E. L. Post. Consistency is of fundamental importance. An inconsistent deductive system is worthless, for all of its formulas are provable as theorems, including those which are explicit negations of others. When the undefined terms are assigned meanings, these contradictory formulas become contradictory propositions and cannot possibly all be true. Since they cannot possibly be true, they cannot serve as a systematization of knowledge—for knowledge is expressed in true propositions only.

If one succeeds in deriving both a formula and its negation as theorems of a system, that proves the system to be inconsistent. But if one tries and does not succeed in deriving both a formula and its negation as theorems, that does *not* prove that the system is consistent. It may only reflect a lack of ingenuity in making proofs on the part of the investigator. How, then, can the consistency of a deductive system be established? One method of proving the consistency of a formal deductive system is to find an interpretation of the system in which all its axioms and theorems are true propositions. Since its theorems are logical consequences of its axioms, any interpretation that makes its axioms true will make its theorems true, also. Hence it is sufficient for the purpose of proving that a system is consistent to find an interpretation that makes all of its axioms true.

The axioms of a deductive system are said to be *independent* (or to exhibit *independence*) if no one of them can be derived as a theorem from the others. A deductive system that is not consistent is logically objectionable and utterly worthless. There is no *logical* objection to a deductive system whose axioms are not independent. However, it is often felt that making more assumptions than necessary for the development of a system is extravagant and inelegant and should be avoided. When a formula need not be assumed as an axiom but can be proved as a theorem from the other axioms that are assumed, aesthetic and prudential considerations dictate that it *ought* to be proved and not assumed for the sake of 'economy'. A set of axioms that are not independent is said to be 'redundant'. A redundant set of axioms is aesthetically inelegant, but it is not logically 'bad'.

If one of the axioms of a system *can* be derived from the remaining ones, the set of axioms is thereby shown to be redundant. But if one tries and is not able to derive any of the axioms from the remaining ones, the axioms are *not* thereby shown to be independent. The failure to find a demonstration may be due only to the investigator's lack of ingenuity. To prove any particular axiom independent of the others, it suffices to find an interpretation that makes the axiom in question *false* and the remaining ones all *true*. Such an interpretation will prove that the axiom in question is not derivable as a theorem from the others, for if it were, it would be made true by any assignment of meanings that made the others true. If such an interpretation can be found for each axiom, this will prove the set of axioms to be independent.

The notion of *deductive completeness* is very important. The term 'completeness' is used in various senses. In the least precise sense of the term, we can say that a deductive system is complete if all the *desired* formulas can be proved within it. We may have an extrasystematic criterion for the truth of propositions about the subject matter for which we constructed the deductive system. If we have, then we may call that system complete when all of its formulas that become true propositions on the intended interpretation are provable formulas or theorems of the system. (In any sense of the term, an *inconsistent* system will be *complete*, but in view of the worthlessness of inconsistent systems, we shall confine our attention here to *consistent* systems.)

There is another conception of *completeness* that can be explained as follows. Any formal deductive system will have a certain collection of special undefined or primitive terms. Any terms that are definable within the system are theoretically eliminable, that is, they are replaceable in any formula in which they occur. They can be so replaced by the sequence of undefined terms by means of which they were first defined. We shall ignore defined terms for the present. All formulas that contain no terms other than these special undefined terms (and logical terms) are expressible within the system. We may speak of the totality of undefined terms as the *base* of the system. The formulas expressible in the system are all formulas constructed *on that base*. In general, the totality of formulas constructed on the base of a given system can be divided into three groups. First is the group of all formulas that are provable as theorems within the system. Second is the group of all formulas whose negations are provable within the system and third, the group of all formulas such that neither they nor their negations are provable within the system. For *consistent* systems, the first and second groups are *disjoint* or *mutually exclusive*, that is, they have no formulas in common. Any system whose third group is empty, that is, contains no formulas at all, is said to be *deductively complete*. An alternative way of phrasing this sense of completeness is to say that every formula of the system is such that either it or its negation is provable as a theorem.

Another definition of 'completeness', that is entailed by, but not equivalent to, the preceding one, is that a deductive system is complete when every formula constructed on its base is either a theorem or else its addition as an axiom would make the system inconsistent.

An example of an incomplete deductive system would be Euclidean geometry minus the parallel postulate. For the parallel postulate is, itself, a formula constructible on the base of the Euclidean system. Yet, neither it, nor its negation, is deducible from the other postulates. It is clear that, although completeness is an important attribute, an incomplete deductive system may be very interesting and valuable. By investigating the incomplete system of Euclidean geometry without the parallel postulate, we can discover those properties possessed by space independently of the question of whether it is Euclidean or non-Euclidean. Perhaps a more cautious and less misleading formulation of the same point is to say that by investigating the incomplete system, we can discover the *common features* of Euclidean and non-Euclidean geometries. Yet, for many purposes, a complete system is to be preferred.

6.5 Logistic Systems

The most important attribute that a deductive system should possess is that of rigor. A system has rigor when no formula is asserted to be a theorem unless it is logically entailed by the axioms. For the sake of rigor, arbitrary rather than familiar symbols are taken as undefined or primitive terms and the system is developed *formally*. Listing clearly all the undefined terms and explicitly stating all the axioms used as premisses for the theorems will help specify precisely which formulas are to be esteemed as theorems and which are not. With the increased emphasis on rigor that characterizes the modern period, critical mathematicians have seen that this procedure is not enough. To achieve rigor, more is required.

A system is rigorous only when its theorems are proved logically, or derived logically from its axioms. However clearly its axioms are stated, a formal system will lack rigor unless the concept of *logical proof* or *logical derivation* is also specified precisely. All deductive systems of the sort that have been mentioned, even formal deductive systems that contain logical terms in addition to their own special uninterpreted symbols, depend upon 'ordinary logic' for their development. They *assume* logic, in the sense that their theorems are supposed to follow *logically* from their axioms. But they do not specify what this 'logic' is. Hence all earlier deductive systems, for geometry, or physics, or psychology, or the like, contain concealed assumptions that are not explicitly stated. These hidden assumptions are the rules or principles of logic to which one appeals in constructing proofs or derivations of theorems. Hence all those deductive systems fall short of complete rigor, since not all of their presuppositions are acknowledged. Therefore, their developments are not entirely rigorous, but are more or less loose.

The question naturally arises: How can this looseness be eliminated and greater rigor be achieved? The answer is obvious enough. A deductive system will be developed more rigorously when it is specified not only what axioms are assumed as premisses in deriving the theorems, but also what principles of inference are to be used in the derivations. The axioms must be supplemented by a list of valid argument forms or principles of valid inference.

The demand for rigor and for system does not stop even here, however. For the sake of rigor, in addition to its own special axioms, a deductive system must specify explicitly what forms of inference are to be accepted as valid. It would be unsystematic—and probably impossible—simply to list or catalog *all* required rules of logic or valid modes of inference. A deductive system of logic, itself, must be set up. Such a deductive system will have deduction, itself, as its subject matter. A system of this type, often referred to as a *logistic system,* must differ from the ordinary, less formal varieties in several important respects. Since the subject matter is deduction itself, the logical terms 'if . . . then . . .', 'and', 'or', 'not', and so on, cannot occur in it with their ordinary meanings simply assumed. Uninterpreted symbols must be used in their stead. The logical principles or rules of inference that *it* assumes for the sake of deducing logical theorems from logical axioms must be few in number and explicitly stated.

A second fundamental difference between logistic systems and other formal deductive systems is that in the latter, the concept of a significant or 'well formed' formula need not be specified. It is absolutely required in a logistic system, however. In an ordinary (nonformal) deductive system, it will be obvious which sequences of its words are significant propositions of English (or of whatever the natural language is in which the system is expressed). In a formal but nonlogistical deductive system, the sequences of its symbols are easily divided into those which 'make sense' and those which do not, for they will contain such ordinary logical words as 'if . . . then . . .', 'and', 'or', or 'not', by whose disposition in it the sequence can be recognized as significant or otherwise. An example will make this clear. In a formal deductive system that contains '*A*', '*B*', and '*C*' as uninterpreted primitive symbols, the sequence of symbols 'If any *A* is a *B*, then it is a *C*' is clearly a complete and 'significant' formula that may or may not be provable as a theorem. The sequence of symbols 'If any *A* is a *B*' is obviously *incomplete,* while the sequence 'And or or *A B* not not if' is clearly nonsense. These are recognized as 'complete' or 'well formed', as 'incomplete' or 'ill formed' by the presence in them of *some* symbols whose meanings are understood. In a logistic system, however, *all* symbols are uninterpreted. There are no familiar words within its formulas (or sequences of symbols) to indicate which are 'well formed' and which are not. Where the symbols '*A*', '*B*', '$\sim$', and '$\supset$' are uninterpreted, there must be some method of distinguishing between a well formed formula like '$A \supset \sim B$' and one like '$AB \supset \sim$', which is not well formed. By our knowledge of the normal interpretations of these symbols, we can recognize the difference and classify them correctly, but for the *rigorous* development of our system, we must be able to do this in abstraction from the (intended) meanings of the symbols involved.

The matter may be expressed in the following terms. As ordinarily conceived, a nonformal deductive system (interpreted, like Euclidean geometry) is an arrangement or organization of propositions about some specified subject matter. Consisting of propositions, it is a *language* in which the subject matter may be discussed. Understanding the language, we can divide all sequences of its words into those which are meaningful statements and those which are meaningless or nonsensical. This division is effected in terms of meanings and is thus done *nonformally.* In a logistic system, the situation is different because prior to the extrasystematic assignment of meanings or interpretation, *all* sequences of symbols are without meaning. Yet we want, prior to and independent of its interpretation, a comparable division of all of its formulas into two groups. When meanings are assigned to the primitive symbols of a logistic system, some of its formulas will express propositions, while others will not. We may informally characterize a formula which on the intended interpretation becomes a significant statement as a 'well formed formula' (customarily abbreviated '*wff*'). Any formulas which on the intended interpretation do *not* become significant statements, are *not* well formed formulas. In a logistic system, there must be a *purely formal* criterion for distinguishing well formed formulas from all others. To characterize the criterion as 'purely formal' is to say that it is *syntactical* rather than *semantical,* pertaining to the formal characteristics and arrangements of the symbols in abstraction from their meanings. Thus a logistic system must contain only uninterpreted symbols and must provide a criterion for dividing sequences of these symbols into two groups. The first group will contain all well formed formulas, the second group will contain all others. Of the well formed formulas, some will be designated as Axioms (or Postulates) of the logistic system.

All well formed formulas that are not axioms are divided into two groups, those which are theorems and those which are not. The former are those which are derivable from the axioms or postulates, *within the system.* Although uninterpreted, the well formed formulas of a logistic system constitute a 'language' in which derivations or proofs can be set down. Some well formed formulas are assumed as axioms, and other well formed formulas are derived from them as theorems. It might be proposed to define 'theorem' as any *wff* that is the conclusion of a valid argument whose premisses include only axioms of the system. This proposed definition of 'theorem' will be acceptable only if the concept of a *valid* argument within the logistic system can be formally defined. Because all *wffs* of the system are uninterpreted, the ordinary conception of validity cannot be used to characterize arguments within the system. This is the case because the usual concept of validity is *semantical.* An argument is regarded as valid if and only if the *truth* of its premisses entails the *truth* of its conclusion. Consequently, a purely formal or syntactical *criterion of validity* must be provided for arguments that are expressed *within the system.* 'Valid' arguments within the system may have not only postulates or already established theorems as premisses, with new theorems as conclusions. They may also have as premisses *any wffs,* even those which are neither postulates nor theorems, and they may have as conclusions *wffs* that are not theorems. Of course, it is desired that any argument within the system which is syntactically 'valid' will become, when given its intended or 'normal' interpretation, a semantically valid argument.

Any logistic system, then, will contain the following elements: (1) a list of primitive symbols which, together with any symbols defined in terms of them, are the only symbols that occur within the system; (2) a purely formal or syntactical criterion for dividing sequences of these symbols into formulas that are well formed (*wffs*) and those that are not; (3) a list of *wffs* assumed as postulates or axioms; (4) a purely formal or syntactical criterion for dividing sequences of well formed formulas into 'valid' and 'invalid' arguments; and (5), derivatively from (3) and (4), a purely formal criterion for distinguishing between theorems and nontheorems of the system.

Different logistic systems may be constructed as systematic theories of different parts of logic. The simplest logistic systems are those which formalize the logic of truth-functional compound statements. These systems are called *propositional calculi* or, less frequently, *sentential calculi.* One particular propositional calculus will be presented and discussed in Chapter 8.

7

Set Theory

7.1 The Algebra of Classes

The four traditional types of subject-predicate propositions discussed in Section 4.1 can be understood as being about classes. On this interpretation, these categorical propositions (as they have been traditionally called) are understood to affirm or to deny that one class is included in another, either in whole or in part. Thus the ***A*** proposition 'All humans are mortal' is taken to assert that the whole of the class of humans is included in the class of mortal beings, and the ***I*** proposition 'Some humans are mortal' asserts that *part* of the class of humans is included in the class of mortals. Their respective negations, the ***O*** and ***E*** propositions, simply deny these inclusions.[1]

The notion of a *class* is too basic to be defined in terms of more fundamental concepts. A variety of synonyms are available: collection, aggregate, totality, set, and so on. Although mentioning them may help to make the notion of class more intuitive, they can scarcely provide a useful definition. In the following sections, where axiomatic treatments are set forth, 'class' is an undefined or primitive term.

On the class interpretation of categorical propositions, both their subject and predicate terms designate classes. What has been said so far may suggest that the relation of class inclusion is fundamental. But in developing an algebraic treatment of classes, statements about them are most conveniently expressed as equations and inequalities. To permit this kind of formulation, three *operations* on classes are taken as fundamental, and in terms of these operations the relation of class inclusion can readily be defined.

In this section and the next, we use lower-case Greek letters to symbolize classes. Given the classes α, β, γ, . . . additional classes can be defined in terms of them. We can define the class of all objects that belong either to α or to β; this class is called the *sum* or *union* of α and β and symbolized by '$\alpha \cup \beta$'. We can define the class of all objects that belong to both α and β; this class is called the *product* or *intersection* of α and β and is symbolized by '$\alpha \cap \beta$' or simply

[1] See Chapters 5 and 6, Irving M. Copi, *Introduction to Logic*, Macmillan Publishing Co., Inc., New York and London, 1978.

'$\alpha\beta$'. And we define the class of all objects that do *not* belong to α; this class is called the complement of α and is symbolized either by $-\alpha$ or by $\bar{\alpha}$.

Many statements about classes can be formulated using the ordinary equals sign. The equation '$\alpha = \beta$' asserts that all members of α, if any, are also members of β, *and* that all members of β, if any, are also members of α. Many of the properties of the sum, product, and complement of classes can be expressed by means of equations. It is clear, for example, that the operations of forming the sum and forming the product of two classes are commutative. For example, symbolically, we have

$$(\alpha \cup \beta) = (\beta \cup \alpha) \quad \text{and} \quad \alpha\beta = \beta\alpha$$

They are also associative

$$(\alpha \cup \beta) \cup \gamma = \alpha \cup (\beta \cup \gamma) \quad \text{and} \quad (\alpha\beta)\gamma = \alpha(\beta\gamma)$$

Two principles of distribution also hold for class sums and products. Any object that belongs either to α or to both β and γ must belong either to α or to β and must also belong either to α or to γ, and conversely. In symbols,

$$\alpha \cup (\beta\gamma) = (\alpha \cup \beta)(\alpha \cup \gamma)$$

Moreover, any object that belongs both to α and to either β or γ must belong either to both α and β or to both α and γ. In symbols,

$$\alpha(\beta \cup \gamma) = \alpha\beta \cup \alpha\gamma$$

Two other principles express the *idempotence* of $\cup$ and $\cap$; they resemble the tautology principle for statements (see page 40) and are immediate consequences of the definitions of the sum and the product of classes:

$$\alpha = \alpha \cup \alpha \quad \text{and} \quad \alpha = \alpha\alpha$$

Another immediate consequence of those definitions is the principle of absorption:

$$\alpha = \alpha \cup \alpha\beta$$

Turning now to the concept of class complement, we observe that since any object belongs to a given class if and only if it does not belong to the class of all objects that do not belong to the given class, the complement of the complement of a class is the class itself. We thus have a sort of double negative rule for complementation, which can be expressed in symbols as

$$\alpha = \bar{\bar{\alpha}}$$

An object that does not belong to the sum of two classes belongs to neither of them and must, therefore, belong to both of their complements. And an object that does not belong to the product of two classes must belong to the class complement of at least one of them. These two propositions, and their converses, which are also true, can be expressed symbolically as

$$\overline{\alpha \cup \beta} = \bar{\alpha}\bar{\beta} \quad \text{and} \quad \overline{\alpha\beta} = \bar{\alpha} \cup \bar{\beta}$$

which are versions of De Morgan's theorems applying to classes.

Two special classes are the *empty* class, which has no members, and the *universal* class, to which all objects belong. The empty class is symbolized as '0' (sometimes as 'Λ', and usually in set theory proper as '$\varnothing$') and the universal class is symbolized as '1' (sometimes as 'V'). It is clear that the empty class is the complement of the universal class:

$$\bar{1} = 0$$

The following two equations are immediate consequences of the preceding definitions:

$$\alpha \cup \bar{\alpha} = 1 \quad \text{and} \quad \alpha\bar{\alpha} = 0$$

Further immediate consequences are these:

$$\alpha \cup 0 = \alpha,\ \alpha 1 = \alpha,\ \alpha 0 = 0, \quad \text{and} \quad \alpha \cup 1 = 1$$

It is easily shown that any class can be designated by infinitely many different *class-expressions*. Thus the class designated by 'α' can also be designated by '$\alpha(\beta \cup \bar{\beta})$' (since $\beta \cup \bar{\beta} = 1$ and $\alpha 1 = \alpha$), and by '$[\alpha(\beta \cup \bar{\beta})](\gamma \cup \bar{\gamma})$', and so on. By this 'Law of Expansion' we can always introduce any class symbol we choose into a given class expression in such a way that the original and the expanded class expressions designate the same class.

By the principle of distribution, the class $\alpha(\beta \cup \bar{\beta})$ is the same as $\alpha\beta \cup \alpha\bar{\beta}$. To aid in describing the form of the latter expression, let us use the phrase 'simple class term' to refer to the class symbols 'α', 'β', 'γ', . . . , in contrast to other class expressions such as sums and products. Now we can describe the expression '$\alpha\beta \cup \alpha\bar{\beta}$' as a sum of distinct products, such that in each product only simple class terms or their complements appear and such that any simple class term which appears anywhere in the entire expression appears exactly once in every product.[2] By means of the equations presented thus far, any class expression can be transformed into another, perhaps simpler, expression that designates the same class. Thus $\overline{\alpha(\bar{\alpha} \cup \beta)}$ is equal by De Morgan's Theorem to $\bar{\alpha} \cup (\overline{\bar{\alpha} \cup \beta})$ which is equal, again by De Morgan's Theorem, to $\bar{\alpha} \cup \bar{\bar{\alpha}}\bar{\beta}$. By

[2]Compare these with the Boolean Expansions for statements discussed in Appendix B.

double negation this result is equal to $\bar{\alpha} \cup \alpha\bar{\beta}$, which is equal by expansion to $\bar{\alpha}(\beta \cup \bar{\beta}) \cup \alpha\bar{\beta}$, which by distribution is equal to $\bar{\alpha}\beta \cup \bar{\alpha}\bar{\beta} \cup \alpha\bar{\beta}$. (Our association principle, $(\alpha \cup \beta) \cup \gamma = \alpha \cup (\beta \cup \gamma)$, permits us to drop parentheses and write either simply as $\alpha \cup \beta \cup \gamma$.)

Any class, whatever, will divide the universal class into two subdivisions or subclasses which are mutually exclusive and jointly exhaustive. That is, for any class α: $1 = \alpha \cup \bar{\alpha}$ and $\alpha\bar{\alpha} = 0$. Any two classes will divide the universal class into *four* subclasses that are exclusive and exhaustive. Thus for any classes α and β: $1 = \alpha\beta \cup \alpha\bar{\beta} \cup \bar{\alpha}\beta \cup \bar{\alpha}\bar{\beta}$, and the product of any two of those four products is the empty class. Similarly, any n classes will divide the universal class into 2^n subclasses which are exclusive and exhaustive.

The notations introduced thus far permit symbolizing the **A** and **E** subject-predicate propositions. The **E** proposition, *No* α *is* β, asserts that the classes α and β have no members in common, which means that their product is empty. The **E** proposition is, therefore, symbolized as

$$\alpha\beta = 0$$

The **A** proposition, *All* α *is* β, asserts that there is nothing which belongs to α but not to β, which means that the product of α and the complement of β is empty. The **A** proposition is, therefore, symbolized as

$$\alpha\bar{\beta} = 0$$

To symbolize the ***I*** and ***O*** categorical propositions, we must introduce the inequality sign '$\neq$', where '$\alpha \neq \beta$' asserts that either α contains an object that is not a member of β or β contains an object that is not a member of α. The ***I*** proposition: *Some* α *is* β, asserts that there is at least one member of α which is also a member of β, that is, that the product of α and β is not empty. In symbols, the ***I*** proposition appears as

$$\alpha\beta \neq 0$$

The ***O*** proposition: *Some* α *is not* β, asserts that there is at least one member of α that is not a member of β, that is, the product of α and $\bar{\beta}$ is not empty. In symbols, the ***O*** proposition is expressed as

$$\alpha\bar{\beta} \neq 0$$

When formulated in our class notation, it is completely obvious that the **A** and ***O*** propositions are contradictories, as are the **E** and ***I*** propositions.

Some of the traditional 'immediate inferences' involving categorical propositions are already contained in the notation of the class algebra. Thus every proposition has exactly the same symbolization as its obverse, for example, '$\alpha\bar{\beta} = 0$' symbolizes both *All* α *is* β and *No* α *is non-*β. And conversion, where it is valid, is an immediate consequence of the principle of commutation, for example, *Some* α *is* β and *Some* β *is* α are symbolized by '$\alpha\beta \neq 0$' and '$\beta\alpha \neq 0$', respectively, which are obviously equivalent, since $\alpha\beta = \beta\alpha$ by commutation.

When we turn to the 'mediate inferences' of the traditional categorical syllogism, we can divide all categorical syllogisms into two kinds: those which contain only universal propositions (**A** and **E**), and those which contain at least one existential proposition (**I** or **O**). It is easily shown that all valid syllogisms of the first kind have the form

$$\alpha\bar{\beta} = 0,\ \beta\bar{\gamma} = 0 \therefore \alpha\bar{\gamma} = 0$$

The validity of this form can be derived within the algebra of classes by appealing to the results already set forth in this section. Since $\bar{\gamma}0 = 0$, and $\alpha\bar{\beta} = 0$ is a premiss, we have $\bar{\gamma}(\alpha\bar{\beta}) = 0$, which by association and commutation yields $(\alpha\bar{\gamma})\bar{\beta} = 0$. Now $\alpha 0 = 0$, and $\beta\bar{\gamma} = 0$ is a premiss, so we have $\alpha(\beta\bar{\gamma}) = 0$, which by association and commutation yields $(\alpha\bar{\gamma})\beta = 0$. Hence $(\alpha\bar{\gamma})\beta \cup (\alpha\bar{\gamma})\bar{\beta} = 0$, which by distribution yields $(\alpha\bar{\gamma})(\beta \cup \bar{\beta}) = 0$. Since $\beta \cup \bar{\beta} = 1$ and $(\alpha\bar{\gamma})1 = \alpha\bar{\gamma}$, we have $\alpha\bar{\gamma} = 0$, the syllogism's conclusion.

It can also be shown that all valid syllogisms of the second kind have the form

$$\alpha\beta \neq 0,\ \beta\bar{\gamma} = 0 \therefore \alpha\gamma \neq 0$$

To establish the validity of this form, we first observe that since $\alpha 0 = 0$, if $\alpha\beta \neq 0$ then $\alpha \neq 0$ and $\beta \neq 0$. Since $\alpha 0 = 0$, and $\beta\bar{\gamma} = 0$ as a premiss, $\alpha(\beta\bar{\gamma}) = 0$. By association, we have $(\alpha\beta)\bar{\gamma} = 0$. Now $\alpha\beta = (\alpha\beta)1$ and $\gamma \cup \bar{\gamma} = 1$, hence $\alpha\beta = (\alpha\beta)(\gamma \cup \bar{\gamma})$, and by distribution, $\alpha\beta = (\alpha\beta)\gamma \cup (\alpha\beta)\bar{\gamma}$. But $(\alpha\beta)\gamma \cup 0 = (\alpha\beta)\gamma$, and we have already shown that $(\alpha\beta)\bar{\gamma} = 0$, hence $\alpha\beta = (\alpha\beta)\gamma$. Since $\alpha\beta \neq 0$ is a premiss, we know that $(\alpha\beta)\gamma \neq 0$. By association and commutation, we obtain $(\alpha\gamma)\beta \neq 0$ from which it follows that $\alpha\gamma \neq 0$, which is the syllogism's conclusion. Hence the algebra of classes is adequate not only to validate immediate inferences involving categorical propositions, but is capable of validating categorical syllogisms also.

The symbol '$\subset$' is introduced to symbolize *class inclusion*. The expression '$\alpha \subset \beta$' asserts that all members of α, if any, are also members of β, and is used as an alternative formulation of the **A** proposition, *All* α *is* β. It can be defined in terms of the symbols already introduced in various ways: either as $\alpha\bar{\beta} = 0$ or as $\alpha\beta = \alpha$ or as $\alpha \cup \beta = \beta$ or as $\bar{\alpha} \cup \beta = 1$, all of which are obviously equivalent. The relation $\subset$ is reflexive and transitive (see pages 135–136) and has the (transposition) property that if $\alpha \subset \beta$ then $\bar{\beta} \subset \bar{\alpha}$. The latter is an immediate consequence of double negation and commutation when '$\alpha \subset \beta$' is rewritten as '$\alpha\bar{\beta} = 0$' and '$\bar{\beta} \subset \bar{\alpha}$' is rewritten as '$\bar{\beta}\bar{\bar{\alpha}} = 0$'. Its reflexiveness is obvious when '$\alpha \subset \alpha$' is rewritten as '$\alpha\bar{\alpha} = 0$', and its transitivity has already been established in our algebraic proof of validity for categorical syllogisms containing only universal propositions.

7.2 Axioms for Class Algebra

The algebra of classes can be set up as a formal deductive system. Such a system is called a Boolean Algebra, and a vast number of alternative postulate sets for Boolean Algebra have been devised. One of them can be set forth as follows.

Special undefined primitive symbols:

$$\mathbf{C},\ \cap,\ \cup,\ -,\ \alpha,\ \beta,\ \gamma,\ \ldots$$

AXIOMS:

Ax. 1. If α and β are in $\mathbf{C}$, then $\alpha \cup \beta$ is in $\mathbf{C}$.
Ax. 2. If α and β are in $\mathbf{C}$, then $\alpha \cap \beta$ is in $\mathbf{C}$.
Ax. 3. There is an entity 0 in $\mathbf{C}$ such that $\alpha \cup 0 = \alpha$ for any α in $\mathbf{C}$.
Ax. 4. There is an entity 1 in $\mathbf{C}$ such that $\alpha \cap 1 = \alpha$ for any α in $\mathbf{C}$.
Ax. 5. If α and β are in $\mathbf{C}$, then $\alpha \cup \beta = \beta \cup \alpha$.
Ax. 6. If α and β are in $\mathbf{C}$, then $\alpha \cap \beta = \beta \cap \alpha$.
Ax. 7. If α, β, γ are in $\mathbf{C}$, then $\alpha \cup (\beta \cap \gamma) = (\alpha \cup \beta) \cap (\alpha \cup \gamma)$.
Ax. 8. If α, β, γ are in $\mathbf{C}$, then $\alpha \cap (\beta \cup \gamma) = (\alpha \cap \beta) \cup (\alpha \cap \gamma)$.
Ax. 9. If there are unique entities 0 and 1 satisfying Axioms 3 and 4, then for every α in $\mathbf{C}$ there is an $-\alpha$ in $\mathbf{C}$ such that

$$\alpha \cup -\alpha = 1 \quad \text{and} \quad \alpha \cap -\alpha = 0$$

Ax. 10. There is an α in $\mathbf{C}$ and a β in $\mathbf{C}$ such that $\alpha \neq \beta$.

The present system[3] is a formal deductive system rather than a logistic system. On its intended interpretation, of course, $\mathbf{C}$ is the collection of all classes, 0 and 1 are the empty and universal classes, respectively, and the symbols $\cup$, $\cap$, and $-$ represent class addition, multiplication, and complementation, respectively.

The reader who is interested in deducing some theorems from these axioms will find the following fairly easy to derive:

Th. 1. There is at most one entity 0 in $\mathbf{C}$ such that $\alpha \cup 0 = \alpha$.
*Th. 2. There is at most one entity 1 in $\mathbf{C}$ such that $\alpha \cap 1 = \alpha$.
Th. 3. $\alpha \cup \alpha = \alpha$.
Th. 4. $\alpha \cap \alpha = \alpha$.
Th. 5. $\alpha \cup 1 = 1$.
Th. 6. $\alpha \cap 0 = 0$.
*Th. 7. $0 \neq 1$.
Th. 8. If $\alpha = -\beta$ then $\beta = -\alpha$.

[3] From E. V. Huntington's 'Sets of Independent Postulates for the Algebra of Logic', *Transactions of the American Mathematical Society*, vol. 5 (1904), p. 288.

TH. 9. $\alpha = --\alpha$.
TH. 10. If $\alpha \cap \beta \neq 0$, then $\alpha \neq 0$.
TH. 11. $\alpha = (\alpha \cap \beta) \cup (\alpha \cap -\beta)$.
*TH. 12. $\alpha \cup (\beta \cup \gamma) = (\alpha \cup \beta) \cup \gamma$.
TH. 13. $\alpha \cap (\beta \cap \gamma) = (\alpha \cap \beta) \cap \gamma$.
TH. 14. $0 = -1$.
TH. 15. $\alpha \cup (\alpha \cap \beta) = \alpha$.
TH. 16. $\alpha \neq -\alpha$.
TH. 17. $-(\alpha \cap \beta) = -\alpha \cup -\beta$.
TH. 18. $-(\alpha \cup \beta) = -\alpha \cap -\beta$.
TH. 19. If $\alpha \cap -\beta = 0$ and $\beta \cap -\gamma = 0$, then $\alpha \cap -\gamma = 0$.
*TH. 20. If $\alpha \cap \beta \neq 0$ and $\beta \cap -\gamma = 0$, then $\alpha \cap \gamma \neq 0$.

The methods of proof proceed largely by the substitution of equals for equals. For example, Th. 1 is proved by considering any entities 0_1 and 0_2 in $\boldsymbol{C}$ such that $\alpha \cup 0_1 = \alpha$ and $\alpha \cup 0_2 = \alpha$. Since α is any member of $\boldsymbol{C}$, we have both $0_1 \cup 0_2 = 0_1$ and $0_2 \cup 0_1 = 0_2$. Since $0_1 \cup 0_2 = 0_2 \cup 0_1$ by Ax. 5, we obtain, by substitution, first $0_1 \cup 0_2 = 0_2$ and then $0_1 = 0_2$, which establishes the theorem.

Since a Boolean Algebra is a formal deductive system in its own right, it is susceptible of various interpretations. One of them, of course, is the algebra of classes. But we can give our Boolean Algebra a propositional rather than a class interpretation. Suppose we interpret '$\boldsymbol{C}$' as the collection of all propositions, and 'α', 'β', 'γ', . . . as symbolizing propositions, and interpret '$\cap$', '$\cup$', and '$-$' as symbolizing conjunction, (weak) disjunction, and negation. Then if we further interpret the equals sign as symbolizing material equivalence, all axioms and theorems of the Boolean Algebra become logically true propositions of the propositional calculus. Hence we can say that the propositional calculus is *a* Boolean Algebra.[4]

7.3 Zermelo-Fraenkel Set Theory (ZF)—The First Six Axioms

Abstract set theory was developed in the nineteenth century by Georg Cantor (1845–1918) in connection with his mathematical researches.[5] Unfortunately, Cantor's set theory gave rise to a number of contradictions, some of which will be discussed later in this section and also in Sections 7.4 and 7.7. The occurrence of these contradictions motivated mathematicians and logicians to search for an improved version of set theory that would be consistent. It was suspected that some unrecognized assumption was responsible for the

[4]For more details see Robert R. Stoll, *Set Theory and Logic*, W. H. Freeman and Company, Publishers, San Francisco and London, 1963, pp. 273–274; or Robert R. Stoll, *Sets, Logic, and Axiomatic Theory*, 2nd ed., W. H. Freeman and Company, Publishers, San Francisco and London, 1974, pp. 201–202.

[5]The mathematical background is described at length by P. E. B. Jourdain in his Introduction to Georg Cantor, *Contributions to the Founding of the Theory of Transfinite Numbers*, P. E. B. Jourdain, tr., Dover Publications, Inc., New York, 1915.

contradictions. So it was natural to attempt to construct a deductive system for set theory, in which every assumption would be explicitly stated as an axiom.

Cantor conceived a set to be 'any collection into a whole . . . of definite and separate objects . . . The objects are called the *elements* [members] of the set'.[6] Of the ideas involved in this conception the following seems clear. First, what is meant by 'definite' is that there is some criterion for membership in the set, so that any object can be cognized, at least in theory, to be a member or not to be a member of the set in question. Second, what is meant by 'separate' (or 'distinct') is that any member can be recognized, at least in theory, as different from any other member, so that no member gets counted more than once: thus the set containing just the objects 1, 2, and 2–1, is to be counted as having just *two* members rather than three. Finally, that a set is 'a collection into a whole' indicates that sets themselves are objects and, therefore, are eligible to be members of other sets.

The fundamental relation in abstract set theory is that of a membership. The symbol for 'is a member of' or 'belongs to' is the Greek letter '$\in$'. To assert that a is a member of the set A, we write '$a \in A$'. And to deny that b belongs to A, we write '$b \notin A$'. If we regard the nation, France, as the set of all its citizens, and Pierre is a citizen of France, it is true that

$$\text{Pierre} \in \text{France}$$

And since France is a member of the United Nations,

$$\text{France} \in \text{U.N.}$$

But because the U.N. has only nations as members, it is clear that

$$\text{Pierre} \notin \text{U.N.}$$

which shows that $\in$ is not a transitive relation.

It is obvious that different attributes may belong to exactly the same things. Thus (apart from the plucked chicken of Diogenes) the attribute of being a featherless biped belongs to all and only those things that have the (different) attribute of being a rational animal. A set, however, is determined by its members. Thus sets that contain exactly the same members are the same set. Thus the set of all featherless bipeds, F, is equal to or identical with, the set of all rational animals, R. We write this as

$$F = R$$

And to deny that the set A is the same as the set B, we write

$$A \neq B$$

[6] Op. cit., p. 85.

In Chapter 5 Leibniz's principle of the Identity of Indiscernibles was stated as

> $x = y$ if and only if every attribute of x is an attribute of y, and conversely

When quantification over predicate variables was introduced, the identity relation was defined, in accord with Leibniz's principle:

$$x = y = \text{df } (F)(Fx \equiv Fy)$$

From this definition, it was remarked, there follow all of the characteristics of the identity relation that are embodied in the rules of inference that were labeled 'Principle of Identity'. This principle is included in the underlying logic that is used in set theory in deriving conclusions from premisses, and in deriving theorems from axioms in axiomatic set theory.

The difference between sets and attributes was remarked above. Although two attributes may belong to exactly the same objects and yet be distinct, this is not true of sets. If the set A contains exactly the same members as the set B, then $A = B$. This is a fact about sets. It can be regarded as the most basic fact about sets. And so it is stated as the first axiom of the **ZF** system: not as a logical truth about the identity relation but as an assumption about sets.

ZF-1 $$A = B = \text{df } (x)(x \in A \equiv x \in B)$$

This is the Principle or Axiom of Extensionality and it states that a set is determined or defined by its members.

A set C, all of whose members are elements of a set D, is called a *subset* of D and if C is a subset of D, then C is said to be *included* in D. The subset relation '$\subset$' is defined in our notation by

$$C \subset D = \text{df } (x)(x \in C \supset x \in D)$$

This definition implies that every set is a subset of itself; in other words, set inclusion is a reflexive relation. It is obvious that if $C \subset D$ and $D \subset E$, then $C \subset E$. Thus set inclusion is a transitive relation. If $C \subset D$ and $D \subset C$, then by the Principle of Extensionality, $C = D$. Where $C \subset D$ but $C \neq D$, C is said to be a *proper subset* of D.

In the case of sets containing only a small number of members, a set may be symbolized by putting braces around an enumeration of the names of those members. Thus the set containing just the two objects a and b can be symbolized as '$\{a, b\}$' or as '$\{b, a\}$'. The order in which the names of the members is written is indifferent, since sets containing the same members are identical. A set containing just two members is called a 'pair set' or 'doubleton'.

For sets containing many members, another notation is available. The new notation involves, in addition to braces, a variable (such as 'x'), a colon (representing the concept of *such that*), and a propositional function that expresses some condition that is satisfied or fulfilled by all those objects and only those objects that belong to the set. Thus the set of all citizens of France would be symbolized by '$\{x$: x is a citizen of France$\}$'. More generally, where Fx is satisfied by just those objects belonging to the set S, $S = \{x: Fx\}$. So for any object y, $y \in \{x: Fx\} \equiv Fy$. This more powerful way of symbolizing sets can be seen to subsume the first method, for the set $\{a, b, c\}$ can also be symbolized as '$\{x: x = a \vee x = b \vee x = c\}$'. This notation is in the spirit of Bertrand Russell's remark that '. . . a class [set] may be defined as all the terms [objects] satisfying some propositional function.'[7]

Cantor apparently believed—for a time at least[8]—that *any* condition on objects could determine or define a set, the set containing or comprehending just those objects satisfying that condition. Although Cantor did not formulate this belief as a principle, it could be stated as

$$(\exists S)(x)(x \in S \equiv \varphi x)$$

where 'φ' represents any predicate or condition and there is no free occurrence of 'S' in φx. Unfortunately, this unrestricted principle of comprehension leads to a contradiction, the one generally known as the Russell paradox. If we take as φx the propositional function $x \notin x$, we have the set $R = \{x: x \notin x\}$, whence $y \in R \equiv y \notin y$, and putting R for y yields $R \in R \equiv R \notin R$, which is a contradiction.[9]

To avoid that contradiction, some restriction on the comprehension principle is needed. The one most commonly used today was first proposed in 1908 by Ernst Zermelo (1871–1953),[10] and the axiomatization of set theory proposed by him then was later (1922) augmented by Abraham A. Fraenkel[11] (b. 1891) to produce the system usually referred to as '**ZF**'.[12] In **ZF**, instead of assuming that *any* condition on *any* objects defines the set of all such objects, it is stated that *given* any set A, any condition on members of A defines the *subset* of A that contains just those members of A which satisfy that condition. This more restricted principle of comprehension (called the *Aussonderung Axiom* by

[7] Bertrand Russell, *The Principles of Mathematics*, Cambridge University Press, Cambridge, England, 1903; 2nd ed., W. W. Norton & Company, Inc., New York, 1938.

[8] See Cantor's letter to Dedekind in Jean van Heijenoort, ed., *From Frege to Godël*, Harvard University Press, Cambridge, Mass., 1967, pp. 113–117.

[9] Cf. pp. 159 ff. supra, and exercise 1 on p. 343.

[10] Ernst Zermelo, 'Investigations in the Foundations of Set Theory I', Stefan Bauer-Mengelberg, tr., in Jean van Heijenoort, *From Frege to Gödel*, op. cit., pp. 199–215.

[11] Abraham A. Fraenkel, 'Zu den Grundlagen der Cantor-Zermeloschen Mengenlehre', *Math. Annalen*, vol. 86, (1922), pp. 230–237.

[12] Because of Thoralf A. Skolem's contribution to the theory it has been suggested that '. . . it would seem historically more appropriate to call it Zermelo-Fraenkel-Skolem set theory'. See Patrick Suppes, *Axiomatic Set Theory*, D. Van Nostrand Company, New York, 1960, p. 8.

Zermelo) can be stated as

ZF-2 $$(\exists S)(x)(x \in S \equiv x \in A \cdot \varphi x)$$

where, again, it is understood that the variable 'S' has no free occurrence in the propositional function φx. Thus, given any set A and any condition φ, there is a set S which is that subset of A which contains just those members of A that satisfy the condition φ. (Since the expression 'φx' in the Axiom ranges over all kinds of propositional functions, the Axiom is more than a single axiom and is usually referred to as an 'Axiom Schema'.) This *Aussonderung* principle has been referred to, also, as *Abstraction,*[13] as Comprehension,[14] as Separation,[15] as Specification,[16] and as the Axiom of Subsets.[17] How the Axiom of Separation avoids the Russell paradox can be seen as follows. Given any set A, if in the Axiom of Separation we take, as φx, the propositional function $x \notin x$, we have the set $R = \{x: x \in A \cdot x \notin x\}$, whence $y \in R \equiv y \in A \cdot y \notin y$. Now putting R for y yields $R \in R \equiv R \in A \cdot R \notin R$, which is *not* a contradiction but can be simplified to $R \notin R \cdot R \notin A$.[18] From this last formula it immediately follows that $R \notin A$. But since A was any set whatever, and R's existence follows from the Axiom of Separation, it has been shown that, given any set, there is something that is not a member of it. Halmos has described this result somewhat whimsically as the conclusion "that *nothing contains everything,* or, more spectacularly, *there is no universe.* 'Universe' here is used in the sense of 'universe of discourse,' meaning in any particular discussion, a set that contains all the objects that enter into that discussion."[19] Here we have a fundamental difference between the Algebra of Classes, in which there is posited a universal class (1 or V), and **ZF** set theory, in which—as has just been proved—there is *no* universal set.

For the Axiom of Separation to produce a set S, there must already be given a set A of which the new set S is a subset. As Halmos remarked, "All the basic principles of set theory, except only the axiom of extension, are designed to make new sets out of old ones."[20] If we were to make a minimal assumption that there is at least one set A (as we assumed in Chapter 4 that there is at least one individual), then we can prove by the Axiom of Separation that there is an *empty set,* that is, a set that has no members. We already have at hand the propositional function $x \neq x$ that nothing satisfies, so by Separation, given

[13] Patrick Suppes, *Axiomatic Set Theory,* D. Van Nostrand Company, New York, 1960, p. 5.

[14] Abraham A. Fraenkel and Yehoshua Bar-Hillel, *Foundations of Set Theory,* North-Holland Publishing Company, Amsterdam, 1958, pp. 137 ff.

[15] Ernst Zermelo, op. cit., p. 202; Patrick Suppes, op. cit., p. 6; and Paul J. Cohen, *Set Theory and the Continuum Hypothesis,* W. A. Benjamin, Inc., Menlo Park, Calif., 1966, p. 55.

[16] Paul R. Halmos, *Naïve Set Theory,* D. Van Nostrand Company, New York, 1960, p. 4.

[17] Abraham A. Fraenkel, *Abstract Set Theory,* North-Holland Publishing Company, Amsterdam, 1966, p. 16.

[18] See exercise 5 on p. 343.

[19] Halmos, op. cit., pp. 6 ff.

[20] Halmos, op. cit., p. 4.

the set A, we have

$$(\exists B)(x)(x \in B \equiv x \in A \cdot x \neq x)$$

Thus $B = \{x: x \in A \cdot x \neq x\}$. Since no y is such that $y \in A \cdot y \neq y$, it follows that for every y, $y \notin B$, and by Extensionality there cannot be two empty sets. It is customary to symbolize the empty set as '$\varnothing$' (sometimes as 'Λ' or as '0').

Objections have been raised against the very idea of an empty set. Zermelo referred to it as a 'fictitious' set in his original formulation. If a set is determined or defined by its members, no set would seem to be defined if there are no members to define it. As Russell observed long ago: '. . . there is a grave logical difficulty in trying to interpret extensionally a class [set] which has no extension.'[21]

He went on to say both that 'Great difficulties are associated with the null-class [empty set] . . .'[22] and that '. . . with the strictly extensional view of classes . . . a class which has no terms [members] fails to be anything at all: what is merely and solely a collection of terms cannot subsist when all the terms are removed.'[23] Yet there are reasons for countenancing an empty set. For one thing, we may not *know* when what would otherwise be a perfectly acceptable set might be empty. That is so even in mathematics, for it is not yet known whether there are any members in the set of all integers n greater than 2, such that there exist integers x, y, z all greater than 0 where $x^n + y^n = z^n$. For another, it is convenient to have a set that is the product or intersection of two sets—even where it might not be known whether the two sets have any members in common. And then, if there are any sets at all, the Axiom of Separation entails the existence of an empty set. Near the very end of his *Principles of Mathematics*, Russell finally concluded: 'It is quite essential . . . that there should be a null-class . . .'[24]

Instead of making the minimal assumption that there is at least one set, from which Separation gives us the empty set, the **ZF** system makes a much stronger existence assumption as its Axiom VII. That stronger assumption, however, appears to presuppose the existence of the empty set, as will be seen. Hence it is really simpler to postulate the existence of the empty set directly. That is the procedure followed by Zermelo in his original formulation, as (part of) Axiom II, directly following Extensionality, which was his first axiom. We assume then, as another axiom (the third, in the sequence presented here) the Empty Set Axiom:

ZF-3 *There is a set $\varnothing$ such that for any object x, $x \notin \varnothing$. It is an immediate consequence that for any set S, $\varnothing \subset$ S.*

[21] Bertrand Russell, *Principles of Mathematics,* Cambridge University Press, Cambridge, England, 1903, p. 32.

[22] Ibid., p. 73.

[23] Ibid., p. 74.

[24] Ibid., p. 514.

We can now see another fundamental difference between the Algebra of Classes and set theory. In the former, every class has an absolute complement. But in set theory this cannot be true, for the absolute complement of Ø would contain everything and we have already shown that there can be no such set.

At this point, it might be well to avow the general practice of those who develop one or another variant of **ZF** set theory. The objects we talk about here are limited to sets exclusively. Mathematicians work with sets all the time: sets of points, sets of lines, sets of numbers of various kinds. In fact, all of the familiar mathematical objects such as numbers and functions can be construed as sets, so for mathematical purposes no other kinds of objects need be considered.[25] Apart from truth-functional operators, quantifiers, the special relational symbols '$\in$', '$\notin$', '$=$', '$\neq$', the colon, and our general notation 'φx' for conditions on x, all of our symbols here, 'a', 'b', 'c', . . . as well as 'A', 'B', 'C', . . . will range over sets exclusively. There are no individuals in **ZF** set theory in its usual formulations. To allow for individuals would complicate matters more than is necessary for the mathematical purposes for which set theory is typically developed.[26]

Another useful principle is the Axiom of Pairing:

ZF-4 *Given any sets a and b, there is a set S having as members just a and b. In symbols:* $(\exists S)(x)(x \in S \equiv x = a \vee x = b)$.

So for any sets x and y, there is the *pair* set $\{x, y\}$, which is sometimes called the *unordered pair*—in contrast with another concept to be introduced in the next section. No distinctness assumption was made in the Pairing Axiom. Thus, given any set a, we can form the unordered pair $\{a, a\}$ which can also be written as '$\{a\}$,' and is called a *unit set* or *singleton*.

Russell puzzled over the question whether $\{a\}$ was the same as a, itself, initially being inclined to say that it appeared to be 'self-evident' that 'A whole composed of one term only is that one term.'[27] But he subsequently decided that 'it is necessary to distinguish a single term from the class [set] whose only member it is'.[28] He was persuaded by an argument he attributed to Frege: Let u be a set having more than one member; let $\{u\}$ be the set whose only member is u; then $\{u\}$ has one member, u has many members, hence $u \neq \{u\}$. Had Russell not also had misgivings about the empty set, he might have been equally well persuaded by the following argument suggested by Stoll. Because Ø has no members and $\{$Ø$\}$ has a member, it follows that Ø $\neq$ $\{$Ø$\}$, which proves that $u \neq \{u\}$.

Apart from the nonexistence of a universal set and the absence of absolute complements, we want our sets to conform to the rules governing classes set

[25] Even points can be construed as sets, as explained in Alfred North Whitehead, *Process and Reality*, Cambridge University Press, Cambridge, England, 1929, Part 4.

[26] A set theory containing individuals can be found in Suppes, op. cit.

[27] B. Russell, *Principles of Mathematics*, op. cit., p. 77. See also pp. 68, 73.

[28] Ibid., p. 106. See also pp. 514 and 517.

forth in the preceding sections. It is easy to see that any two sets a and b have an intersection or product $a \cap b$. This is assured by the Separation Axiom. If we let '$x \in b$' be the φx in that axiom, we have

$$(\exists S)(x)(x \in S \equiv x \in a \cdot x \in b).$$

Since S is unique (by Extensionality), we can introduce the connective '$\cap$' by the following definition:

$$a \cap b = \text{df } \{x: x \in a \cdot x \in b\}$$

This process can be used repeatedly to obtain $(a \cap b) \cap c$ if c is also a set, $[(a \cap b) \cap c] \cap d$ if d is also a set, and so on. It is easily shown that $a \cap \varnothing = \varnothing$, and that $\cap$ is commutative, associative, and idempotent. There is a more general intersection principle than the preceding would indicate. If A is a nonempty collection of sets $a_1, a_2, a_3, \ldots$, there is a set B that is the intersection of *all* the members of A. This is given by the Separation Axiom, if we let a_i be any particular set in A (there is one because A is nonempty) and let φx be the condition that x belongs to every set in A:

$$(\exists B)(x)[x \in B \equiv x \in a_i \cdot (a)(a \in A \supset x \in a)]$$

Here B is in no way dependent on the choice of one a_i rather than another; indeed $B = \{x: (a)(a \in A \supset x \in a)\}$. Since B is unique (by Extensionality), we can define the general intersection symbol '$\bigcap$' by the following:

$$\bigcap A = \text{df } \{x: (a)(a \in A \supset x \in a)\}$$

Other notations for $\bigcap A$ are '$\bigcap_i a_i$', '$\bigcap\{a \in A\}$', '$\bigcap\{x: x \in A\}$', and '$\bigcap_{x \in A} x$'. It is obvious that $\bigcap\{a, b\} = a \cap b$.

Our definition of $\bigcap A$ was conditional on the set A's being nonempty. If we did not impose that condition, there would be no set a_i whose membership in A is needed for using the Separation Axiom. So '$\bigcap \varnothing$' is not yet defined. We cannot define it as $\{x: (a)(a \in \varnothing \supset x \in a)\}$ because every x satisfies the condition that $a \in \varnothing \supset x \in a$, since $a \notin \varnothing$ for every a. Thus the suggested definition would make $\bigcap \varnothing$ contain everything, whereas we have already shown that there is no universal set in **ZF**. We *could* leave it undefined (as division by zero is left undefined in arithmetic), or we could define $\bigcap \varnothing$ to be $\varnothing$ itself, in order to make $\bigcap S$ defined for any set S whatever.

We also want any two sets a and b to have a union or sum $a \cup b$ which contains all (and only) those things that are either members of a or members of b. The Separation Axiom will not help us here. We *could* introduce a new axiom

$$(\exists B)(x)(x \in B \equiv x \in a \vee x \in b)$$

where, by Extensionality, B is the unique set $\{x: x \in a \vee x \in b\}$ for which we introduce the notation '$a \cup b$'. Then we could use it repeatedly to obtain $(a \cup b) \cup c$ if c is also a set, and $[(a \cup b) \cup c] \cup d$ if d is also a set, and so on. Were we to do so, it would be clear that $a \cup \varnothing = a$, that $\cup$ is commutative, associative, and idempotent, and also that $\cap$ and $\cup$ are each distributive with respect to the other. But such a 'one at a time' method of forming unions of given sets is not really enough, because we want a set that contains every object that belongs to any set a_i of a collection A of sets, $\{a_1, a_2, a_3, \ldots, a_i, \ldots\}$, regardless of whether the collection A is finite or infinite. (A collection is *infinite* if there is no natural number 0, 1, 2, . . . that is the number of members of that collection. The concepts of finite and infinite will be discussed at length in later sections of this chapter.)

Given a collection A of sets $a_1, a_2, a_3, \ldots$ we want an axiom that will produce a set B containing anything that belongs to any member of A. This is stated as a more general Union Axiom:

ZF-5 $$(A)(\exists B)(x)[x \in B \equiv (\exists a)(x \in a \cdot a \in A)]$$

As usual the Extensionality Axiom guarantees the uniqueness of the set B, so we can use any of a number of available alternative notations for it. Where $A = \{a_1, a_2, a_3, \ldots\}$, the set B is symbolized as '$\bigcup A$' or as '$\bigcup_i a_i$' or as '$\{x: (\exists a)(x \in a \cdot a \in A)\}$' or as '$\bigcup\{a: a \in A\}$' or as '$\bigcup\{x: x \in A\}$' or as '$\bigcup_{x \in A} x$'.

From this general Union Axiom and the Axiom of Pairing, we can obtain the more modest $a \cup b$ from sets a and b as follows. By the Axiom of Pairing, we have $\{a, b\}$. The set $\bigcup\{a, b\}$ given by the Union Axiom is $\{x: (\exists z)(x \in z \cdot z \in \{a, b\})\}$. But the only z's belonging to $\{a, b\}$ are a and b, themselves, so $\bigcup\{a, b\} = \{x: x \in a \vee x \in b\}$, which is the desired $a \cup b$. It is obvious that $\{x: x \in \varnothing\} = \varnothing$, so $\bigcup \varnothing = \varnothing$, and that $\bigcup\{x: x \in \{a\}\} = a$, so $\bigcup\{a\} = a$. It is also clear that $\bigcup\{x: x \in \{a, b\}\} = a \cup b$, so $\bigcup\{a, b\} = a \cup b$, and that $\{a\} \cup \{b\} = \{a, b\}$.

We have already remarked that there is no *absolute* complement of any set in **ZF**. But the concept of *relative complement* is available. The *relative complement* $A - B$ *of* B *in* A, whose existence and uniqueness is given by Separation and Extensionality:

$$A - B = \text{df}\ \{x: x \in A \cdot x \notin B\}$$

which is the set of all members of a that are not members of B. It is obvious that $A - \varnothing = A$, that $A - A = \varnothing - A = \varnothing$, and that if $A \cap B = \varnothing$ then $A - B = A$. The relative complement is more useful in mathematics than an absolute complement would be. For example, one would want the complement of the set of even numbers to be the set of odd numbers only rather than a set containing, in addition to odd numbers, points and circles and functions and everything else. The relative complement of B in A is sometimes written as '$A \backslash B$'. If we make the assumption in a particular discussion that all of the sets $a, b, c, \ldots$ being considered are subsets of some specified set V, then we can symbolize the various relative complements of a in V, of b in V, of c in V, and so on, as '$\overline{a}$', '$\overline{b}$', '$\overline{c}$', and so on. Then in the context of that discussion, the sets $a, b, c, \ldots, \overline{a}, \overline{b}, \overline{c}, \ldots$ will obey *all* the laws of the Algebra of Classes presented in Sections 7.1 and 7.2.

The notion of a subset of a set has already been defined. There are occasions when one wants to consider the totality of subsets of a set. That the totality of subsets of a given set E form a set is asserted by the Power Set Axiom:

ZF-6 $$(E)(\exists S)(x)(x \in S \equiv x \subset E)$$

Extensionality insures uniqueness, and we define the power set of a set E, symbolized as '$\mathcal{P}E$', by

$$\mathcal{P}E = \text{df}\,\{x\colon x \subset E\}$$

Where E is a finite set, it is easy to see that $\mathcal{P}E$ contains more members than E does. For example: $\mathcal{P}\varnothing = \{\varnothing\}$, $\mathcal{P}\{a\} = \{\varnothing, \{a\}\}$, $\mathcal{P}\{a, b\} = \{\varnothing, \{a\}, \{b\}, \{a, b\}\}$. It is easy to prove by mathematical induction[29] that if E contains n members, then $\mathcal{P}E$ contains 2^n members, which explains the origin of the term 'power set'. For any set E, we know that $E \subset E$ and $\varnothing \subset E$, so for any set E, we have both $E \in \mathcal{P}E$ and $\varnothing \in \mathcal{P}E$. From this last, it is obvious that $\bigcap\{x\colon x \in \mathcal{P}E\} = \varnothing$. Other immediate consequences of our definitions and axioms are that $E \subset F$ if and only if $\mathcal{P}E \subset \mathcal{P}F$, that $\mathcal{P}E \cup \mathcal{P}F \subset \mathcal{P}(E \cup F)$, that $\mathcal{P}(E \cap F) = \mathcal{P}E \cap \mathcal{P}F$, and that, although $E = \bigcup \mathcal{P}E$, in general we have only that $E \subset \mathcal{P} \bigcup E$.

7.4 Relations and Functions

Before exploring further implications of the Power Set Axiom, it will be useful to discuss relations and functions. Relations were discussed in Chapter 5, where many of the attributes of binary relations were defined and their interrelations examined. Because not every binary relation is symmetrical, it is necessary to distinguish between *Rab* and *Rba*. Because *Rab* might be true while *Rba* was false, the *order* of the arguments a and b is essential. We could express this by saying that the relation R holds for (or the propositional function *Rxy* is satisfied by) the *ordered pair* $\langle a, b\rangle$ but not the ordered pair $\langle b, a\rangle$.

Early in the preceding section we remarked that sets can be regarded as the extensions of predicates: subject to the restrictions of the Separation Axiom, the

[29]The method of proof by mathematical induction is reviewed on pp. 223 ff.

set F consists of all objects that satisfy the predicate or condition φx. By analogy, we focus our attention on the collection of ordered pairs of objects, written $\langle x, y \rangle$, between the first and second of which the relation Rxy holds. To develop this notion within set theory, the idea of 'order' must somehow be defined in terms of sets. Although its members determine a unique set independently of any order in which they are arranged or in which they may be named, it is, nevertheless, possible to define the notion of order in purely set theoretic terms.[30] The essence of the notion of *order* is that the ordered pair $\langle x, y \rangle$ is the same as the ordered pair $\langle a, b \rangle$ if, and only if, both $x = a$ and $y = b$. This essence is captured by the Wiener-Kuratowski definition of ordered pair:

$$\langle x, y \rangle = \text{df } \{\{x\}, \{x, y\}\}$$

Given any x and y, we use the Axiom of Pairing to produce the set in question and it is unique by Extensionality. It is immediately obvious that if $x = a$ and $y = b$, then $\langle x, y \rangle = \langle a, b \rangle$. To establish the converse, we assume that $\langle x, y \rangle = \langle a, b \rangle$, that is, $\{\{x\}, \{x, y\}\} = \{\{a\}, \{a, b\}\}$. Since $\{x\} \in \{\{x\}, \{x, y\}\}$, we have $\{x\} \in \{\{a\}, \{a, b\}\}$. Hence $\{x\} = \{a\}$ or $\{x\} = \{a, b\}$. In either case $x = a$. Since $\{a, b\} \in \{\{a\}, \{a, b\}\}$, we have $\{a, b\} \in \{\{x\}, \{x, y\}\}$. Hence either (1) $\{a, b\} = \{x\}$ or (2) $\{a, b\} = \{x, y\}$. By a similar argument, we also have either (3) $\{x, y\} = \{a\}$ or (2) $\{x, y\} = \{a, b\}$. If (1) and (3) are both true, then $x = a = y = b$, which includes the desired result. If either (1) or (3) is not true, then $\{x, y\} = \{a, b\}$, whence by our previous result that $x = a$, we have that $\{a, y\} = \{a, b\}$. Here, if $a \neq b$ then $y = b$, but if $a = b$, then $y = b$. So in every case $y = b$.

Given two sets A and B, we can form ordered pairs $\langle x, y \rangle$ with $x \in A$ and $y \in B$. The collection of *all* such pairs is called the *Cartesian Product* $A \times B$ of A and B, and for each ordered pair $\langle a, b \rangle$ in it, a is called the *first coordinate* and b is called the *second coordinate*. (An example is the following: given the Euclidean plane with x and y axes established, if A is the set of all real numbers identified with points on the x-axis and B is the set of all real numbers identified with points on the y-axis, then the set of all ordered pairs $\langle x, y \rangle$ represents the Cartesian plane itself—which accounts for the name). That this collection of ordered pairs *is* a set, follows from the Power Set, the Union, and the Separation Axioms. Given sets A and B, we form their union $A \cup B$, which contains every $x \in A$ and every $y \in B$. Next, we form the power set $\mathcal{P}(A \cup B)$ which contains every set $\{x\}$ and every set $\{x, y\}$ as members—along with other members. Then, we form *that* set's power set $\mathcal{P}\mathcal{P}(A \cup B)$ which contains every set $\{\{x\}, \{x, y\}\}$ as a member—along with other members. Now, the Separation Axiom tells us that there is a subset that is exactly the set wanted.

[30] The method for doing so was first devised by N. Wiener and subsequently improved by C. Kuratowski. See 'A Simplification of the Logic of Relations' by N. Wiener, *Proc. of the Cambridge Philosophical Soc.*, vol. 17 (1914), pp. 387–390, and, 'Sur la notion de l'ordre dans la théorie des ensembles', C. Kuratowski, *Fundamenta Mathematica*, vol. 2, (1921), pp. 161–171.

$$(\exists S)(z)[z \in S \equiv [(z \in \mathcal{P}\mathcal{P}(A \cup B)) \cdot (\exists x)(\exists y)(x \in A \cdot y \in B \cdot z = \langle x, y \rangle)]]$$

So for any sets A and B, there is a set that contains all and only those ordered pairs $\langle x, y \rangle$ with $x \in A$ and $y \in B$. Extensionality insures uniqueness, so '$A \times B$' can symbolize the set in question. Much the same sort of argument, but using unions instead of powers, will show that if $\langle x, y \rangle \in A$, then both x and y belong to $\bigcup \bigcup A$. One point should be made clear: The Cartesian product of A and B does not require $A \neq B$. The product $A \times A$ is a perfectly good Cartesian product, that of A with itself.

We can now define a (binary) relation quite simply as any set of ordered pairs. R is a relation if and only if

$$(x)(x \in R \supset (\exists u)(\exists v)(x = \langle u, v \rangle))$$

Obvious examples of relations are $\varnothing$, and the Cartesian product of any sets A and B, and any subset of a Cartesian product.

The *domain* of a relation R, (dom R), is the set of all first coordinates of R. The *range* of a relation R, (ran R), is the set of all second coordinates of R, and the *field* of a relation R (fld R) is the set of all coordinates of R. Given any relation R, the axioms permit us to specify both its domain and its range. We first use the Union Axiom twice to produce the set $\bigcup \bigcup R$, and then by Separation

$$(\exists S)(x)[x \in S \equiv [x \in \bigcup \bigcup R \cdot (\exists y)(\langle x, y \rangle \in R)]]$$

gives us the unique dom R, and

$$(\exists T)(x)[x \in T \equiv [x \in \bigcup \bigcup R \cdot (\exists z)(\langle z, x \rangle \in R)]]$$

gives us the unique ran R. Examples are plentiful: dom $\varnothing$ = ran $\varnothing$ = $\varnothing$, dom $(A \times B) = A$ and ran $(A \times B) = B$, and if R is equality of members of E, then dom R = ran R = E.

Except for the empty relation $\varnothing$, one must always specify a set within, for, or over which, the relation is defined. If we tried to define the identity relation $X = Y$ without limiting its domain (or range, or field) we should get into trouble—for the aggregate Z of all ordered pairs $\langle x, y \rangle$ such that $x = y$ would contain every set $z = \{\{x\}, \{x, x\}\}$, and the union of its union, $\bigcup \bigcup Z$ would be the universal set, which has been proved not to exist. The same problem would arise in connection with set membership $(x \in y)$ and set inclusion $(x \subset y)$.

The notion of a function is a familiar one. The pressure of a fixed volume of gas is a function of its temperature, and the cost of sending a letter by mail from New York to San Francisco is a function of its weight. We define a function in set theoretic terms as a many-one relation, that is, a set of ordered pairs no two of which have the same first coordinate. If X and Y are sets, a function *from* (or *on* X) *to* (or *into*) Y is a relation, f, such that dom $f = X$ and such that for each x in X, there is a unique y in Y with $\langle x, y \rangle \in f$. That f is a function from X to Y is stated as '$f{:}X \rightarrow Y$', and is defined as

$$f \subset X \times Y \cdot (x)[x \in X \supset (\exists y)[y \in Y \cdot \langle x, y \rangle \in f \cdot (z)(\langle x, z \rangle \in f \supset y = z)]]$$

For each x in X, the unique y in Y such that $\langle x, y \rangle \in f$ is represented by '$f(x)$'; so $f(x) = y$. The object y is the *value* that the function f *assumes* or *takes on at* or *for* the *argument x*. Another locution commonly used is that *f maps x onto y*. The domain of a function f from X into Y is X, but its range can be a proper subset of Y. If the range of f is equal to Y, then f is said to map X onto Y.

A function f from X to Y that maps distinct elements onto distinct elements is called *one-to-one*. If f: $X \rightarrow Y$ is one-to-one and onto Y, then f effects a pairing of the elements of X with those of Y, and, as such, is called a *one-to-one correspondence* between X and Y. If there exists a one-to-one correspondence between X and Y, these sets are said to be *equivalent* (or *equinumerous*); this is symbolized as '$X \sim Y$'.

The familiar 0, 1, 2, 3, . . . are called *natural numbers* or *counting numbers*. They will be defined in set theoretic terms in the following section. A *finite set* is defined as a set onto which a one-to-one function maps the set of all natural numbers that are less than some natural number n. Such a set is said to contain n elements. Thus the one-to-one function that maps 0 onto the letter a, 1 onto b, 2 onto c, . . . , 24 onto y, and 25 onto z, maps the set of all natural numbers less than 26 onto the set of all letters of the alphabet. An *infinite set* is then defined (negatively) as any set that is not finite.

No finite set is equivalent to any of its proper subsets. But the matter is otherwise with infinite sets. Galileo remarked long ago that, although not all numbers are perfect squares, yet 'there are as many squares as there are numbers, because they are just as numerous as their roots and all the numbers are roots.'[31] Some equivalences are even more surprising. There are as many whole numbers as there are rational fractions. The one-to-one correspondence between these two sets is shown by the following enumeration of all fractions: of two fractions, we enumerate first that fraction, the sum of whose numerator and denominator is smaller. Of two fractions whose numerator and denominator have the same sum, we count first the one with the smaller numerator. (Of course, we do not repeat equal fractions in our enumeration: If 1/2 has already been counted, we do not also include 2/4 or 3/6 or · · · .) With this ordering of the fractions, the one-to-one correspondence appears as follows:

1/1	1/2	2/1	1/3	3/1	1/4	2/3	3/2	4/1	1/5	5/1	1/6	· · ·
0	1	2	3	4	5	6	7	8	9	10	11	· · ·

[31] Galileo Galilei, *Dialogues Concerning Two New Sciences,* Northwestern University Press, Evanston and Chicago, 1946, p. 31.

An 'enumeration' is of course a one-to-one function from the natural numbers (0, 1, 2, 3 . . .) onto the set being enumerated. Any set, all of whose members can be mapped one-to-one onto the natural numbers, is said to be *denumerable* (or *countable*, or *enumerable*). (Of course, we have not yet shown that the collection of all natural numbers is a *set*. This will be discussed in the next section.)

If there is a one-to-one correspondence between X and a proper subset of Y but no one-to-one correspondence between X and Y, then Y is said to be *larger than* X, or to have *more elements* than X, and X is said to be *smaller than* Y or to have *fewer elements than* Y. Cantor proved that there are more real numbers than there are integers by showing that in any enumeration of real numbers there must be some real number that gets left out. (Of course, the integers can be mapped one-to-one onto a subset of the real numbers by the function f where $f(n) = n$.) With no loss in generality, we confine our attention to real numbers in the interval from 0 to 1, and consider *any* enumeration of real numbers from that interval:

$$\begin{array}{lllllll}
0. & a_{11} & a_{12} & a_{13} & \cdots & a_{1j} & \cdots \\
0. & a_{21} & a_{22} & a_{23} & \cdots & a_{2j} & \cdots \\
0. & a_{31} & a_{32} & a_{33} & \cdots & a_{3j} & \cdots \\
\cdots & \cdots & \cdots & \cdots & \cdots & \cdots & \cdots \\
0. & a_{i1} & a_{i2} & a_{i3} & \cdots & a_{ij} & \cdots \\
\cdots & \cdots & \cdots & \cdots & \cdots & \cdots & \cdots
\end{array}$$

where a_{ij} is the jth digit of the ith decimal fraction enumerated. Given any such array representing a one-to-one mapping of natural numbers into real numbers, a 'missing' real number can be specified by the following diagonal procedure. One missing real number is the decimal fraction $d = 0.d_1\, d_2\, d_3 \ldots d_k \ldots$ whose kth digit d_k is defined to be 7 if $a_{kk} \leqq 5$, or 3 if $a_{kk} > 5$. This 'diagonal' number is clearly in the interval from 0 to 1, but it cannot be any one of the numbers in the given array because it differs from each of them in at least one digit. For any k, the diagonal number differs in its kth decimal place from the kth decimal fraction of the array. Hence no enumeration of real numbers can contain all of them—which proves that the real numbers are nondenumerable, that is, that there are more of them than there are integers.

Now, we are in a position to consider another implication of the Power Set Axiom. The Cantor diagonal proof just described is a special form of the proof of a more general theorem, which states that for any set A, its power set $\mathcal{P}A$ is larger than A, itself. There is obviously a one-to-one mapping from A *onto* the proper subset of $\mathcal{P}A$ that consists of all singletons in $\mathcal{P}A$. (For every a in A, its singleton $\{a\}$ is in $\mathcal{P}A$, and $s(a) = \{a\}$ is obviously one-to-one.) Now consider any one-to-one function f from A into $\mathcal{P}A$, and form the subset A' of $\mathcal{P}A$ that consists of just those members a of A which are not members of $f(a)$, the element of $\mathcal{P}A$ onto which they are mapped: $A' = \{x\colon\ x \notin f(x)\}$. Now, consider any a in A. If $a \in A'$, then $a \notin f(a)$ and so $f(a) \neq A'$, but if $a \notin A'$, then $a \in f(a)$ and so again, $f(a) \neq A'$. But a is *any* member of A, so *no* member of A is mapped by f onto A', which is, therefore, left out of the mapping f. But f was *any* one-to-one function from A to $\mathcal{P}A$, so $\mathcal{P}A$ is larger than A.

Because Cantor's principle of Comprehension was not restricted in the way that the **ZF** Axiom of Separation is, Cantor's set theory contained the set S of all sets. According to the result derived in the preceding paragraph from the Power Set Axiom, it follows that the power set $\mathcal{P}$S of the set S of all sets is larger than S. But this is impossible, because every subset of S, being a set, is a member of S. The Cantor paradox, then, is that given any set s there is a still larger set $\mathcal{P}s$, yet there is a largest possible set S. Cantor was aware of this problem and communicated it to Dedekind in 1899.[32] But Cantor did not work out a satisfactory resolution of this difficulty. In the **ZF** set theory, the Cantor paradox does not arise. The problem is solved by the more limited Axiom of Separation that takes the place of Cantor's too generous principle of Comprehension. Although larger and ever larger sets can be produced by repeated use of the Power Set Axiom, no 'largest' set is ever reached.

There are other definitions for 'finite' and 'infinite'. One can define an *infinite set* as a set that is equivalent to one of its proper subsets. Given this definition of infinite, a *finite set* can then be defined as a set that is not an infinite set. This pair of definitions can be proved to be equivalent to those first mentioned on page 184, in the presence of the Choice Axiom, which will be introduced in Section 7.6.

Still another definition of finite set, attributed to Dedekind, is 'A set S is called *finite* if there exists a mapping of S into itself such that no proper subset of S is mapped into itself.'[33]

7.5 Natural Numbers and the Axiom of Infinity

Numbers are useful in many ways. Early in childhood we learn to use the cardinal numbers 'one', 'two', 'three', . . . to count and to answer the question 'How many?' And we learn also that the ordinal numbers 'first', 'second', 'third', . . . can be used in ordering a collection of objects. Our task is to develop or define these numbers within **ZF** set theory. We postpone our discussion of ordinal numbers until Section 7.7. In Section 7.6 we shall consider cardinal numbers in general. In this section we are concerned with the 'natural numbers': 0, 1, 2, 3, ...

Mathematicians have two ways of characterizing a collection of (mathematical) objects that they propose to study. One way is to set forth an axiom system that will provide an 'implicit definition' of those objects, from which—it

[32] See Cantor's letter in van Heijenoort, op. cit., pp. 113–117.

[33] See Raymond L. Wilder, *Introduction to the Foundations of Mathematics,* 2nd ed., John Wiley & Sons, Inc., New York-London-Sydney, 1965, p. 66.

is hoped—all of the other facts concerning those objects can be derived as theorems. The other way is to *construct* (or explicitly define) the new objects in terms of old ones that are already at hand.

During the nineteenth century mathematicians carried out what has come to be called 'the arithmetization of analysis'. That involved the construction of complex numbers, real numbers, algebraic numbers, rational fractions, and negative numbers, on the basis of—or in terms of—the natural numbers 0, 1, 2, 3, . . . using notions of intuitive set theory.[34] But the natural numbers, themselves, were developed axiomatically, by Dedekind and subsequently by Peano.[35] Peano's formulation of an axiom system for the natural numbers has long been accepted as an admirably compendious characterization (or implicit definition) of natural numbers, from which all their essential mathematical attributes can be derived.

Peano's system is based on three primitive or undefined concepts: *zero*, *number*, and *successor*. These are not defined explicitly *in* the system, but rather are implicitly defined *by* the axioms that make statements about them. It is convenient to represent the successor of x as x': What Peano intended here can be explained informally (and, of course, extrasystematically) by saying that the successor of a number is obtained by adding 1 to it. The five axioms are

P 1. 0 is a number.[36]
P 2. If x is a number, then x' is a number.
P 3. If x and y are numbers and $x' = y'$, then $x = y$.
P 4. There is no number x such that $x' = 0$.
P 5. If $\varphi 0$ and for every number x, if φx then $\varphi x'$, then for every number x, φx.

Peano himself acknowledged that not only the natural numbers satisfy these axioms. For example, the sequence 1, 1/2, 1/4, 1/8, . . . also satisfy them, if 'zero' is taken to denote 1, 'number' is taken to denote the set of positive fractions with numerator 1 and denominator any power of 2, and 'successor of

[34] A lucid and detailed account of these constructions is found in Edmund Landau, *Foundations of Analysis*, Chelsea Publishing Company, New York, 1951. Or see H. B. Enderton, *Elements of Set Theory*, Academic Press, Inc., New York, 1977, or Suppes, op. cit.

[35] R. Dedekind, '*Was sind und was sollen die Zahlen?*', Brunswick, 1888, W. W. Beman, tr., in *Essays on the Theory of Numbers*, The Open Court Publishing Company, Chicago and London, 1901. See also his 'Letter to Kefferstein' (1890) H. Wang and S. Bauer-Mengelberg, tr., in van Heijenoort, op. cit. G. Peano, *Arithmetices Principia, Nova Methodo Exposita*, Turin, 1889, tr., in van Heijenoort, op. cit.

[36] Some people find it objectionable to call zero a number, on the grounds, for example, that if there are no apples in a box, it is misleading to say that the box contains a number of apples, since the number is zero. One should less misleadingly say that there are no apples in the box or no number of apples in the box. Indeed Peano's own original formulation of his system had the natural numbers begin with 1 rather than with 0. However, the number 0 is so useful in our notations and computations that it is accepted in mathematics in spite of possibly misleading locutions to which it may give rise.

x' is taken to denote $x/2$. But systems other then the natural numbers that satisfy the axioms can be shown to be 'just like' the natural numbers, and could be taken to be just different expressions for them.

It is clearly desirable to characterize the natural numbers in a more univocal way than Peano's. However, Peano's axioms are very useful: If we manage to *construct* the natural numbers, somehow, we can check our constructions by verifying that they satisfy the Peano axioms, for the latter do yield all of the strictly formal arithmetic truths that mathematicians prize.

Frege, and later independently Russell,[37] proposed a brilliant 'constructive' definition of number, based on the following considerations. If A and B are equinumerous classes, that is, classes between whose members there is a one-to-one correspondence, we say that A and B have *the same number* of members. Every pair set, for example, has exactly two members. Is 'two', then, the name for some characteristic shared by all pair sets? Does 'two' signify *the* common characteristic of all pairs? It may be that all pair sets share *several* characteristics, so the number 'two' cannot be defined in quite this way. Frege and Russell avoided the 'positing' of a unique shared characteristic by defining 'two' to be the *class* (or *set*) of all pairs—which is unique by Extensionality. The Frege-Russell definition of 'zero' is the class whose only member is the empty set. The definition of 'one' is the class of all singletons and the definition of 'three' is the class of all triples, and so on. There is much wisdom here and deep insight as well as ingenuity. But it cannot be incorporated into **ZF** set theory directly, because if there *were* a set of all singletons (or all doubletons, or etc.), then its Union would be the universal set, which has been proved not to exist. Zermelo proposed to identify the natural numbers 0, 1, 2, 3, . . . with the sets $\varnothing, \{\varnothing\}, \{\{\varnothing\}\}, \{\{\{\varnothing\}\}\}, \ldots$ Later, John von Neumann[38] proposed the use of an alternative and more convenient sequence of sets to define the natural numbers. By the Axioms of the Empty Set, Pairing, and Union, we know that each of the following sets exist: $\varnothing$, $\{\varnothing\}$, $\{\varnothing, \{\varnothing\}\}$, $\{\varnothing, \{\varnothing\}, \{\varnothing, \{\varnothing\}\}\}$, . . . where each set u is followed by the set $u \cup \{u\}$. It is convenient to define the successor set u^+ of the set u to be the sum of u and $\{u\}$. Then, 0 is defined to be the empty set; 1 is defined to be the successor set 0^+ which is $\varnothing \cup \{\varnothing\}$ or simply $\{\varnothing\}$; 2 is defined to be 1^+, which is $\{\varnothing\} \cup \{\{\varnothing\}\}$ or simply $\{\varnothing, \{\varnothing\}\}$, and so on. On this approach, instead of trying to define n with Frege and Russell as the set of all n-membered sets, which we have seen to be impossible in **ZF**, we define n to be a particular and special n-membered set, a *representative* n-membered set, that the **ZF** axioms guarantee to exist:

[37] G. Frege, *Die Grundlagen der Arithmetik*, Verlag von Wilhelm Koebner, Breslau, 1884, J. L. Austin, tr., *The Foundations of Arithmetic*, Basil Blackwell, Oxford, 1950. B. Russell, *Principles of Mathematics*, op. cit. There is some difference between these authors' formulations. Frege regarded a number as the number of a concept, whereas Russell regarded a number as the number of the extension of a concept, that is, of the class of objects satisfying the concept.

[38] J. von Neumann 'Zur Einführung der transfiniten Zahlen', *Acta litterarum ac scientiarum Regiae Universitatis Hungaricae Francisco-Josephinae, Sectio scientiarum mathematicarum 1*, 1923, tr., in van Heijenoort, op. cit.

Now we have the following:

$$
\begin{aligned}
0 &= \varnothing \\
1 &= \{0\} \\
2 &= \{0, 1\} \\
3 &= \{0, 1, 2\} \\
&\vdots \\
n^+ &= \{0, 1, 2, \ldots, n\} \\
&\vdots
\end{aligned}
$$

These infinitely many sets exist, but is there a set containing all of them? Another axiom is required to insure that there is.

ZF-7 Axiom of Infinity. *There exists a set containing* 0 and containing the successor of each of its elements.

By itself, as here stated, the Axiom of Infinity does not provide us with a unique set whose *only* members are 0, 1, 2, 3, . . . , because other objects *might* belong to a set, S, guaranteed to exist by the Axiom of Infinity. We shall call any set an *inductive set* if it contains 0 and the successor of each of its members. It is clear that 0, 1, 2, 3, . . . all belong to every inductive set. The Axiom of Infinity states that there exists *an* inductive set, S. Every inductive subset of S contains 0, 1, 2, 3, . . . , so the intersection of all inductive subsets of S, which we shall call ω, exists by the Separation Axiom, is unique by Extensionality, and contains *just* the natural numbers.

$$
\begin{aligned}
\omega &= \text{df} \bigcap \{S_i\colon S_i \subset S \cdot S_i \text{ is an inductive set}\} \\
&= \{0, 1, 2, 3, \ldots\}
\end{aligned}
$$

To verify that ω satisfies the Peano axioms is fairly easy. That $0 \in \omega$ follows from the fact that ω is an inductive set, as does also the fact that if $u \in \omega$, then $u^+ \in \omega$, so P1 and P2 hold for ω. Axiom P4, which states that no u in ω is such that $u^+ = 0$, follows immediately from the fact that u^+ always contain u, whereas 0 is the empty set $\varnothing$ and does *not* contain any u. Because ω is the minimal inductive set, it follows that if any subset S of ω is an inductive set, then $S = \omega$. That is, if $S \subset \omega$, if $0 \in S$, and if $x^+ \in S$ whenever $x \in S$, then $S = \omega$. This is all that is stated by P5, which is a formulation of the principle of mathematical induction.

To prove that P3 holds for every n in ω, we introduce a new concept and then prove two lemmas concerning it (a *lemma* is a subsidiary or preliminary theorem introduced for the purpose of proving some further theorem). The new concept is that of a *transitive* set, defined to be a set that contains every member of any of its members. Formally,

$$a \text{ is a } \textit{transitive} \text{ set} = \text{df } (x)(y)[(x \in y \cdot y \in a) \supset x \in a]$$

Presumably, a transitive set is so-called because $\in$ between its members and their members is a transitive relation. Alternative equivalent ways in which 'a is a transitive set' can be defined are

$$\bigcup a \subset a \qquad \text{or} \qquad (x)(x \in a \supset x \subset a) \qquad \text{or} \qquad a \subset \mathcal{P}a$$

Next we establish the two lemmas.

LEMMA 1. *If a is a transitive set, then $\bigcup(a^+) = a$.*

Proof:
$$\begin{aligned} \bigcup(a^+) &= \bigcup(a \cup \{a\}) \text{ (by definition of }{}^+) \\ &= \bigcup a \cup \bigcup\{a\} \\ &= \bigcup a \cup a \\ &= a \qquad \text{(because } \bigcup a \subset a \text{ for a transitive set } a) \end{aligned}$$

LEMMA 2. *Every natural number n is a transitive set.*

Proof: (by induction, appealing to P5) Let T be the set of natural numbers that are transitive sets, that is, $T = \{x\colon x \in \omega \cdot \bigcup x \subset x\}$. First we observe that $0 \in T$, since for every x, $x \notin 0$, whence $x \in 0 \supset x \in T$.

Next we consider any natural number n and show that $n \in T \supset n^+ \in T$. Suppose $n \in T$. By Lemma 1, $\bigcup(n^+) = n$, and since $n^+ = n \cup \{n\}$, $n \in n^+$, and so $\bigcup(n^+) \in n^+$. Therefore, $n^+ \in T$. Now, by induction, $T = \omega$, that is, every natural number is a transitive set.

Now we can prove P3, that for any n and m in ω, if $n^+ = m^+$, then $n = m$. If $n^+ = m^+$, then $\bigcup(n^+) = \bigcup(m^+)$. Since both n and m are in ω, both are transitive sets by Lemma 2. And by Lemma 1, $\bigcup(n^+) = n$ and $\bigcup(m^+) = m$, whence $n = m$.

This finally verifies that the von Neumann sets satisfy all of the Peano axioms and, therefore, have all the usual arithmetic properties of natural numbers. They are not the philosophically more plausible Frege-Russell numbers (which cannot be sets in **ZF**) but are *instances* of the Frege-Russell numbers. Each number n is itself equivalent to every set containing just that number n of members. Hence they are admirably suited for *counting* the number of members in any finite set. They have, indeed, been called 'counter sets'.[39] Because each natural number n is a set containing all of the natural numbers less than n, the present definition is exactly in accord with what was stated on page 184 in

[39] W. V. O. Quine, *Mathematical Logic*, Harvard University Press, Cambridge, Mass., 1947, p. 247.

defining a finite set. Each natural number, then, can serve the function of a cardinal number and be used to answer the question 'How many?' when asked about a finite set.[40] Some of their peculiar set theoretic features are surprising. For example, each natural number n 'belongs to' or 'is a member of' every larger number, which may seem repugnant to common sense. But to a greater or lesser degree, every scientific advance marks some departure from the common sense that preceded it.[41]

7.6 Cardinal Numbers and the Choice Axiom

In Section 7.5 the natural numbers 0, 1, 2, 3, . . . were given a set theoretic definition. Each of them is the number of members in some finite set or other and *is* a finite set with that number of members. The Axiom of Infinity stated in that section guarantees the existence of the infinite set ω. In Section 7.4, it was proved that the set ω has fewer members than its power set $\mathcal{P}\omega$, and more generally that any set A has fewer members than its power set $\mathcal{P}A$. There are, then, infinitely many sets, each containing a different finite number of members, and also infinitely many more sets, each containing infinitely many members. Thus there is an unending ascension of sets, each larger than any that precedes it.

Cantor symbolized the cardinal number of a set A by '$\bar{\bar{A}}$', where the two lines over the symbol for the set signify a double abstraction from the concrete nature of the set, first, abstracting from the particular nature of its members and secondly, abstracting from the particular order in which those members are arranged. It is more usual today to symbolize the cardinal number of a set A by 'card A'. The essential nature of a cardinal number is given by

$$\text{card } A = \text{card } B \equiv A \sim B$$

The Frege-Russell concept of a cardinal number as a set of all equivalent or equinumerous sets cannot be used in **ZF**, so we took for our finite cardinal numbers *representative* sets from among sets equivalent to each other. It is customary to do the same for infinite cardinal numbers. We associate with each set, whether finite or infinite, the concept of its *size* or *magnitude*. And we conceive a cardinal number to be a 'measure' of the size of any set whose cardinal number it is. From each plurality of equivalent sets we select one representative set to serve as the cardinal number of any of them. So any pair of equivalent sets has the same cardinal number. For definiteness we should specify *which* set from among a plurality of equivalent sets is selected as the

[40] 'The natural numbers in the role of cardinal numbers are the *finite cardinals*, and sets which have these cardinals are *finite sets*.' Robert R. Stoll, *Set Theory and Logic*, W. H. Freeman and Company, Publishers, San Francisco and London, 1963, p. 83.

[41] See the brilliant and provocative discussion by Paul Benacerraf, 'What Numbers Could Not Be', *The Philosophical Review*, vol. 74 (1965), pp. 47–73.

representative, or cardinal number. This has already been done for the finite case. It will be convenient to delay this specification for the infinite case until Section 7.7. In that section it will also be remarked that the cardinal numbers are *serially* (*totally*, or *simply*) ordered by the relation $<$ (is less than). That is, for any cardinal numbers m and n, if $m \neq n$ then either $m < n$ or $n < m$, where the relation $<$ is asymmetric, irreflexive, and transitive (these attributes of relations were defined on pages 134–136). This feature of cardinal numbers is, of course, necessary if they are to fulfill their intended function, which is to measure the size of sets.

It was remarked in Section 7.4 that the set of all natural numbers, ω, is equivalent to various other sets: the set of all perfect squares and the set of all rational fractions. There is a single cardinal number, card ω, that is the number of members in each of these equivalent sets. We define the symbol for it:

$$\aleph_0 =_{\text{df}} \text{card } \omega$$

this is the Hebrew letter aleph with subscript zero, and is read 'aleph-null'. It is the smallest infinite cardinal number.

Because the Power Set Axiom produces ever larger sets, there are infinitely many cardinal numbers. Each is the unique representative of sets equivalent to each other and to itself. Infinite cardinal numbers have an arithmetic that, in some respects, is very similar to the familiar arithmetic of finite numbers, although, in other respects, it is very different. The definitions of the fundamental arithmetic operations on cardinal numbers are the same for both finite and infinite cardinals.

Intuitively, the sum of two numbers, m and n, is the number p if and only if there are disjoint sets M and N with $m = \text{card } M$ and $n = \text{card } N$ and $p = \text{card } (M \cup N)$. We define $m + n$ to be card $(M \cup N)$ where M and N satisfy the conditions just stated. In other words, the arithmetic sum of two cardinal numbers x and y is defined to be the cardinal number of the set-theoretic sum (or union) of any two disjoint sets X and Y that have those cardinal numbers, respectively. That addition, so defined, is independent of the particular disjoint sets M and N, can be seen from the fact that for any other disjoint sets M' and N', if $M \sim M'$ and $N \sim N'$, then $(M \cup N) \sim (M' \cup N')$. The familiar commutation and association laws for $+$ follow from the commutation and association of $\cup$.

The differences between arithmetic addition for finite and infinite cardinals can be brought out clearly by the following contrasts. For finite numbers m and n, $m + n = m$ only if $n = 0$. But for infinite cardinal numbers, that is not true. The set of natural numbers ω can be decomposed into a finite and an infinite part. For any finite number n, $\omega = \{0, 1, 2, \ldots, n - 1\} \cup \{n, n + 1, n + 2, \ldots\}$, where the first *summand* has cardinality n and the second summand has cardinality $\aleph_0$. Hence

$$n + \aleph_0 = \aleph_0 + n = \aleph_0 \tag{1}$$

for any finite number n, even where $n \neq 0$. Another contrast is the following. The sum of any two finite numbers different from 0 is larger than either summand. Because the set of all odd numbers $\{1, 3, 5, 7, \ldots\}$ and the set of all even numbers $\{0, 2, 4, 6, \ldots\}$ each has the cardinality $\aleph_0$ and their union is the set of all natural numbers, we have

$$\aleph_0 + \aleph_0 = \aleph_0 \tag{2}$$

Finally, it should be remarked that for any finite cardinal numbers m and n, where $m \leq n$, the result of subtracting m from n, $(n - m)$ is unique. The result of subtracting m from n, where m and n are infinite cardinals, however, is not unique. (1) and (2) above show that $\aleph_0 - \aleph_0$ can equal any cardinal number from 0 to $\aleph_0$ itself.

Arithmetic addition is defined also for infinitely many summands. If $\{m_i\}$ is a set of infinitely many cardinal numbers, then where $\{M_i\}$ is a set of infinitely many sets, such that $m_i = \text{card } M_i$ and $i \neq j \supset M_i \cap M_j = \varnothing$ for every i and j, then by definition

$$\sum_i m_i = \text{card } (\bigcup_i M_i)$$

For example, the denumerable set of all natural numbers can be decomposed into denumerably many sets, each of which contains denumerably many natural numbers, as in the following (diagonal) array:

```
 1  2  4  7...........
 3  5  8..............
 6  9.................
10....................
......................
```

which shows that

$$\aleph_0 + \aleph_0 + \aleph_0 + \cdots = \aleph_0.$$

Intuitively, the product $m \cdot n$ (or mn) of two numbers m and n is the result of adding m to itself n times. An alternative (equivalent) way to define the product of m and n is to produce sets M and N where $n = \text{card } M$ and $n = \text{card } N$ and then use the Cartesian product to define

$$m \cdot n = \text{card } (M \times N).$$

It is easily seen that this product is unique (i.e., independent of the particular sets M and N whose Cartesian product is formed), that it is equal to 0 if, and only if, either $M = \varnothing$ or $N = \varnothing$. It is also evident that the product of m and n

obeys the familiar commutation and association laws for multiplication and that, for the infinite case as well as for the finite one, multiplication is distributive with respect to addition.

There is, however, a complication in the case of multiplication that involves infinitely many factors—even where there are 'only' $\aleph_0$ factors and even where each factor is finite. Suppose $\mathfrak{M}$ to be a set of infinitely many disjoint nonempty finite sets M_i where m_i = card M_i, and we wish to define the product $\prod_i m_i$. Intuitively, this number is the number of distinct selection sets, μ, where each μ contains exactly one element from each of the sets M_i in $\mathfrak{M}$. In the finite case where $\mathfrak{M} = \{\{a, b\}, \{c, d, e\}\}$, that is, where $M_1 = \{a, b\}$ and $M_2 = \{c, d, e\}$, the distinct sets μ are $\{a, c\}$, $\{a, d\}$, $\{a, e\}$, $\{b, c\}$, $\{b, d\}$, $\{b, e\}$. There are six such sets, which is the expected product since 2 = card M_1 and 3 = card M_2. The actual process of forming distinct sets μ is simple. Select one member from M_1, and select one member from M_2. Form their pair set, which is our first μ. Then select one member of M_1 and select a different member of M_2. Next, form their pair set, the second μ, and so on. Now, the problem in the infinite case is this: an arbitrary choice must be made from among the members of each of the infinitely many sets M_i of the collection $\mathfrak{M}$—unless each set M_i (that is not a singleton) has its members already arranged alphabetically, numerically, by size, or in any other order that permits the statement of a rule that dictates which should be chosen first for inclusion in the first μ, which next for inclusion in the next μ, and so on. Now for any single finite set M_i, or even for an infinite one, there is no great problem of making an arbitrary selection. But the notion of making an infinite sequence of arbitrary choices is somewhat fanciful. What is needed here is another axiom, one that asserts the existence of a selection set, μ, that contains exactly one member from each of any number of sets, M_i. Zermelo stated this axiom in approximately the following form:

ZF-8 AXIOM OF CHOICE (*Axiom der Auswahl*): *If $\mathfrak{M}$ is a set whose elements are all sets that are different from $\varnothing$ and mutually disjoint, its union $\bigcup\mathfrak{M}$ includes at least one subset μ having one and only one element in common with each element of $\mathfrak{M}$.*[42]

Bertrand Russell called it the 'multiplicative axiom', because of its essential role in defining multiplication for infinitely many cardinal numbers as factors. One of his ingenious illustrations to clarify its role and need concerned a mythical millionaire who accumulated $\aleph_0$ pairs of shoes and $\aleph_0$ pairs of socks. The questions 'How many shoes did he have? And how many socks?' require different methods of answering. Russell wrote:

> The reason for the difference is this: Among boots we can distinguish right and left, and therefore we can make a selection of one out of each pair, namely, we can choose all the right boots or all the left boots; but with socks no such principle of selection suggests itself, and we cannot be

[42] Zermelo, op. cit., in van Heijenoort, op. cit.

> sure, unless we assume the multiplicative axiom, that there is any class consisting of one sock out of each pair. Hence the problem.
>
> We may put the matter in another way. To prove that a class has $\aleph_0$ terms, it is necessary and sufficient to find some way of arranging its terms in a progression. There is no difficulty in doing this with the boots. The *pairs* are given as forming an $\aleph_0$, and therefore as the field of a progression. Within each pair, take the left boot first and the right second, keeping the order of the pair unchanged; in this way we obtain a progression of all the boots. But with the socks we shall have to choose arbitrarily, with each pair, which to put first; and an infinite number of arbitrary choices is an impossibility. Unless we can find a *rule* for selecting, *i.e.* a relation which is a selector, we do not know that a selection is even theoretically possible.[43]

Clearly, if we have one subset μ, we can get another by taking the relative complement of each set $M_i - \mu$, in turn, and using the Choice Axiom to get another μ, and so on. Hence the new axiom guarantees the existence of all the selection sets μ in $\bigcup \mathfrak{M}$, and legitimizes the definition:

If $\{m_i\}$ is a set of cardinal numbers, and if $\{M_i\}$ is a corresponding set of sets, such that card $M_i = m_i$ for each i, then

$$\prod_i m_i = \text{card}\ (\bigtimes_i M_i)$$

The same definition in more familiar symbols would be written:

$$\begin{aligned}\prod_i m_i =\text{df card}\ \{\mu\colon (m_i)(m_j)(\exists M_i)(\exists M_j)(m_i &= \text{card}\ M_i \cdot \\ m_j = \text{card}\ M_j \cdot (i)(j)(i \neq j \supset M_i \cap M_j &= \varnothing)\cdot \\ (\exists x)(x \in M_i \cdot x \in \mu \cdot (y)((y \in M_i \cdot y \in \mu) &\supset y = x)))\}\end{aligned}$$

Another formulation of this useful axiom is

Axiom of Choice (Alternative version): *For any set M there is a is a function f (a 'choice function' for M) such that the domain of f is the set of nonempty subsets of M and $f(S) \in S$ for every nonempty $S \subset M$.*

Here, in the case of defining the product of cardinal numbers m_i, the choice function f for the set $\bigcup \mathfrak{M}$ will select an element x_i from each nonempty subset M_i of $\bigcup \mathfrak{M}$; each determines a singleton $\{x_i\}$, and the union of all of them will be a selection set μ.

There are several theorems equivalent to the Choice Axiom, one of which plays an important role in the next section.[44] Much has been written about the

[43] Bertrand Russell, *Introduction to Mathematical Philosophy*, George Allen & Unwin Ltd., London, 1919, p. 126.

[44] See Suppes, op. cit., pp. 243–250. A very full discussion of various applications of the Choice Axiom in different branches of mathematics can be found in Fraenkel and Bar-Hillel, op. cit., pp. 59–74.

Choice Axiom. Efforts were made to derive it from the other axioms of set theory. None succeeded. Some people worried that a contradiction might arise from adding the Choice Axiom to the other axioms. Those worries proved groundless. In 1938 Gödel proved that if the other axioms of set theory are consistent, then the result of adding the Choice Axiom to them is also consistent.[45] In 1963 Paul J. Cohen proved that the Choice Axiom cannot be derived from the other axioms.[46]

Exponentiation of cardinal numbers follows the same pattern. In the finite case, just as multiplication involves the addition of equal summands, so exponentiation involves the multiplication of equal factors. We recall that for finite nonzero cardinal numbers a, b, c, d, the term b^a is the result of multiplying b by itself a times, whence $b^{a+c} = b^a \cdot b^c$, $(b \cdot d)^a = b^a \cdot d^a$, $(b^a)^c = (b^c)^a = b^{a \cdot c}$, and where $b \neq 0$, $b^0 = 1$, and where $a \neq 0$, $0^a = 0$.

It is useful at this point to introduce a new notation for certain sets of functions:

$$B^A = \text{df } \{f\colon f \text{ is a function on } A \text{ to } B\}$$

This notation is intended to reveal and exploit the analogy with the familiar notation for exponentiation. That there is an analogy is shown by the following examples. Suppose that $B = \{b_1, b_2, \ldots, b_n\}$, so $n = \text{card } B$; and let $A = \{a_1\}$, so card $A = 1$. In this case there are exactly n functions in B^A: $f_1(a_1) = b_1$, $f_2(a_1) = b_2, \ldots, f_n(a_1) = b_n$. In this case, card $(B^A) = (\text{card } B)^{\text{card } A}$. Now for the same B, let $A = \{a_1, a_2\}$; here card $A = 2$. In this case there are n^2 functions in B^A:

$$\begin{array}{lll}
f_{11}(a_1) = b_1 & f_{12}(a_1) = b_1 & f_{13}(a_1) = b_1 \ldots f_{1n}(a_1) = b_1 \\
f_{11}(a_2) = b_1 & f_{12}(a_2) = b_2 & f_{13}(a_2) = b_3 \ldots f_{1n}(a_2) = b_n \\
\\
f_{21}(a_1) = b_2 & f_{22}(a_1) = b_2 & f_{23}(a_1) = b_2 \ldots f_{2n}(a_1) = b_2 \\
f_{21}(a_2) = b_1 & f_{22}(a_2) = b_2 & f_{23}(a_2) = b_3 \ldots f_{2n}(a_2) = b_n \\
\cdots\cdots\cdots & \cdots\cdots\cdots & \cdots\cdots\cdots\cdots\cdots\cdots \\
f_{n1}(a_1) = b_n & f_{n2}(a_1) = b_n & f_{n3}(a_1) = b_n \ldots f_{nn}(a_1) = b_n \\
f_{n1}(a_2) = b_1 & f_{n2}(a_2) = b_2 & f_{n3}(a_2) = b_3 \ldots f_{nn}(a_2) = b_n
\end{array}$$

In this case also card $(B^A) = (\text{card } B)^2 = (\text{card } B)^{\text{card } A}$. In general, for any finite nonempty sets A and B, card $(B^A) = (\text{card } B)^{\text{card } A}$. So it is plausible to extend the definition of exponentiation to cover all cardinal numbers a and b,

[45] K. Gödel, 'The Consistency of the Axiom of Choice and of the Generalized Continuum Hypothesis', *Proceedings of the National Academy of Sciences, U.S.A.*, vol. 24 (1938), pp. 556–557. See also *The Consistency of the Axiom of Choice and of the Generalized Continuum Hypothesis with the Axioms of Set Theory*, Annals of Mathematics Studies, No. 3, Princeton, N.J., 1940.

[46] Paul J. Cohen, 'The Independence of the Continuum Hypothesis', I and II, *Proceedings of the National Academy of Sciences, U.S.A.*, vol. 50 (1963), pp. 1143–1158, and vol. 51 (1964), pp. 105–110. See also Paul J. Cohen, *Set Theory and the Continuum Hypothesis*, W. A. Benjamin, Inc., Menlo Park, Calif., 1966.

where a and b can be either finite or infinite:

$$b^a = \text{df card } (B^A) \qquad \text{where } a = \text{card } A \text{ and } b = \text{card } B.$$

Just as cardinal addition is based on the union of disjoint sets that have the cardinal numbers being added, and just as cardinal multiplication is based on the Cartesian product of sets having the cardinal numbers being multiplied, so cardinal exponentiation is based on mappings of one set onto another, where the range has the cardinal number that is raised to a power and the domain has the cardinal number that is the exponent. It is easily shown that for any nonzero infinite cardinal numbers a, b, c, d the same equalities obtain as for finite cardinals. So we have $\aleph_0^0 = 1$, and for any finite cardinal $a > 0$,

$$\aleph_0^a = \aleph_0 \cdot \aleph_0 \cdot \aleph_0 \cdot \ldots \cdot \aleph_0 = \aleph_0.$$

If the set B in B^A is equal to $2\ (= \{0, 1\})$, we have the set of all *characteristic functions* of subsets of A. The definition is this:

For any subset $A' \subset A$ the characteristic function

$$\begin{aligned} F_{A'}\colon A \rightarrow 2 &\text{ is such that} \\ &(x)(x \in A' \supset f_{A'}(x) = 1) \\ \text{and } &(x)(x \in (A - A') \supset f_{A'}(x) = 0). \end{aligned}$$

It is useful to think of the characteristic function $f_{A'}$, for each $A' \subset A$, as a *condition* on members of A, such that every member x of A for which $f_{A'}(x) = 1$ *satisfies* the condition, and such that every member x of A for which $f_{A'}(x) = 0$ does *not* satisfy the condition. The Separation Axiom states that any such condition $f_{A'}(x)$ on members of A determines a unique subset $A' \subset A$. Here we have the converse of that statement, namely, that any subset A' of A determines a characteristic function or condition $f_{A'}(x)$. It is obvious, then, that the set of all characteristic functions of subsets of A is equivalent or equinumerous to the set of all subsets of A. That is, for any set A, $2^A \sim \mathcal{P}A$, and in particular, $2^\omega \sim \mathcal{P}\omega$. Since $\{A'\colon A' \subset \omega\} = \mathcal{P}\omega$ and is a set by the Power Set Axiom, we know that $2^\omega \sim \{A'\colon A' \subset \omega\}$. Cantor proved that for any set S, $S < \mathcal{P}S$, and in particular, $\omega < \mathcal{P}\omega$. We also know, by Cantor's diagonal proof, that $\omega < \{x\colon x \text{ is a real number such that } 0 \leq x \leq 1\}$, so $\omega < \mathbb{R}$, where $\mathbb{R}$ is the set of all real numbers.

Both $\mathcal{P}\omega$ and $\mathbb{R}$ are greater than ω. It is not difficult to prove that $\mathcal{P}\omega \sim \mathbb{R}$. There are as many real numbers between 0 and 1 as there are real numbers altogether so that without any loss in generality, we can confine our attention to real numbers in that closed interval. Every such real number is representable as a binary fraction, $0.d_1d_2d_3\ldots$, where each d_i is one of the two binary digits 0 and 1. (1 is $0.111\ldots1\ldots$, 0 is $0.000\ldots0\ldots$, 1/2 is represented both by $0.100\ldots0\ldots$ and by $0.011\ldots1\ldots$). Each binary fraction determines a unique subset of ω, namely, the subset that contains the natural number n if and only if $d_n = 1$. Thus the binary fraction 0.111 . . . 1 . . . determines the set ω, itself; the binary fraction 0.000 . . . 0 . . . determines the empty subset of ω; and the binary fraction 0.101 . . . 1 . . . determines the subset of ω that contains every natural number except 2. The binary fraction 0.0101 . . . 0101 . . . determines the set of all even numbers, where only d_n's with even-numbered subscripts are 1's, and so on. Conversely, every subset A' of ω determines a binary fraction $0.d_1d_2d_3 \ldots$, in which $d_n = 1$ if, and only if, $n \in A'$. Thus we have a one-to-one correspondence between $\mathcal{P}\omega$ and $\{x\colon x$ is a real number such that $0 \leq x \leq 1\}$ and hence between $\mathcal{P}\omega$ and $\mathbb{R}$, so $\mathcal{P}\omega \sim \mathbb{R}$, and card $(\mathcal{P}\omega) =$ card $\mathbb{R}$. The usual symbol for card $\mathbb{R}$ is $\aleph$, the cardinal number of the continuum. Since $2^\omega \sim \mathcal{P}\omega$, card $(2^\omega) =$ card $(\mathcal{P}\omega)$, and since card $(2^\omega) =$ (card $2)^{\text{card}\,\omega}$ which is $2^{\aleph_0}$, we have established that $2^{\aleph_0} = \aleph$. As for $\aleph_0$, it is also true for $\aleph$ that $\aleph + \aleph = \aleph \cdot \aleph = \aleph$. More infinite cardinals exist by the Power Set Axiom. Some of them will be discussed in the next section.

7.7 Ordinal Numbers and the Axioms of Replacement and Regularity

The elements of a set may be related to each other in a variety of ways: some may be smaller than others, some may be elements of others, some may be subsets of others, and so on. Any such relation among its members can be said to impose some kind of order, even if only partial, on the set. It may be useful to consider some examples.

1. Let S be the power set $\mathcal{P}\{a, b, c\}$ with the ordering relation that of *set inclusion*, $\subset$.
2. Let P be the set of positive integers and let the ordering relation be *divides without remainder, d*.
3. Let ω be the set of natural numbers with the ordering relation *less than or equal to*, $\leq$.

Each of these sets is *partially ordered* by the relation mentioned for it, where we define *partial order in a set X* to be a reflexive, antisymmetric, and transitive relation in X. These notions were discussed and defined in Chapter 5 (pages 135–136 and 143). Their definitions in the context of **ZF** set theory are

1. R is reflexive in $X = df(x)(x \in X \supset xRx)$
2. R is antisymmetric in $X = df(x)(y)((x \in X \cdot y \in X) \supset ((xRy \cdot yRx) \supset x = y))$
3. R is transitive in $X = df(x)(y)(z)((x \in X \cdot y \in X \cdot z \in X) \supset ((xRy \cdot yRz) \supset xRz))$

A little thought will confirm that $\subset$ is a partial order in $\mathcal{P}\{a, b, c\}$, d is a partial order in P, and $\leq$ is a partial order in ω. The first two sets are *only* partially ordered by the relations mentioned for them, because there are distinct members of each set that are not related to each other by the ordering relation. The third set is *totally* ordered, or *simply* ordered, which—as the term is generally used—includes or entails partial order.

A very useful kind of ordered set to work with is a *well ordered set*, which is defined to be a partially ordered set, every nonempty subset of which has a first (or least, or smallest) element, that is, an element that is related by the ordering relation to every other element in the subset. The third of the preceding examples is a well ordered set under the ordering relation mentioned for it. But neither of the first two examples is well ordered under the relation specified for it. (The subset of $\mathcal{P}\{a, b, c\}$ containing just $\{a\}$ and $\{b\}$ does not contain a first element, since neither $\{a\} \subset \{b\}$ nor $\{b\} \subset \{a\}$. And the subset of P containing just 2 and 3 does not contain a first element, since neither divides the other without remainder.) Each of the first two examples is well ordered, however, under *another* relation of its members. If we consider the relation of alphabetical precedence, π, among a, b, and c, $a\pi b$, $b\pi c$, and $a\pi c$, we can then define a derived alphabetical precedence relation Π among the members of $\mathcal{P}\{a, b, c\}$ in the obvious way. Our definition is first, for any subset α of $\{a, b, c\}$, if $\alpha \neq \varnothing$ then $\varnothing \Pi \alpha$, and if $\alpha \neq \{a, b, c\}$ then $\alpha \Pi \{a, b, c\}$, and then $\{x_1\} \Pi \{x_2\}$ if and only if $x_1 \pi x_2$, $\{x_1, y_1\} \Pi \{x_2, y_2\}$ if and only if either $x_1 \pi x_2$ or $x_1 = x_2 \cdot y_1 \pi y_2$. It is obvious that $\mathcal{P}\{a, b, c\}$ is well ordered under this lexicographic ordering. And the set P of the second example is well ordered by $\leq$. In summary, a set may be well ordered by one relation, but not by another.

In developing the theory of well ordered sets, it is convenient to work with the narrower relation of precedence $<$, or 'less than', rather than the disjunctive 'less than or equal to', $\leq$. The precedence relation is irreflexive, and asymmetric rather than antisymmetric, but like $\leq$, $<$ is transitive. The first remark to make about well ordered sets is that they are *connected* by the relation that well orders them, that is, for any two distinct members $x_1 \neq x_2$ of a well ordered set, either $x_1 < x_2$ or $x_2 < x_1$. This follows immediately from the well ordering condition, for any doubleton $\{x_1, x_2\}$ is a nonempty subset and, therefore, contains a least element.

Two partially ordered sets are said to be *similar* if there is a one-to-one correspondence between them that preserves order. That is, if X and Y are partially ordered sets and there is a one-to-one correspondence f from X onto Y, such that for any x_1 and x_2 in X, $x_1 \leq x_2$ (in X) if, and only if, $f(x_1) \leq f(x_2)$ (in Y), then X is similar to Y, symbolized as $X \cong Y$ (sometimes as $X \approx Y$). It is obvious that $X \cong Y$ entails that $X \sim Y$. An infinite well ordered set can be similar to a proper subset of itself. For example, the function $m = 2n$ is a one-to-one correspondence between the set of all natural numbers and the set of all even numbers which preserves order in the two sets.

One important feature of any such mapping f, from a well ordered set X onto a subset of itself, is that $x \leq f(x)$, for every x in X. This is proved as follows. Suppose that there were an x in X, such that $f(x) < x$. Then since X is well ordered, there must be a least such x, call it x_0. It could not be the least element in X, because $f(x_0) \in X$ and $f(x_0) < x_0$. For any y in X, if $y < x_0$ then $y \leq f(y)$. But $f(x_0)$ is such a y, so $f(x_0) \leq f(f(x_0))$. But since $f(x_0) < x_0$, by the order preserving nature of f, $f(f(x_0)) < f(x_0)$. Now by transitivity, $f(x_0) < f(x_0)$, which contradicts the irreflexiveness of $<$. So the supposition that there is an x in X, such that $f(x) < x$, must have been false.

There is an important consequence of the result just obtained. To state it we must define a new term. If X is a well ordered set, and $a \in X$, then the subset $\{x: x \in X \cdot x < a\}$ is called the *initial segment* determined by a, and is usually denoted by $s(a)$. Now we can prove that no well ordered set can be similar to one of its initial segments. For if X were a well ordered set containing a as an element, and f were a similarity function from X onto $s(a)$, then we would have $f(a) \in s(a)$, whence $f(a) < a$, which was proved to be impossible.

Many truths about well ordered sets can be proved by what is called the *principle of transfinite induction*. It is analogous to the principle of mathematical induction discussed in Section 8.5, and, indeed, for finite sets the two are equivalent. In the case of infinite well ordered sets, only the new principle is applicable. It can be stated as follows:

> If S is a subset of a well ordered set X, and if an element x of X belongs to S whenever the initial segment determined by x is included in S, then S = X.

Its proof in the general case is simple. If $X - S$ were not empty, then it would contain a least element, say x. That implies that every element of the initial segment $s(x)$ must belong to S and, by the induction hypothesis, $x \in S$, also. But this would be impossible, because x cannot belong both to S and to $X - S$. Thus $X - S$ must be empty and since $S \subset X$, $S = X$. This principle, as applied to natural numbers, is called 'strong induction,' and its use will be illustrated in Chapters 8, 9, and 10.

If two simply ordered sets are similar, they are said to be *isomorphic* (with each other). Ordered sets that are isomorphic with each other are said to have the same *order type*. The most important order types are those of well ordered sets. Some writers define 'ordinal numbers' as the order types of well ordered sets.[47] An order type of a given nonempty well ordered set, however, cannot be defined in **ZF** set theory as the set of all well ordered sets that are isomorphic with the given well ordered set, because the union of such a set of all similar well ordered sets would be the universal set—which we know does not exist. This situation is the same as that which made the Frege-Russell definition of cardinal number incompatible with **ZF** set theory. In the case of cardinal numbers, discussed in Section 7.6, we let an unspecified representative set from

[47] Raymond L. Wilder, op. cit., p. 127.

among a plurality of equivalent sets serve as the cardinal number of any of the sets equivalent to it. *Which* set would serve as *the* cardinal number was specified only for the finite case. In dealing with ordinal numbers, we *can* specify which well ordered set will serve as the ordinal number of any well ordered set isomorphic with it.

Every natural number n is a subset of the well ordered set ω and is a well ordered set that contains all of the natural numbers less than n. For any natural numbers m and n, as they were defined in Section 7.5, $m < n$ if and only if $m \in n$. Each $n = \{m: m \in \omega \cdot m < n\}$. The ordinal number of any *finite* well ordered set is defined to be the natural number that is the number of elements in the well ordered set in question. Given any natural numbers m and n with $m \in n$, the initial segment $s(m)$ of n determined by m is $\{x: x \in n \cdot x < m\}$. That is, every natural number is a well ordered set such that the initial segment determined by each element in it is the same as that element. This characterization of natural numbers as ordinal numbers can be generalized to give a definition of ordinal number that includes the infinite case.[48] That definition is

α is an ordinal number $=$ df α is a well ordered set such that $s(\xi) = \xi$ for every ξ in α.

The first and smallest of the infinite ordinal numbers covered by this definition is ω. (Infinite numbers are also called *transfinite* numbers.) Using the same definition of successor set that was given in Section 7.5, $u^+ = u \cup \{u\}$, but now allowing u to range over ordinal numbers, we moved beyond ω to ω^+, then to $(\omega^+)^+$, and so on. These are easily seen to be ordinal numbers also. For if α is an ordinal number, then so is α^+, which is $\alpha \cup \{\alpha\}$. For if $\xi \in \alpha^+$, then either $\xi \in \alpha$ or $\xi \in \{\alpha\}$. In case $\xi \in \alpha$, then since α is an ordinal number, $s(\xi) = \xi$; and in case $\xi \in \{\alpha\}$, $\xi = \alpha$, in which case $s(\xi) = \alpha$, that is, $s(\alpha) = \alpha$. So in either case $s(\xi) = \xi$ for any $\xi \in \alpha^+$, and α^+ is an ordinal number.

Before stating the next **ZF** axiom, which will permit the construction of transfinite ordinal numbers out to any distance that might be desired, it will be convenient to define the various arithmetical operations that can be performed with ordinal numbers.

Well ordered sets can be combined by putting one directly after the other. For this to be done, of course, the sets must be disjoint. But if the set A and B that we want to combine in this way are not disjoint, we can easily obtain other sets A' and B' that *are* disjoint, and such that $A \cong A'$ and $B \cong B'$. A' can be the set of all ordered pairs $\langle a, 0\rangle$ with a in A and B' can be the set of all ordered pairs $\langle b, 1\rangle$ with b in B. The obvious one-to-one correspondences are $a \to \langle a, 0\rangle$ and $b \to \langle b, 1\rangle$, with the order within A' and B' simply borrowed from A and B. Therefore, given that A and B are disjoint well ordered sets, we

[48] See John von Neumann, 'Uber die Definition durch transfinite Induktion und verwandte Fragen der allgemeinen Mengenlehre', *Mathematische Annalen*, vol. 99 (1928), pp. 373–391. See also R. M. Robinson, 'The Theory of Classes. A Modification of von Neumann's System', *The Journal of Symbolic Logic*, vol. 2 (1937), pp. 29–36.

define the order in $A \cup B$ so that pairs of elements in A and pairs of elements in B keep the order they had, and let each element of A precede each element of B. The fact that A and B were well ordered entails that $A \cup B$ is well ordered also and it is said to be the *ordinal sum* of A and B.

Now, we can define addition for ordinal numbers. Given ordinal numbers α and β, let A and B be disjoint well ordered sets such that $\alpha = \text{ord } A$ (i.e., the unique ordinal number similar to A) and $\beta = \text{ord } B$, and let C be the ordinal sum of A and B. The sum $\alpha + \beta$ is now defined to be ord C; so ord A + ord B = ord $(A \cup B)$. It is clear that the sum $\alpha + \beta$ is independent of the particular choice of sets A and B. Any disjoint pair of similar sets would give the same result.

Some of the properties of addition for ordinal numbers are quite similar to those for cardinal addition. Some examples are $\alpha + 0 = \alpha = 0 + \alpha$, $\alpha + 1 = \alpha^+$, and the Association Law $\alpha + (\beta + \gamma) = (\alpha + \beta) + \gamma$. Another property is the fact that $\alpha < \beta$ if, and only if, there is an ordinal number γ $(\neq 0)$, such that $\beta = \alpha + \gamma$. But the Commutative Law fails for the addition of ordinals, where at least one is infinite. Thus $1 + \omega = \omega$ but $\omega + 1 \neq \omega$, for if we put a new element in front of an infinite sequence (of type ω), the result is similar to the original sequence. If we put a new element at the back of the sequence, however, the similarity is gone because the new set has a last element and the original set did not.

By analogy with multiplication for natural numbers, we conceive the product of well ordered sets A and B to be the result of adding A to itself B times. To do so, we must have B disjoint sets, each familiar to A. These can be produced as well ordered sets $A_b = A \times \{b\}$ for each b in B. Then, the set of all these disjoint A_b's, $\{A_b: b \in B\}$, has, as its sum, the union $\bigcup \{A_b: b \in B\}$, ordered in the following way: $\langle a, b \rangle < \langle a', b' \rangle$ if and only if either $b < b'$ or $(b = b'$ and $a < a')$. We define the *ordinal product* of two well ordered sets A and B as the Cartesian product $A \times B$, ordered as explained above. To define the product of ordinal numbers α and β, we introduce well ordered sets A and B such that ord $A = \alpha$ and ord $B = \beta$, and let C be the ordinal product of A and B. Then the product of the ordinal numbers α and β, written $\alpha\beta$, is defined to be ord C. The easiest well ordered sets to introduce and use here as A and B are the ordinal numbers α and β themselves, since each is a well ordered set whose ordinal number is itself.

Some of the properties of ordinal multiplication are quite similar to those for cardinal multiplication: $\alpha 0 = 0 = 0\alpha$, $\alpha 1 = \alpha = 1\alpha$, the associative law $\alpha(\beta\gamma) = (\alpha\beta)\gamma$, the left distributive law $\alpha(\beta + \gamma) = \alpha\beta + \alpha\gamma$, and the fact that if the product of two ordinal numbers is zero, then one of the factors must be zero. But the commutative law for multiplication fails where at least one of the ordinal numbers is infinite. For example, $2\omega = \omega$, for the product is an infinite sequence of ordered pairs. However, $\omega 2 \neq \omega$ because this product is an ordered pair of infinite sequences. The right distributive law fails too: $(\alpha + \beta)\gamma$ is, in general, different from $\alpha\gamma + \beta\gamma$. For example, $(1 + 1)\omega = 2\omega = \omega$, but $1\omega + 1\omega = \omega + \omega = \omega 2 \neq \omega$.

Just as ordinal products were defined in terms of repeated addition, so ordinal exponentiation can be defined in terms of repeated multiplication. The result of careful definition along this line will have the following desired properties: $0^\alpha = 0$ for $\alpha \geq 1$, $1^\alpha = 1$, $\alpha^{\beta+\gamma} = \alpha^\beta\alpha^\gamma$, and $\alpha^{\beta\gamma} = (\alpha^\beta)^\gamma$. But not all the familiar laws of exponents hold: $(\alpha\beta)^\gamma$ is, in general, different from $\alpha^\gamma\beta^\gamma$, since $(2 \cdot 2)^\omega = 4^\omega = \omega$, but $2^\omega 2^\omega = \omega\omega = \omega^2$. A few more words on this topic might be in order. If m is finite and greater than 1, m^ω is the limit of the expanding product $m \cdot m \cdot m \ldots$ which is an infinite sequence of m-tuples and thus, ω. Alternatively conceived, m^ω is the limit of the set of all finite ordinals m^n for $n < \omega$, so again $m^\omega = \omega$. Similarly, ω^ω is the limit of the set of all powers ω^n for $n < \omega$. Since $1 + \omega + \omega^2 + \cdots + \omega^n = \omega^n$, we may also write $\omega^\omega = 1 + \omega + \omega^2 + \cdots + \omega^k + \cdots$.

We turn next to the problem of whether all of the ordinal numbers discussed in the preceding paragraphs really exist in the **ZF** system. Since ω is a set, we can use the Axiom of Pairing and the Union Axiom to form successor sets on the 'far' side of ω. The successor of ω, ω^+, is $\omega \cup \{\omega\}$, and is written—using the notation for ordinal addition already introduced—as $\omega + 1$. Next, $(\omega^+)^+$ is $\omega^+ \cup \{\omega^+\}$, and is written $\omega + 2$, and so on. It is obvious that each of them exists as a set and is an ordinal number. But is there a set containing them all? Just as we needed an additional axiom, the Axiom of Infinity, to prove that there was a set of all the natural numbers, so here we need still another axiom (or set of axioms presented as an axiom schema) to prove that sets of ordinal numbers beyond ω exist. The needed axioms here are

ZF-9 Axiom Schema of Replacement. *If $\varphi(x, y)$ is a formula such that for each member x of the set X, $\varphi(x, y)$ and $\varphi(x, z)$ imply that $y = z$, then there exists a set S such that $y \in S$ if and only if there is an x in X such that $\varphi(x, y)$.*

Now, if we specify X to be ω and $\varphi(x, y)$, as $y = \omega + x$, we have an Axiom of Replacement that asserts the existence of a set containing

$$\omega, \omega + 1, \omega + 2, \ldots$$

The union of this set and ω we denote by $\omega 2$. It is easily seen to be an ordinal number. After $\omega 2$ come further ordinal numbers: $\omega 2 + 1$, $\omega 2 + 2$, and so on. After all of them comes $\omega 3$, whose existence as a set is asserted by another specification of Replacement. Continuing, we get $\omega 3 + 1$, $\omega 3 + 2$, and so on, and then $\omega 4$. In the same way, we obtain further ordinal numbers. Thus we get, successively, $\omega, \omega 2, \omega 3, \ldots$. By Replacement, again, we obtain an ordinal number, ω^2, that is related to these other ordinal numbers in the series in the way that ω is to the natural numbers. It is convenient to introduce new symbols for some of the transfinite ordinal numbers produced by the Axiom of Replacement. For example, the 'next' ordinal after the sequence ω^ω, $\omega^{(\omega^\omega)}$, $\omega^{\omega^{(\omega^\omega)}}$,

... is usually written as ε_0. Also, there is an ordinal number ω_1, which will be discussed later.

It might be helpful at this point to look at some particular transfinite well ordered sets composed of familiar elements. Three examples follow.

1. The set of all fractions $\frac{2^n - 1}{2^n}$, together with 1, ordered by the familiar '$<$'. This set looks like

$$0,\ 1/2,\ 3/4,\ 7/8,\ \ldots;\ 1$$

in which the initial segment $s(1)$ is clearly similar to ω. The whole well ordered set must have an ordinal number greater than ω. (In fact, its ordinal number is $\omega + 1$.)

2. The set of all odd numbers in order of magnitude, followed by the set of all even numbers in order of magnitude. This set looks like 1, 3, 5, . . . ; 2, 4, 6, . . . and must have an ordinal number twice as large as ω. (In fact, its ordinal number is $\omega + \omega$ or $\omega \cdot 2$.)

3. The set of all positive integer powers of prime numbers p^n, ordered according to the relation $<$ defined as follows: $p_m^i < p_n^j$ if either $p_m < p_n$ or both $p_m = p_n$ and $i < j$. This set looks like

$$2,\ 4,\ 8,\ 16,\ \ldots;\ 3,\ 9,\ 27,\ 81,\ \ldots;\ 5,\ 25,\ 125,\ 625,\ \ldots;\ \ldots$$

and must have an even larger ordinal number. (In fact, its ordinal number is $\omega \cdot \omega$ or ω^2.)

Some remarks about the Axioms of Replacement are appropriate at this point. First, the condition in the hypothesis that $\varphi(x, y)$ and $\varphi(x, z)$ implies that $y = z$ serves to prevent taking as a specification of $\varphi(x, y)$, the formula $x \subset y$. If that were allowed, then—since $\varnothing$ is a subset of every set—taking X as $\{\varnothing\}$ would make S contain all sets, which has been shown to be impossible. Perhaps it might occur to the reader that the condition of being single-valued could more simply be expressed by specifying a function f such that $f(x) = y$, rather than invoking an expression like $\varphi(x, y)$. A function is well defined, however, only if its domain is a set and a set can be specified that contains its range. The Replacement Axioms are needed precisely to produce a set that *is* the range of f. The Replacement Axioms are enormously powerful. Indeed, the Separation Axioms can be derived from Replacement. Given any set A and any condition $\varphi(x)$, if we specify the X in the Replacement Axiom to be A and we specify $\varphi(x, y)$ to be ($x = y$ and $\varphi(x)$), then the '$\varphi(x, y)$ and $\varphi(x, z)$ imply $y = z$' part of the hypothesis of the Replacement Axiom is satisfied. Its conclusion then asserts (leaving out the trivially true '$x = x$')

$$(\exists S)(x)(x \in S \equiv x \in A \cdot \varphi x)$$

which is the Axiom Scheme of Separation. Finally, it should be remarked that the Axiom of Pairing is derivable from Replacement plus the Power Set Axiom.

Here, we specify the set A in Replacement to be the power set $\mathcal{P}\mathcal{P}(\varnothing)$, which is $\{\varnothing, \{\varnothing\}\}$. If a and b are the two objects whose pair set is wanted, we specify $\varphi(x, y)$ to be $(x = \varnothing \cdot y = a) \vee (x = \{\varnothing\} \cdot y = b)$. Then, for each x in $\mathcal{P}\mathcal{P}(\varnothing)$, there is exactly one y such that $\varphi(x, y)$: for $x = \varnothing$, $y = a$, and for $x = \{\varnothing\}$, $y = b$. With these specifications, the hypothesis of the Replacement Axiom is satisfied and its conclusion then states that

$$(\exists S)(x)(x \in S \equiv x = a \vee x = b)$$

which is the Axiom of Pairing.

The unending ascension of ordinal numbers can be indicated by the following:

$$0, 1, 2, \ldots, \omega, \omega + 1, \omega + 2, \ldots, \omega 2, \omega 2 + 1, \omega 2 + 2, \ldots, \omega 3, \ldots, \omega 4, \ldots, \omega^2, \omega^2 + 1, \omega^2 + 2, \ldots, \omega^2 + \omega, \omega^2 + \omega + 1, \omega^2 + \omega + 2, \ldots, \omega^2 + \omega 2, \omega^2 + \omega 2 + 1, \ldots, \omega^2 + \omega 3, \ldots, \omega^2 + \omega 4, \ldots, \omega^2 2, \ldots, \omega^2 3, \ldots, \omega^3, \ldots, \omega^4, \ldots, \omega^\omega, \ldots, \omega^{(\omega^\omega)}, \ldots, \omega^{\omega^{(\omega^\omega)}}, \ldots, \varepsilon_0, \varepsilon_0 + 1, \varepsilon_0 + 2, \ldots, \varepsilon_0 + \omega, \ldots, \varepsilon_0 + \omega 2, \ldots, \varepsilon_0 + \omega^2, \ldots, \varepsilon_0 + \omega^\omega, \ldots, \varepsilon_0 2, \ldots, \varepsilon_0 \omega, \ldots, \varepsilon_0 \omega^\omega, \ldots, \varepsilon_0{}^2, \ldots$$

Up to any point in this ascension, the set of all ordinal numbers, to that point, is a well ordered set, W, which has an ordinal number greater than any member of W. If there *were* a set of *all* ordinal numbers, Ω, it would have an ordinal number greater than any ordinal number *in* Ω, that is, an ordinal number greater than any ordinal. This contradiction is called the Burali-Forti Paradox, and shows that there can be no set containing *all* ordinal numbers.[49]

In Section 7.6 much was said about transfinite cardinal numbers, but their definition was postponed to the present section. We did say that a cardinal number was a representative set, serving as the cardinal number of any set equivalent to it. We want now to specify which sets are to serve as these representatives. To do so we make use of an important theorem proved by Zermelo, using the Axiom of Choice, that *every* set can be well ordered.[50] We know that repeated uses of the Power Set Axiom establish the existence of an endless series of sets with ever larger cardinality. By Zermelo's theorem, every set is equivalent to some ordinal number. This suggests that we look to ordinal numbers for a specification of the representative sets to serve as cardinal numbers.

An infinite set, of course, can be equivalent to a great many different ordinal numbers. In fact, all of the transfinite ordinal numbers indicated on this page have cardinality of 'only' $\aleph_0$. But all of the ordinal numbers equivalent to a given set S form a set. The argument here is quite simple. The power set of S,

[49] See C. Burali-Forti, 'Una Questione Sui Numeri Transfiniti', *Rendiconti del Circolo Matematico di Palermo*, vol. 11 (1897), pp. 154–64, tr. in van Heijenoort, op. cit., pp. 104–11. See also I. Copi, 'The Burali-Forti Paradox', *Philosophy of Science*, vol. 25 (1958), pp. 281–86.

[50] E. Zermelo, 'Beweis, das jede Menge wohlgeordnet werden kann', *Mathematische Annalen*, vol. 59 (1904), pp. 514–516. Tr. in van Heijenoort, op. cit., pp. 139–141.

$\mathcal{P}S$, has higher cardinality than S. Any ordinal number, say σ, that is equivalent to $\mathcal{P}S$, has cardinality greater than the cardinality of S, and σ is therefore larger than all of the ordinal numbers equivalent to S. Since every ordinal number β less than σ is a member of σ, σ is a set that contains every ordinal number equivalent to S. Now the Separation Axiom gives us a subset σ' of σ containing all and only ordinal numbers equivalent to S. Since σ' is an ordered set, it contains a smallest or least element α, which can then be specified as *the* representative set equivalent to S which is selected to serve as the cardinal number of S. We can define *cardinal number* now as an ordinal number α such that if β is another ordinal number equivalent to α, then $\alpha < \beta$. It is clear that this definition accords with the essential characteristic of cardinal numbers, that card A = card B if and only if $A \sim B$.

It should be realized that powers and products of ordinals must not be interpreted cardinally. If A and B are well ordered sets, then, in general, card $((\text{ord } B)^{\text{ord} A}) < (\text{card } B)^{\text{card} A}$. For example, in the case of 2^ω we have card $((\text{ord } 2)^\omega) = \text{card } \omega = \aleph_0$, but $(\text{card } 2)^{\text{card } \omega} = 2^{\aleph_0} = \aleph$, and $\aleph_0 < \aleph$. And in the case of ω^ω, we have $(\text{card } \omega)^{\text{card } \omega} = \aleph_0^{\aleph_0} = \aleph$, but card $((\text{ord } \omega)^{(\text{ord } \omega)}) = \aleph_0$ because all positive integers can be arranged into a well ordered set whose ordinal number is ω^ω. And the set of all positive integers has cardinality $\aleph_0$.[51] The nondenumerable set $\mathbb{R}$ of all real numbers with cardinal $\aleph$ has an ordinal number located beyond all of the ordinal numbers indicated on page 209. There are ordinal numbers with nondenumerably many members. The set of all ordinal numbers, up to some point beyond the location of the ordinal number of the well ordered set of real numbers is, itself, a well ordered set. Therefore, it has a least member—which is called ω_1, the least ordinal number with nondenumerable cardinality, and its cardinal number is $\aleph_1$. There is, in fact, a whole sequence of alephs with increasing cardinality. Their cardinal numbers are

$$\aleph_0, \aleph_1, \ldots, \aleph_n, \ldots, \aleph_\omega, \aleph_{\omega+1}, \ldots, \aleph_{\omega 2}, \ldots$$

We know that $2^{\aleph_0} = \aleph$ and that $\aleph_1$ is the least cardinal number greater than $\aleph_0$. From this, we know that $\aleph_1 \leq \aleph$. The much discussed *continuum hypothesis* asserts that $\aleph_1 = \aleph$ or $\aleph_1 = 2^{\aleph_0}$. The *generalized continuum hypothesis* states that for every ordinal number α, $\aleph_{\alpha+1} = 2^{\aleph_\alpha}$. Efforts to prove the continuum hypothesis were made by such distinguished mathematicians as Cantor and Hilbert. In 1939, Gödel proved that if the axioms of set theory are consistent, the continuum hypothesis cannot be *disproved*, and in 1963 Paul Cohen proved that the continuum hypothesis cannot be *proved* from the axioms of set theory either.[52]

The nine axioms for **ZF** set theory stated thus far are believed to be sufficient for the development of mathematics. It might seem desirable to add a final

[51] See A. A. Fraenkel, *Abstract Set Theory*, 3rd rev. ed., op. cit., p. 205.

[52] See the references cited in footnotes 45 and 46 on page 200. Gödel's and Cohen's results also apply to the generalized continuum hypothesis.

restrictive axiom stating that there are no sets other than those whose existence is required or entailed by the given axioms. A hint as to the direction that such an axiom might take is given by another paradox, due to Mirimanoff.[53] It involves the concept of a *grounded* set, defined as any set, x, for which there is no sequence of (not necessarily distinct) sets $y_1, y_2, y_3, \ldots$ such that $\ldots \in y_3 \in y_2 \in y_1 \in x$. Let W be the set of all grounded sets. If W is grounded, then $W \in W$, whence $\ldots W \in W \in W$, so W is *not* grounded. But if W is not grounded, then there is a sequence of sets $y_1, y_2, y_3, \ldots,$ such that $\ldots \in y_3 \in y_2 \in y_1 \in W$, whence y_1 is not grounded, and therefore cannot be a member of W. In either case, W involves us in a contradiction. The limited kind of comprehension stated in the Separation Axiom would seem to prevent the formation of such a set as W. But the very idea of a nonempty set with no ultimate members, but containing an infinite descending sequence of sets, seems repugnant. In view of the need to distinguish between a singleton $\{x\}$ and its only member x, the idea of a set's being a member of itself also seems suspect. But for the limitation in the Separation Axiom, the very condition $x \in x$ would lead to the Russell paradox. This paradox can be generalized into infinitely many others involving what might be called '$\in$-cycles': $x \in y \cdot y \in x$, $x \in y \cdot y \in z \cdot z \in x$, and so on.[54] Such ungrounded sets, self-membered sets, or $\in$-cycle sets, all seem both repugnant to our thinking and of no possible utility in mathematics or logic. They are all 'ruled out' by the final **ZF** axiom:

ZF-10 AXIOM OF REGULARITY (*Axiom der Fundierung*). *Every nonempty set A contains an element b such that* $A \cap b = \varnothing$.

That this axiom does rule out the undesirable sets in question is easily shown.

1. If A is a set, then $A \notin A$. If $A = \varnothing$, A has no members whatever so $A \notin A$. If $A \neq \varnothing$, by the Regularity Axiom, there is an x in A, such that $\{A\} \cap x = \varnothing$. Since $\{A\}$ is a singleton, only $A \in \{A\}$ so that $x = A$ and $\{A\} \cap A = \varnothing$. But since $A \in \{A\}$, it follows that $A \notin A$, which was to have been proved.

2. There are no sets A and B such that $A \in B \cdot B \in A$. Suppose that there were such sets: then both

$$A \in \{A, B\} \cap B \quad \text{and} \quad B \in \{A, B\} \cap A \tag{1}$$

By the Regularity Axiom there is an x in $\{A, B\}$, such that $\{A, B\} \cap x = \varnothing$. But $\{A, B\}$ is the doubleton of A and B, so either $x = A$ or $x = B$. Therefore, either $\{A, B\} \cap A = \varnothing$ or $\{A, B\} \cap B = \varnothing$, which contradicts (1), proving the supposition false.

3. There is no ungrounded set y_0, such that there is a sequence of sets $y_1, y_2,$

[53] D. Mirimanoff, 'Les Antinomies de Russell et de Burali–Forti et le Problème Fondamental de la Theorie des Ensembles', *L'Enseignement Mathematique*, vol. 19 (1917), pp. 37–52. See also S. Yuting, 'Paradox of the Class of All Grounded Classes', *Journal of Symbolic Logic*, vol. 18 (1953), p. 114 and R. Montague, 'On the Paradox of Grounded Classes', *ibid.*, vol. 20 (1955), p. 140.

[54] See W. V. O. Quine, *Mathematical Logic*, op. cit., pp. 128 ff.

$y_3, \ldots$, such that $\ldots \in y_3 \in y_2 \in y_1 \in y_0$. For if there were such a set y_0 and sequence of sets $y_1, y_2, y_3, \ldots$, we could use the Axiom of Replacement with X specified to be ω and $\varphi(x, y)$ specified to be $y_{x+} \in y_x$. The condition that for each x in ω, $\varphi(x, y)$ and $\varphi(x, z)$ implies $y = z$ must be satisfied because our second result shows that there can be no $\in$-cycles. Therefore, all of the sets $y_1, y_2, y_3, \ldots$ must be distinct. With its hypothesis thus satisfied, the Axiom of Replacement would guarantee the existence of a set $R = \{x: x = y_n \cdot n \in \omega\}$. Hence there would be a function f with $\omega = \text{dom } f$ and $R = \text{ran } f$, where $f(n^+) \in f(n)$ for each $n \in \omega$. Then, any member r of R would have to equal $f(n)$ for some n, from which it would follow that $f(n^+) \in R \cap r$. This implies that for every r in R, $R \cap r \neq \varnothing$. That, however, would contradict the Axiom of Regularity.

This single chapter was intended to provide a bare introduction to the fascinating and important branch of mathematics known as set theory. Only a few of its topics and problems have been mentioned and, of course, not everything said has been proved. It is hoped that the readers' interest and curiosity have been sufficiently aroused to make them want to learn more about set theory.

SUGGESTED FURTHER READINGS

Useful books that can be consulted with pleasure and profit are listed here.

Cohen, Paul J., *Set Theory and the Continuum Hypothesis*, W. A. Benjamin, Inc., Menlo Park, Calif., 1966.

Enderton, Herbert, *Elements of Set Theory*, Academic Press, Inc., New York, 1977.

Fraenkel, Abraham, *Abstract Set Theory*, 3rd rev. ed., North-Holland Publishing Company, Amsterdam, 1966.

Fraenkel, Abraham, and Yehoshua Bar-Hillel, *Foundations of Set Theory*, North-Holland Publishing Company, Amsterdam, 1958.

Halmos, Paul, *Naïve Set Theory*, D. Van Nostrand Company, New York, 1960.

Hausdorff, Felix, *Set Theory*, 2nd ed., Chelsea Publishing Company, New York, 1957.

Jech, Thomas J., *Lectures in Set Theory*, Springer-Verlag, Berlin-Heidelberg-New York, 1971.

Quine, W. V. O., *Set Theory and Its Logic*, Harvard University Press, Cambridge, Mass., 1963.

Stoll, Robert R., *Set Theory and Logic*, W. H. Freeman and Company, Publishers, San Francisco and London, 1963.

Suppes, Patrick, *Axiomatic Set Theory*, D. Van Nostrand Company, New York, 1960.

8

A Propositional Calculus

8.1 Object Language and Metalanguage

The logistic system to be constructed in this chapter is intended to be adequate for the formulation of arguments whose validity depends upon the ways in which statements are truth functionally compounded. Our logistic system will be a *language,* although for the sake of rigor in developing its theorems, it will be regarded as uninterpreted. We shall *talk about* this language. It will be the *object* of our discussion, and is, therefore, called the 'object language'. Because it is uninterpreted, its symbols and formulas have no meaning, and we cannot *use* it until it is given an interpretation—which will be postponed until after its development. We must, therefore, use a *different* language in order to talk about our object language. The language *used* in talking about a language is called a 'metalanguage'. In any investigation of language, there is an object language that is the *object* of investigation, and there is a metalanguage that is *used* by the investigators in talking about the object language.

An object language may be discussed from alternative points of view. Its relationship to its users may be investigated, as in a study of dialect changes in English usage and pronunciation in various parts of the country. Or the meaning or interpretation of a language may be investigated, as in compiling a dictionary, and in this latter inquiry a *semantical* metalanguage must be used. Finally, the formal structure of a language may be investigated, as in a grammar textbook, or in describing the development of theorems in an uninterpreted logistic system, for which a *syntactical* metalanguage or *Syntax Language* is used. In discussing the object language to be constructed here, we shall sometimes use a semantical metalanguage, and sometimes a syntactical one. In discussing its consistency, completeness, and its adequacy for expressing all truth-functional compound statements, we shall have to use a semantical metalanguage, for these topics involve its intended interpretation. In describing the purely formal criteria for its well formed formulas and the syntactical 'validity' of its arguments, only a Syntax Language need be used, for no references to meanings have to be made in these connections.

The language to be used as metalanguage in discussing our logistic system will be ordinary English, plus elementary arithmetic, with the addition of some special symbolic devices that will be introduced and explained as they are needed. It is assumed, of course, that the reader understands the metalanguage, for the entire discussion occurs within it. Only this one metalanguage will be used. It will function in some parts of our inquiry as a semantical metalanguage, in other parts as a Syntax Language.

It should be emphasized that 'object language' and 'metalanguage' are *relative terms*. Any language, no matter how simple or how complex, is an object language when it is being talked *about*. And any language (which must be an interpreted or meaningful one, of course) is a metalanguage when it is *used* to discuss an object language. Since *our* object language is uninterpreted, we cannot use it as a metalanguage. In another context, however, where the object language is an interpreted or meaningful language, one and the same language can function as both object language and metalanguage. Thus in our first chapter the English language was being discussed (and hence was the object language), and the discussion was carried on in English (which was, therefore, the metalanguage, also). A sufficiently rich or complex language can succeed in formulating the whole of its own syntax, and a good deal of its own semantics. But no consistent language can express the whole of its own semantics; certainly not the truth conditions for all of its own statements. That limitation can easily be shown.

That no language can express exhaustively its own semantics is shown by the following version of the 'Paradox of the Liar'.[1] Consider the following English sentence:

The sentence printed on page 214, line 25 of this book is not true.

Let us abbreviate the preceding sentence by the letter 'S'. Now just as it is obvious that 'Snow is white' is true if, and only if, *snow is white*, so it is equally obvious that

'S' is true if and only if *the sentence printed on page 214, line 25 of this book is not true.*

But counting the lines and looking at the page number verifies that 'S' is identical with *the sentence printed on page 214, line 25 of this book.* Hence

'S' is true if and only if 'S' *is not true.*

which is an explicit contradiction. That such a contradictory result can appear as a consequence of apparently innocent assumptions ought not to be regarded as a joke or a sophistry. It is a serious matter which reveals that the assumptions

[1] Attributed to J. Łukasiewicz.

were not as innocent as they appeared. The source of the trouble is generally agreed to lie in the attempt to formulate the truth conditions for the sentences of a language within that language itself. At least, if we distinguish sharply between object language and metalanguage, and do not try to make an object language serve as its own semantical metalanguage, the contradiction does not arise. An alternative method of avoiding such contradictions is discussed in Appendix C at the end of this book.

8.2 Primitive Symbols and Well Formed Formulas

We now proceed to construct our logistic system. There are two kinds of primitive symbols in our propositional calculus: 'propositional' symbols and 'operator' symbols. We use just four of the latter kind, these being

$$\cdot, \sim, (,).$$

We want infinitely many propositional symbols, for which we use the first four capital letters of the alphabet (in boldface type), with and without subscripts:

$$\begin{array}{lllll} \mathbf{A} & \mathbf{A}_1 & \mathbf{A}_2 & \mathbf{A}_3 & \ldots. \\ \mathbf{B} & \mathbf{B}_1 & \mathbf{B}_2 & \mathbf{B}_3 & \ldots. \\ \mathbf{C} & \mathbf{C}_1 & \mathbf{C}_2 & \mathbf{C}_3 & \ldots. \\ \mathbf{D} & \mathbf{D}_1 & \mathbf{D}_2 & \mathbf{D}_3 & \ldots. \end{array}$$

These are the only symbols that our propositional calculus will contain,[2] and in proving theorems and deriving conclusions from premisses *within the system* they are to be regarded as being completely uninterpreted. They may be thought of, prior to the assignment of meanings, as being repeatable and recognizable *marks* rather than 'symbols' at all—though it will be convenient to refer to them as 'symbols'.

Of course, we are guided in setting up our logistic system by the interpretation we intend eventually to give it. This intended interpretation controls our choice of primitive symbols and also governs our syntactical definition of 'well formed formula'. A *formula* of our system is defined to be any finite sequence of primitive symbols of our system. The set of *all* formulas of our system includes such sequences as the following:

[2] We *could* use just **A** and ′ to provide a finite base for the infinite set of propositional symbols written as

$$\mathbf{A}, \mathbf{A}', \mathbf{A}'', \mathbf{A}''', \mathbf{A}'''', \ldots.$$

As will be shown in Chapter 9, we *could* dispense with the symbols '(' and ')'.

$\mathbf{B}_1$
$(\mathbf{A})\cdot(\mathbf{A})$
$\sim(\mathbf{D})\sim(\sim)$
$\sim((\mathbf{A}_1)\cdot(\mathbf{C}_3))$
$\mathbf{B}_2\mathbf{B}_3\mathbf{A}_7\sim(\ )(\ )\cdot(\ )$
)))((
. . . .

Only some of these are to count as well formed formulas, however. Our definition of 'well formed formula' will be stated, of course, in our Syntax Language. It is convenient to introduce some special symbols into our Syntax Language, as an aid in discussing the logistic system clearly and economically. Although the first four capital letters of the alphabet, with and without subscripts, printed in boldface type, are symbols in our object language, those same letters printed in lightface italic type are symbols in our metalanguage. Their meanings in the metalanguage follow the convention that a lightface italic letter that is an element of our metalanguage denotes or means that same letter printed in boldface, which is an element of our object language. In addition, we introduce the capital letters '*P*', '*Q*', '*R*', '*S*', . . . , with and without subscripts, and refer to them as 'propositional variables'. Whereas the propositional symbols *in* our propositional calculus are uninterpreted and have *no meaning*, the propositional variables '*P*', '*Q*', '*R*', '*S*', . . . *in* our Syntax Language *are interpreted* and *do have meaning*. Every propositional variable of our Syntax Language, until further notice, denotes any formula of our object language—subject to the following restriction. In any sentence or sequence of sentences of our Syntax Language, two distinct propositional variables, say '*P*' and '*Q*', may denote either two distinct formulas of our object language, for example, '$\mathbf{B}_1$' and '$\sim((\mathbf{A}_1)\cdot(\mathbf{C}_3))$', or one and the same formula of our object language. Although a propositional variable may denote *any* formula of the object language in any one context, it must continue to denote that same formula wherever it occurs in that context. Thus the propositional variables '*P*', '*Q*', '*R*', '*S*', . . . of our metalanguage may have substituted for them *any name* in the metalanguage of *any formula* of the object language. At this point, we introduce the symbols '·', '∼', '(', and ')' into the Syntax Language, and explain their meanings. Where any propositional variable, say '*P*', of our Syntax Language denotes in some context a particular formula of our object language, say '**A**', then the symbol '$\sim(P)$' of our Syntax Language will, in that context, denote the formula '$\sim(\mathbf{A})$' of our object language. Also, where any propositional variables, say '*P*' and '*Q*', denote in some context two formulas of our object language, say '**A**' and '$\sim(\mathbf{B}_2)$', respectively, then the symbol '$(P)\cdot(Q)$' of our Syntax Language will, in that context, denote the formula '$(\mathbf{A})\cdot(\sim(\mathbf{B}_2))$' of our object language.

We cannot simply give a list of all well formed formulas of our object language, since there are infinitely many of them. It is necessary to give an inductive or *recursive* definition of 'well formed formula', which can be stated as follows, using the symbolic conventions explained in the preceding paragraph.

Recursive Definition for Well Formed Formula[3]

(a) Any propositional symbol is a *wff*.
(b) If any formula P is a *wff*, then $\sim(P)$ is a *wff*.
(c) If any formulas P and Q are both *wffs*, then $(P)\cdot(Q)$ is a *wff*.

(No formula of the object language shall be considered to be a *wff* unless its being so follows from these rules.)

This definition permits infinitely many well formed formulas, but it provides an *effective* criterion for recognizing them. No matter how long a (finite) sequence of symbols of the object language may be, our recursive definition permits us to decide in a finite number of steps whether or not it is a well formed formula. Let us take a relatively simple example to illustrate this:

$$\sim((\mathbf{A})\cdot(\sim(\mathbf{B})))$$

The question to be decided is whether the foregoing is a *wff* or not. By part (b) of the recursive definition, it is a *wff*, provided that $(A)\cdot(\sim(B))$ is a *wff*. The latter is a *wff* by part (c) of the definition, provided that both A and $\sim(B)$ are *wffs*. The first of these is a *wff* by part (a) of the definition, and by part (b), the second is a *wff* provided that B is a *wff*, which it is by part (a) of the definition. Thus the formula in question *is* a *wff*. From here on we shall use the propositional variables 'P', 'Q', 'R', 'S', and so on, of our Syntax Language to denote only *well formed* formulas of our object language.

EXERCISES

Using the Recursive Definition for Well Formed Formulas, show which of the following are *wffs* of our object language:

1. $(\sim(\mathbf{A}_1))\cdot(\mathbf{A}_1)$
2. $\sim(\sim((\mathbf{B}_1)\cdot(\sim(\mathbf{C}_3))))$
3. $\sim(\sim((\mathbf{B}_2)\cdot(\sim(\mathbf{D}_4)))$
4. $\sim((\sim(\mathbf{D}_3))\cdot(\sim(\mathbf{D}_4)))$
*5. $(\sim((\sim)\cdot(\mathbf{C}_2)))\cdot(\sim((\mathbf{B}_3)\cdot(\mathbf{B}_3)))$
6. $\sim(((\mathbf{A})\cdot(\mathbf{B}))\cdot(\sim(\mathbf{B})))$

[3] Alternatives are available here, as elsewhere, in developing logistic systems. In place of the recursive definition in the text, we might have used the following:

(a) If P is a propositional symbol, then (P) is a *wff*.
(b) If P is a *wff*, then $(\sim P)$ is a *wff*.
(c) If P and Q are *wffs*, then $(P\cdot Q)$ is a *wff*. (No formula is a *wff* unless its being so follows from these rules.)

This alternative definition of *wff* would make different formulas *wffs*, but the resulting logistic system could be developed equally well.

7. $\sim((\mathbf{A})\cdot(\mathbf{B}))\cdot(\sim((\mathbf{A})\cdot(\mathbf{A})))$
8. $(\sim(\sim(\mathbf{A}))\cdot(\mathbf{B})\cdot(\mathbf{C})))\cdot(((\sim(\mathbf{A}))\cdot(\mathbf{B}))\cdot(\sim(\mathbf{C})))$
9. $\sim((\sim((\mathbf{A})\cdot(\sim(\mathbf{B}))))\cdot(\sim((\sim((\mathbf{B})\cdot(\mathbf{C})))\cdot(\sim(\sim((\mathbf{C})\cdot(\mathbf{A}))))))$
*10. $\sim(((\sim((\mathbf{A})\cdot(\sim(\mathbf{B}))))\cdot(\mathbf{A}))\cdot(\sim(\sim((\mathbf{A})\cdot(\sim(\mathbf{B}))))))$

On the intended or 'normal' interpretation of our logistic system, its propositional symbols are to be symbolic translations of English statements that contain no other statements as (truth-functional) component parts. Where P is the symbolic translation of any English statement whatever, $\sim(P)$ is the symbolic translation of its negation. Where P and Q are the symbolic translations of any two English statements whatever, $(P)\cdot(Q)$ is the symbolic translation of their conjunction. The question now arises: Is our logistic system adequate, when normally interpreted, to express all truth-functionally compound statements? A system that is adequate to express all truth functions will be said to be '*functionally complete*'. This is a special case of the 'expressive completeness' mentioned in Chapter 6.

We begin our discussion by showing that our logistic system (which we call 'R.S.', since it is Rosser's System[4]) is adequate to the formulation of some familiar truth functions. Negations and conjunctions have already been mentioned, but there are also disjunctions, conditionals, and equivalences to be considered. Where two English statements translate into P and Q, their *weak* disjunction (which is true if either or both are true) is expressible in R.S. as $\sim((\sim(P))\cdot(\sim(Q)))$. Since it is a very common function, we introduce an abbreviation for it into our Syntax Language, defining '$P \vee Q$' to denote identically the same *wffs* of the object language R.S. as are denoted by '$\sim((\sim(P))\cdot(\sim(Q)))$'.[5] Where two English statements translate into P and Q, their *strong* disjunction (which is true if either one is true but not both) is expressible in R.S. as $(\sim((P)\cdot(Q)))\cdot(\sim((\sim(P))\cdot(\sim(Q))))$.

The symbolic translation of an English truth-functional conditional statement whose antecedent has the symbolic translation P and whose consequent has the symbolic translation Q will be $\sim((P)\cdot(\sim(Q)))$. This concept, too, is quite common, so we introduce the symbol '$\supset$' into our Syntax Language, using '$P \supset Q$' to denote identically the same *wffs* of R.S. as are denoted by

[4] The author is indebted to Professor J. Barkley Rosser of Cornell University for permission to include the following material. The calculi presented here and in Chapter 9 are early versions of logistic systems that appear in revised form in Professor Rosser's *Logic for Mathematicians*, McGraw-Hill Book Company, New York, 1953.

[5] Because '$P \vee Q$' is an abbreviation of '$\sim((\sim(P))\cdot(\sim(Q)))$', it is proper to say that $P \vee Q$ is, or is identical with, $\sim((\sim(P))\cdot(\sim(Q)))$, but *not* that $P \vee Q$ is an abbreviation of the latter, for there are no abbreviations in the object language itself. (The symbol 'v' is a significant symbol of the Syntax Language, but *does not even occur* in the object language, although we could introduce it there if we wished.) A parallel may help make this clear. Since 'U.S.S.R.' is an abbreviation of 'Union of Soviet Socialist Republics', the U.S.S.R. is identical with the Union of Soviet Socialist Republics. The U.S.S.R. is not an abbreviation of anything because it is a nation covering one-sixth of the globe. Since 'U.S.S.R.' denotes a large nation, the U.S.S.R. *is* a large nation and has no *meaning* in any literal or semantical sense. Similarly, since in any context '$P \vee Q$' denotes a *wff* of R.S., in that context $P \vee Q$ *is* a *wff* of R.S. Because R.S. is uninterpreted, its *wffs* are meaningless and $P \vee Q$, in particular, is meaningless.

'$\sim((P)\cdot(\sim(Q)))$'. Where two English statements have the symbolic translations P and Q, the statement that they are (materially) equivalent, that is, that they have the same truth values, or that each (materially) implies the other, has the symbolic translation $(P \supset Q)\cdot(Q \supset P)$. This is identically the same *wff* of R.S. as $(\sim((P)\cdot(\sim(Q))))\cdot(\sim((Q)\cdot(\sim(P))))$. Since material equivalence too is a frequently used notion, we introduce the symbol '$\equiv$' into our Syntax Language, defining '$P \equiv Q$' as an abbreviation for '$(P \supset Q)\cdot(Q \supset P)$'.

At this point, two notational conventions will be adopted for our metalanguage. The first is dispensing with the dot (for conjunction), so that $(P)(Q)$ is identically the same formula of the object language as $(P)\cdot(Q)$. The other notational convention is to replace parentheses by brackets or braces wherever such replacement is conducive to easier reading.

Even using brackets and braces, too many punctuation marks make reading difficult, so two further conventions that permit a minimum use of punctuation will be adopted. The first convention is to assign the following *order of precedence* to the symbols

$$\equiv \quad \supset \quad \mathrm{v} \quad \cdot \quad \sim$$

of our Syntax Language. Of these, each has *greater precedence* or greater scope than any listed to its right. What is intended here can be explained by the following examples. The otherwise ambiguous expression '$\sim PQ$' denotes $(\sim(P))\cdot(Q)$ rather than $\sim((P)\cdot(Q))$, because the connective '$\cdot$' (which we have agreed to represent by juxtaposition) has greater scope than the operator '$\sim$'. The scope of the connective '$\cdot$' extends over that of the '$\sim$' symbol. The otherwise ambiguous expression '$P \mathrm{v} QR$' denotes $P \mathrm{v} (Q\cdot R)$ rather than $(P \mathrm{v} Q)\cdot R$, because the connective 'v' has, by our convention, precedence over the connective '$\cdot$', and its scope extends over that of the latter. The otherwise ambiguous expression '$P \supset Q \mathrm{v} RS$' denotes $P \supset [Q \mathrm{v} (R\cdot S)]$, because '$\supset$' has precedence over both 'v' and '$\cdot$', and 'v' has precedence over '$\cdot$'. And the otherwise ambiguous expression '$P \supset Q \equiv \sim Q \supset \sim P$' denotes $[P \supset Q] \equiv [(\sim Q) \supset (\sim P)]$, because '$\equiv$' has precedence over '$\supset$' and '$\sim$', and '$\supset$' has precedence over '$\sim$'. Our second convention is that of *association to the left*, which means that where the convention of *order of precedence* does not suffice to remove the ambiguity of an expression, its parts should be grouped by parentheses *to the left*. That is, when an expression contains two (or more) occurrences of the same connective, and their relative scopes within the expression are not otherwise indicated, the occurrence to the right shall be understood to have the wider (or widest) scope. This, too, is best explained by means of an example or two. Since all of its connectives have *equal* order of precedence, the ambiguity of the expression '$P \supset Q \supset P \supset P$' cannot be resolved by our first convention. According to our second convention, however, we interpret it as denoting $[(P \supset Q) \supset P] \supset P$. Part, but not all of the ambiguity of the expression '$P \equiv Q \equiv PQ \mathrm{v} \sim P \sim Q$' is resolved by the convention regarding order of precedence. Once this convention has been appealed to, we know that it denotes *either* $[P \equiv Q] \equiv [(P \cdot Q) \vee (\sim P \cdot \sim Q)]$ *or* $P \equiv \{Q \equiv [(P \cdot Q) \vee (\sim P \cdot \sim Q)]\}$. We decide that the former is meant by consulting the second convention, which instructs us to associate to the left.

EXERCISES

Write the following expressions of the Syntax Language in unabbreviated form, complete with parentheses:

1. $P \vee \sim P$
2. $\sim\sim P \supset P$
3. $PQ \supset P$
4. $PQ \vee R$
*5. $P \supset (Q \supset P)$
6. $P \supset Q \supset P$
7. $P \supset (P \vee Q)$
8. $(P \supset PQ) \supset (P \supset Q)$
9. $(P \supset Q)(Q \supset R) \supset (P \supset R)$
*10. $P \supset \sim P \equiv \sim P$

The preceding remarks show how to express some truth functions in R.S. But to prove that R.S. is *functionally complete,* that is, adequate to express *all* possible truth functions of any number of statements, more is required. We must have in our semantical metalanguage a method of expressing all possible truth functions. Then we must prove that all of these, or all their substitution instances, can be expressed in our object language R.S. also, in its normal or standard interpretation. Such a method of expressing all possible truth functions is provided by *truth tables*, which were introduced in Chapter 2. The method and notation of truth tables is, therefore, imported into our semantical metalanguage, and will be used freely in discussing the various semantical properties possessed by R.S. on its standard interpretation.

Truth functions may have one, two, or any number of arguments (in the mathematical sense in which an 'argument' is an independent variable). Thus $f(P)$ is a truth function of P if and only if its truth or falsehood is completely determined by the truth or falsehood of P. Similarly, $f(P, Q)$ is a truth function of P and Q if and only if its truth value is determined solely by the truth values of P and Q.

There are exactly four different truth functions of a single argument, and these may be expressed by the following truth tables:

P	$f_1(P)$
T	F
F	T

P	$f_2(P)$
T	T
F	F

P	$f_3(P)$
T	F
F	F

P	$f_4(P)$
T	T
F	T

The functions $f_1(P)$, $f_2(P)$, $f_3(P)$, and $f_4(P)$ are completely defined by these four truth tables. They are the *only* truth functions of a single argument, and are called 'unary', 'monadic', or 'singulary' functions.[6] That they can be expressed in R.S. in its intended or normal interpretations is easily seen. First we note that

[6] Here we are using 'truth function' in the strict and proper sense of a nonlinguistic correlation between or among truth values.

the intended or normal interpretations of $\sim P$ and PQ are given by the truth tables

P	$\sim P$
T	F
F	T

and

P	Q	PQ
T	T	T
T	F	F
F	T	F
F	F	F

That R.S. is adequate to express $f_1(P)$, $f_2(P)$, $f_3(P)$, and $f_4(P)$ is proved by actually formulating them in R.S. The function $f_2(P)$ is true when P is true and false when P is false, and is, therefore, expressible in R.S. as P itself. The function $f_1(P)$ is false when P is true and true when P is false, and is, therefore, expressible in R.S. as $\sim P$. The function $f_3(P)$ is false no matter which truth value P assumes, and is, therefore, expressible in R.S. as $\sim PP$. The function $f_4(P)$ is true in every case and can therefore be expressed in R.S. as the negation of $f_3(P)$, that is, as $\sim(\sim PP)$. We have thereby shown that all singulary truth functions are expressible in R.S.

There are, of course, more truth functions of two arguments than of one argument. These are defined by the following truth tables:

P	Q	$f_1(P, Q)$
T	T	F
T	F	T
F	T	T
F	F	T

P	Q	$f_2(P, Q)$
T	T	T
T	F	F
F	T	T
F	F	T

P	Q	$f_3(P, Q)$
T	T	T
T	F	T
F	T	F
F	F	T

P	Q	$f_4(P, Q)$
T	T	T
T	F	T
F	T	T
F	F	F

P	Q	$f_5(P, Q)$
T	T	F
T	F	F
F	T	T
F	F	T

P	Q	$f_6(P, Q)$
T	T	F
T	F	T
F	T	F
F	F	T

P	Q	$f_7(P, Q)$
T	T	F
T	F	T
F	T	T
F	F	F

P	Q	$f_8(P, Q)$
T	T	T
T	F	F
F	T	F
F	F	T

P	Q	$f_9(P, Q)$
T	T	T
T	F	F
F	T	T
F	F	F

P	Q	$f_{10}(P, Q)$
T	T	T
T	F	T
F	T	F
F	F	F

P	Q	$f_{11}(P, Q)$
T	T	F
T	F	F
F	T	F
F	F	T

P	Q	$f_{12}(P, Q)$
T	T	F
T	F	F
F	T	T
F	F	F

P	Q	$f_{13}(P, Q)$
T	T	F
T	F	T
F	T	F
F	F	F

P	Q	$f_{14}(P, Q)$
T	T	T
T	F	F
F	T	F
F	F	F

P	Q	$f_{15}(P, Q)$
T	T	F
T	F	F
F	T	F
F	F	F

P	Q	$f_{16}(P, Q)$
T	T	T
T	F	T
F	T	T
F	F	T

These are *all* the truth functions of two arguments, and are called 'binary' or 'dyadic' functions. That they are all expressible in R.S. is easily shown by actually expressing them by means of the $\sim$ and $\cdot$ symbols: for example, $f_{14}(P, Q)$ is expressible as PQ, while $f_1(P, Q)$ is expressible as $\sim(PQ)$.

EXERCISES

1. Express each of the binary truth functions $f_1(P, Q)$, $^*f_2(P, Q), \ldots, {}^*f_5(P, Q), \ldots, {}^*f_{11}(P, Q), \ldots, f_{16}(P, Q)$ as *wffs* of R.S.
2. There are 256 ternary (or triadic) truth functions: $f_1(P, Q, R), f_2(P, Q, R), \ldots, f_{256}(P, Q, R)$, each of which is completely determined (or defined) by a different eight-row truth table. Take any ten of them and express each as a *wff* of R.S.

To prove the functional completeness of R.S., it is necessary to show that *any* truth function of *any* number of arguments is expressible by means of $\sim$ and $\cdot$. Any truth function of n arguments is expressible by means of a truth table having n initial columns and 2^n rows. Thus any truth function $f(P_1, P_2, \ldots, P_{n-1}, P_n)$ is completely specified by writing a '**T**' or an '**F**' in every one of the 2^n places in the last column of the following truth table:

	P_1	P_2	...	P_{n-1}	P_n	$f(P_1, P_2, \ldots, P_{n-1}, P_n)$
Row 1:	**T**	**T**	...	**T**	**T**	
Row 2:	**T**	**T**	...	**T**	**F**	
....						
Row $2^n - 1$:	**F**	**F**	...	**F**	**T**	
Row 2^n:	**F**	**F**	...	**F**	**F**	

The truth function $f(P_1, P_2, \ldots, P_{n-1}, P_n)$ must have either an **F** in just one row of its truth table, or **F**'s is more than one row of its truth table, or **F**'s in no rows of its truth table. In any case the truth function can be represented by a *wff* of R.S.

CASE 1. $f(P_1, P_2, \ldots, P_{n-1}, P_n)$ has an **F** in just one row of its truth table. If the **F** occurs in the first row, then the function is represented in R.S. by the *wff* $\sim(P_1 \cdot P_2 \cdot \ldots \cdot P_{n-1} \cdot P_n)$, which has an **F** in the first row of its truth table and **T**'s in all other rows. If the **F** occurs in the second row, then the function is represented in R.S. by the *wff* $\sim(P_1 \cdot P_2 \cdot \ldots \cdot P_{n-1} \cdot \sim P_n)$. The 2^n distinct truth functions, the ith of which has an **F** in its ith row and **T**'s in all its others, are represented in R.S. by the 2^n *wffs*

$$S_1\text{: } \sim(P_1 \cdot P_2 \cdot \ldots \cdot P_{n-1} \cdot P_n)$$
$$S_2\text{: } \sim(P_1 \cdot P_2 \cdot \ldots \cdot P_{n-1} \cdot \sim P_n)$$
....
$$S_{2^n-1}\text{: } \sim(\sim P_1 \cdot \sim P_2 \cdot \ldots \cdot \sim P_{n-1} \cdot P_n)$$
$$S_{2^n}\text{: } \sim(\sim P_1 \cdot \sim P_2 \cdot \ldots \cdot \sim P_{n-1} \cdot \sim P_n)$$

CASE 2. $f(P_1, P_2, \ldots, P_{n-1}, P_n)$ has **F**'s in more than one row of its truth table. If it has **F**'s in the $k(1 < k \leq 2^n)$ rows $i_1, i_2, \ldots, i_k$, then it is represented in R.S. by the *wff* $S_{i_1} \cdot S_{i_2} \cdot \ldots \cdot S_{i_k}$.

CASE 3. $f(P_1, P_2, \ldots, P_{n-1}, P_n)$ has **F**'s in no row of its truth table. Such a tautologous function is represented in R.S. by the *wff* $\sim(\sim P_1 \cdot P_1 \cdot P_2 \cdot \ldots \cdot P_{n-1} \cdot P_n)$.

The preceding cases exhaust the possibilities, so we have shown how to express any truth function of any number of arguments in R.S. This is not a theorem *of* or *in* R.S., but a theorem *about* it, established in our semantical metalanguage. We may, therefore, list it as

METATHEOREM I. *R.S. is functionally complete.*

A rigorous proof of this metatheorem requires the use of mathematical induction, which we discuss in the following paragraphs.

We could have chosen to present a logistic system other than R.S. In addition to parentheses and the propositional symbols P, Q, R, S and so on, we could have selected as primitive operators any of the following pairs of symbols: '∼' and 'v', '∼' and '⊃', '·' and 'v', '·' and '⊃', or 'v' and '⊃', instead of '∼' and '·'. It is instructive to inquire whether or not any of those alternative choices of primitives would have made the resulting system functionally complete (on the normal interpretation of the symbols '∼', '·', 'v', '⊃', which we shall denote by '∼', '·', 'v', '⊃'). Before addressing ourselves to these problems, it will be well to explain briefly the two kinds of mathematical induction we shall use in establishing results about (not *in*) R.S.

We shall use the term 'weak induction' to refer to the more commonly used type of mathematical induction. The schema for weak induction is

$$\begin{array}{l} f(1) \\ \text{for any arbitrary } m, \text{ if } f(m) \text{ then } f(m+1) \\ \hline \text{there, } f(m) \text{ for every } m \end{array}$$

It is frequently used in proving theorems in elementary algebra. For example, one proves that the sum of the first n odd integers is equal to n^2 by first proving the two premisses of the above schema and then drawing the general conclusion indicated. The first premiss, which we shall call the 'α-case', is here established as the trivial equation $1 = 1^2$. Then, the second premiss, which we shall call the 'β-case', is established by assuming $f(m)$ to be true for an arbitrary integer m and from this assumption, deriving the conclusion that $f(m + 1)$ is true. In this particular proof, the β-case assumption is

$$1 + 3 + 5 + \cdots + (2m - 1) = m^2$$

We derive the desired conclusion by adding $(2m + 1)$ to both sides of the equation and performing an elementary regrouping and factoring of terms:

$$1 + 3 + 5 + \cdots + (2m - 1) + (2m + 1) = m^2 + (2m + 1)$$
$$1 + 3 + 5 + \cdots + (2m - 1) + [2(m + 1) - 1] = (m + 1)^2$$

which shows that for an arbitrary m, if the sum of the first m odd integers is m^2, then the sum of the first $(m + 1)$ odd integers is $(m + 1)^2$. We thus have established the β-case; from it and the α-case, by *weak induction,* we draw the desired conclusion that for every m, the sum of the first m odd integers is equal to m^2. Weak induction may be thought of as summarizing an unending sequence of arguments of the form *Modus Ponens:*

$f(1)$	$f(2)$	$f(m)$
if $f(1)$ then $f(2)$;	if $f(2)$ then $f(3)$; . . . ;	if $f(m)$ then $f(m + 1)$; . . .
$\therefore f(2)$	$\therefore f(3)$	$\therefore f(m + 1)$

A proof of Metatheorem I by weak induction can be stated as follows:

α-CASE. Any truth function $f(P)$ of just one argument can be expressed in R.S., as was shown on pages 220–221.

β-CASE. Here we assume that any truth function of k arguments $f(P_1, P_2, \ldots, P_k)$ can be expressed in R.S. and prove that any truth function of $k + 1$ arguments $g(P_1, P_2, \ldots, P_k, P_{k+1})$ can also be expressed in R.S.

Let $g(P_1, P_2, \ldots, P_k, P_{k+1})$ be any truth function of the $k + 1$ arguments P_1, $P_2, \ldots, P_k, P_{k+1}$. It is defined by the truth table:

P_1	P_2	. . .	P_k	P_{k+1}	$g(P_1, P_2, \ldots, P_k, P_{k+1})$
T	T		T	T	V_1
T	T		T	F	V_2
. . .	. . .	. . .	. . .	. . .	. . .
. . .	. . .	. . .	. . .	. . .	. . .
T	F		F	T	V_{2^k-1}
T	F		F	F	V_{2^k}
F	T		T	T	V_{2^k+1}
F	T		T	F	V_{2^k+2}
. . .	. . .	. . .	. . .	. . .	. . .
. . .	. . .	. . .	. . .	. . .	. . .
F	F		F	T	$V_{2^{k+1}-1}$
F	F		F	F	$V_{2^{k+1}}$

where each V_i $(1 \leq i \leq 2^{k+1})$ is a **T** or an **F**.

A truth function $f_1(P_2, P_3, \ldots, P_k, P_{k+1})$ of the k arguments $P_2, P_3, \ldots, P_k$, P_{k+1} can be defined by the first 2^k rows of the preceding truth table if we delete its first column:

P_2	P_3	$\cdots$	P_k	P_{k+1}	$f_1(P_2, P_3, \ldots, P_k, P_{k+1})$
T	**T**		**T**	**T**	V_1
T	**T**		**T**	**F**	V_2
⋯	⋯	⋯	⋯	⋯	⋯
F	**F**		**F**	**T**	V_{2^k-1}
F	**F**		**F**	**F**	V_{2^k}

and by the β-case assumption can be expressed in R.S. by the *wff* S_1. A truth function $f_2(P_2, P_3, \ldots, P_k, P_{k+1}$ of the k arguments $P_2, P_3, \ldots, P_k, P_{k+1}$ can be defined by the last 2^k rows of the first truth table if we delete its first column:

P_2	P_3	$\cdots$	P_k	P_{k+1}	$f_2(P_2, P_3, \ldots, P_k, P_{k+1})$
T	**T**		**T**	**T**	V_{2^k+1}
T	**T**		**T**	**F**	V_{2^k+2}
⋯	⋯	⋯	⋯	⋯	⋯
F	**F**		**F**	**T**	$V_{2^{k+1}-1}$
F	**F**		**F**	**F**	$V_{2^{k+1}}$

and by the β-case assumption can be expressed in R.S. by the *wff* S_2.

Now the truth table for the *wff* $P_1 \cdot S_1$ has the same entries as the original truth table in its first 2^k rows and **F**'s in its last 2^k rows. The truth table for the *wff* $\sim P_1 \cdot S_2$ has all **F**'s its first 2^k rows and the same entries as the original truth table in its last 2^k rows. From this, it is clear that the truth table for the *wff* $(P_1 \cdot S_1) \vee (\sim P_1 \cdot S_2)$ has exactly the same entries as those in the original truth table, whence the arbitrary truth function $g(P_1, P_2, \ldots, P_k, P_{k+1})$ of $k + 1$ arguments can be expressed in R.S. by the *wff* $(P_1 \cdot S_1) \vee (\sim P_1 \cdot S_2)$ or $\sim(\sim(P_1 \cdot S_1) \cdot \sim(\sim P_1 \cdot S_2))$. This establishes the β-case.

Metatheorem I now follows by weak induction from α and β.

The term 'strong induction' will be used to refer to the somewhat less frequently encountered type of mathematical induction whose schema is

$f(1)$
for any arbitrary m, if $f(k)$ for every $k < m$ then $f(m)$
―――――――――――――――――――――――――
therefore, $f(m)$ for every m

Strong induction may also be thought of as summarizing an unending sequence of arguments of the form *Modus Ponens:*

$f(1)$
if $f(1)$ then $f(2)$;
―――――――――
$\therefore f(2)$

$f(1)$ and $f(2)$
if $f(1)$ and $f(2)$ then $f(3)$; . . . ;
―――――――――――――
$\therefore f(3)$

$f(1)$ and $f(2)$ and . . . and $f(m - 1)$
if $f(1)$ and $f(2)$ and . . . and $f(m - 1)$ then $f(m)$; . . .
―――――――――――――――――――――――――
$\therefore f(m)$

To illustrate the use of strong induction, we prove that a symbolic logic based on the propositional symbols *P*, *Q*, *R*, S, . . . and the operators · and v is *not* functionally complete. We do so by proving that no well formed formula[7] of the system based on ·, v, and *P*, *Q*, *R*, S, . . . can express a truth function that has the value *true* when all of its arguments have the value *false.* In our proof we use strong induction on the number of symbols in the well formed formula g(*P*, *Q*, *R*, . . .), ignoring parentheses, and counting each occurrence of *P*, *Q*, *R*, . . . , ·, and v as one symbol.

α-CASE: If g(*P*, *Q*, *R*, . . .) contains just one symbol, to be well formed it must be either *P* alone, or *Q* alone, or *R* alone, or . . . . Where the arguments *P*, *Q*, *R*, . . . all have the value *false,* g(*P*, *Q*, *R*, . . .) will also have the value *false,* since it is one of them. Hence any well formed formula of the present system that contains exactly one symbol cannot have the value *true* when its arguments all have the value *false.*

β-CASE: Here we assume that any well formed formula g(*P*, *Q*, *R*, . . .) containing *fewer* than *m* symbols cannot have the value *true* when all of its arguments have the value *false.* We shall prove, under this assumption, that any well formed formula containing *exactly m* symbols cannot have the value *true* when all its arguments have the value *false.* Consider any formula g(*P*, *Q*, *R*, . . .) that contains exactly *m* symbols (where $m > 1$). To be well formed, g(*P*, *Q*, *R*, . . .) must be either

$$g_1(P, Q, R, \ldots) \cdot g_2(P, Q, R, \ldots)$$

or

$$g_1(P, Q, R, \ldots) \vee g_2(P, Q, R, \ldots)$$

where $g_1(P, Q, R, \ldots)$ and $g_2(P, Q, R, \ldots)$ are well formed formulas containing fewer than *m* symbols. By the β-case assumption, when *P*, *Q*, *R*, . . . are all *false,* both $g_1(P, Q, R, \ldots)$ and $g_2(P, Q, R, \ldots)$ will have the value *false* also. Now, given any two propositions that are *false,* both their disjunction and their conjunction are *false.* Hence in this case g(*P*, *Q*, *R*, . . .) has the value *false,* also. This establishes the β-case.

Having established both the α and β cases, by strong induction we infer that no well formed formula of the system based on ·, v, *P*, *Q*, *R*, . . . can express a truth function that has the value *true* when all of its arguments have the value *false.* Hence the system is *not* functionally complete.

EXERCISES

Prove the functional completeness or incompleteness of the logistic systems based on the propositional symbols *P*, *Q*, *R*, S, . . . , parentheses, and the operators:

*1. v and ~	3. ⊃ and ~
*2. ⊃ and ·	4. ⊃ and v

[7] We assume a recursive definition that is analogous to the one stated on page 217.

5. $\sim$and $+$, where $+$ is the symbol for exclusive disjunction, defined by the truth table for f_7 on page 221.

*6. $\supset$ and $+$

7. v and $+$

8. $+$ and $\equiv$

9. $+$ and $\cdot$

10. $\equiv$ and $\sim$

11. $\equiv$ and $\supset$

12. $\supset$ and $\not\supset$, where $\not\supset$ is the symbol for (material) nonimplication,[8] defined by the truth table for f_{13} on page 221.

13. $\supset$ and $\not\subset$, where $\not\subset$ is the symbol for converse nonimplication, defined by the truth table for f_{12} on page 221.

14. $\not\supset$ and $\not\subset$

15. Which, if any, of the sixteen binary truth functional operators $f_1, f_2, \ldots, f_{16}$ defined on page 221, can be added to the propositional symbols $P, Q, R, S, \ldots$ and parentheses, to give a functionally complete logistic system with a single operator symbol?

8.3 Axioms and Demonstrations

The rules for our system and the proofs of the theorems within it will be much simplified if we assume infinitely many well formed formulas as axioms or postulates. Of course, we cannot actually write out an infinite list of axioms *within our object language*, but we can use our Syntax Language to specify exactly which *wffs* of R.S. are axioms and which are not. There can be no objection to having an infinite number of axioms if there is an *effective* process for determining whether or not any given *wff* is an axiom. The infinite list of axioms of R.S. may be written as

Axiom 1. $P \supset (P \cdot P)$
Axiom 2. $(P \cdot Q) \supset P$
Axiom 3. $(P \supset Q) \supset [\sim(Q \cdot R) \supset \sim(R \cdot P)]$

Each of these syntactical formulas (formulas within our Syntax Language) represents or designates an infinite list of *wffs* of R.S. Thus Axiom 1 designates all the following:

$$\sim((\mathbf{A})\cdot(\sim((\mathbf{A})\cdot(\mathbf{A}))))$$
$$\sim((\mathbf{B})\cdot(\sim((\mathbf{B})\cdot(\mathbf{B}))))$$
$$\sim((\mathbf{C})\cdot(\sim((\mathbf{C})\cdot(\mathbf{C}))))$$
$$\sim((\mathbf{D})\cdot(\sim((\mathbf{D})\cdot(\mathbf{D}))))$$
$$\sim((\mathbf{A}_1)\cdot(\sim((\mathbf{A}_1)\cdot(\mathbf{A}_1))))$$
$$\cdots\cdots\cdots\cdots\cdots\cdots$$
$$\sim((\sim(\mathbf{A}))\cdot(\sim((\sim(\mathbf{A}))\cdot(\sim(\mathbf{A})))))$$
$$\sim((\sim(\mathbf{B}))\cdot(\sim((\sim(\mathbf{B}))\cdot(\sim(\mathbf{B})))))$$
$$\cdots\cdots\cdots\cdots\cdots\cdots$$
$$\sim(((\mathbf{A})\cdot(\mathbf{D}))\cdot(\sim(((\mathbf{A})\cdot(\mathbf{D}))\cdot((\mathbf{A})\cdot(\mathbf{D})))))$$
$$\sim(((\mathbf{A}_3)\cdot(\mathbf{B}_7))\cdot(\sim(((\mathbf{A}_3)\cdot(\mathbf{B}_7))\cdot((\mathbf{A}_3)\cdot(\mathbf{B}_7)))))$$
$$\cdots\cdots\cdots\cdots\cdots\cdots$$
$$\cdots\cdots\cdots\cdots\cdots\cdots$$

[8] So named by Alonzo Church in *Introduction to Mathematical Logic*, vol. 1, Princeton University Press, Princeton, N.J., 1956, p. 37.

and infinitely many more *wffs* of R.S. It designates, in fact, every *wff* of R.S. that is of the indicated pattern. It is effectively decidable whether any given finite sequence of symbols of R.S. is of this pattern or not. Three patterns are set forth, and every *wff* of R.S. that exemplifies any one of these patterns is assumed as an axiom.

That these axioms are reasonable assumptions is evidenced by the fact that on the normal interpretation of the symbols ~ and ·, they are all tautologies.

Our logistic system is intended, upon interpretation, to be adequate for the formulation of arguments. Arguments, as we know, consist of premisses and conclusions, all of which are expressed in statements. Corresponding to such arguments, we have in our logistic system sequences of well formed formulas of which the last formula is the 'conclusion'. We shall set up a criterion that will enable us to distinguish formally between two kinds of sequences of *wffs* within R.S.: those which become valid arguments when interpreted normally and those which do not.

In Chapter 3 a valid argument was characterized as one for which a formal proof or demonstration could be given. A formal proof of an argument's validity was defined as a sequence of statements, each of which was either a premiss or which followed from preceding statements by an elementary valid argument, and whose last statement was the conclusion of the argument that was being proved valid. Thus the question of the validity of *any* argument was 'reduced' to the question of the validity of certain elementary argument forms or rules of inference. Something like that will be proposed for R.S., but it is not desirable to make such a wholesale set of assumptions about valid elementary inferences as was made in Chapter 3, where nineteen rules of inference were postulated, plus rules of Conditional Proof and Indirect Proof. However, *some* rule of inference must be assumed for our logistic system or there will be no inference legitimized within it at all. One rule of inference will suffice for the validation of all arguments and (hence) for the proof of all theorems within R.S. We assume *Modus Ponens*, and state it (in our Syntax Language) as

RULE 1. From P and $P \supset Q$ to infer Q.

Examples of arguments *in* R.S. that are legitimized as *valid* by the assumption of this rule of inference are

$\sim((\mathbf{A})\cdot(\sim\mathbf{B})))$		$\sim(((\mathbf{A}_1)\cdot(\mathbf{A}_2))\cdot(\sim(\sim(\mathbf{C}))))$
$\mathbf{A}$	and	$(\mathbf{A}_1)\cdot(\mathbf{A}_2)$
$\mathbf{B}$		$\sim(\mathbf{C})$

The assumption of *Modus Ponens* as Rule 1 (abbreviated as R 1) of our logistic system, *by itself*, permits only a special kind of argument having two premisses to be regarded as valid. R.S., however, is adequate for the expression of *all* formally valid arguments that are certifiable by means of truth tables, including extended ones that contain any number of premisses. So it is desirable to introduce a method for validating extended arguments in R.S. by showing how their conclusions follow from their premisses by *repeated* applications of R 1. Here our treatment is roughly analogous to that for extended arguments presented in Chapter 3, although with some important differences. We define a *demonstration* of the validity of an argument that has as premisses the formulas $P_1, P_2, \ldots, P_n$ and as conclusion the formula Q, to be a sequence of well formed formulas $S_1, S_2, \ldots, S_k$ such that every S_i is either one of the premisses $P_1, P_2, \ldots, P_n$, or is one of the axioms of R.S., or follows from two preceding S's by R 1 and such that S_k is Q. As before, an argument is to be regarded as valid if and only if there exists a *demonstration* of its validity. A special notation is introduced (into the Syntax Language) to represent this idea. We introduce the special symbol '$\vdash$' (which may be read 'yields') so that

$$P_1, P_2, \ldots, P_n \vdash Q$$

asserts that there is a demonstration of the validity of the argument that has P_1, $P_2, \ldots, P_n$ as premisses and Q as conclusion. For example, all of the infinitely many arguments denoted by

$$P \supset Q$$
$$\sim(QR)$$
$$\sim(RP)$$

are valid because $P \supset Q, \sim(QR) \vdash \sim(RP)$. The demonstration consists of the following sequence of S's:

$$S_1\text{: } (P \supset Q) \supset [\sim(QR) \supset \sim(RP)]$$
$$S_2\text{: } P \supset Q$$
$$S_3\text{: } \sim(QR) \supset \sim(RP)$$
$$S_4\text{: } \sim(QR)$$
$$S_5\text{: } \sim(RP)$$

in which S_1 is Axiom 3, S_2 is the first premiss of the argument, S_3 follows from S_1 and S_2 by R 1, S_4 is the second premiss, S_5 follows from S_3 and S_4 by R 1, and S_5 is the conclusion. Any such sequence of S's will be called a *demonstration* that $P_1, P_2, \ldots, P_n \vdash Q$. Each S_i will be called a *line* of the demonstration. Where a demonstration is given for the validity of all arguments in R.S. of a particular form, as in the example cited, any particular argument of that form may be regarded as having been validated by that demonstration and the general form may be regarded as a *derived rule of inference.*

The definition of *demonstration* given in the preceding paragraphs should have made it clear that the axioms of R.S. function as 'understood' premisses of every argument formulated within the system. Where those axioms are the *only* premisses of a valid argument, then the conclusion is a *theorem* of the system. That a given formula Q is a theorem, is expressed by writing '$\vdash Q$'. The notation '$\vdash Q$' is defined more strictly as the assertion that there exists a demonstration of Q, which is a sequence of well formed formulas $S_1, S_2, \ldots, S_k$ such that every S_i is either an axiom of R.S. or follows from two preceding S's by R 1 and S_k is Q.

It should be noted that *demonstration*, either of a theorem or of the validity of an argument, is an *effective* concept. Given any sequence of S's, however long, it can be decided quite mechanically, in a finite number of operations, whether or not it is a *demonstration*. It can be decided effectively of any S whether or not it is an axiom (and whether or not it is one of the premisses, if an argument is involved). And if some S, say S_j, is neither an axiom nor a premiss, then since only a finite number of S's precede S_j in the sequence, a finite number of inspections will reveal whether or not two of the preceding lines are S_i and $S_i \supset S_j$. Only if these occur earlier in the sequence of S's will S_j follow from two preceding lines by R 1.

With this definition of *theorem*, the three axioms and the one rule of R.S. can be regarded as a sort of symbolic machine for generating well formed formulas. Each axiom itself generates an infinite number of *wffs*, and by repeated applications of R 1 to them infinitely many more *wffs* are produced as theorems. Two questions quite naturally arise at this point. First, is the system consistent? And second, are all the theorems (on the normal interpretation of the primitive symbols) tautologies?

A logistic system that is a propositional calculus (such as R.S.) will be called *analytic* if and only if all of its theorems become tautologies on their normal interpretations. R.S. is analytic, by this definition, provided that $\vdash P$ implies that P is tautologous. We shall now prove that R.S. is analytic.

METATHEOREM II. *R.S. is analytic (that is, if $\vdash P$ then P is a tautology).*

Proof: We use strong induction on the number of uses of R 1 in the demonstration that $\vdash P$.

α-CASE: Suppose P results from a single use of R 1. Then R 1 must be applied to the axioms. The axioms are all tautologies, as is easily verified by actually constructing truth tables for them. Hence P results from applying R 1 to S and $S \supset P$, where S and $S \supset P$ are both tautologies. Clearly, P must be a tautology in this case, for if it were not, there would be an **F** in at least one row of its truth table, say row j. But since S is a tautology, it has all **T**'s in its truth table, including of course a **T** in row j. Hence in the jth row, S would have a **T** and P an **F**, so that $S \supset P$ would have an **F**—contrary to the fact that it is tautologous and has only **T**'s in its truth table. Hence the Metatheorem is true for m uses of R 1, where $m = 1$.

β-CASE: Here we assume the Metatheorem to be true for any number $k < m$ uses of R 1. That is, if S is a theorem whose demonstration involves fewer than m uses of R 1, then S is a tautology. Consider the status of any P

that has a demonstration which involves m uses of R 1. The formula P is either an axiom (in which case it is clearly a tautology) or must result from earlier lines S and $S \supset P$ by the mth use of R 1. The earlier line S is either an axiom (in which case it is clearly a tautology) or is obtained by $k < m$ uses of R 1, and hence is a tautology by the assumption of the β-case. Similarly, the earlier line $S \supset P$ must also be a tautology. Now, because S and $S \supset P$ are tautologies, P must be one also, by the argument of the α-case.

From α and β, by strong induction, we conclude that if $\vdash P$ (by *any* number of applications of R 1), then P is a tautology, which means that R.S. is analytic.

The analyticity of R.S. having been established, its consistency follows immediately, and may be regarded as a corollary of Metatheorem II. In our proof we use the Post criterion for consistency, which states that a deductive system is consistent if it contains a formula that is not provable within it as a theorem.

COROLLARY. *R.S. is consistent.*

Proof: The formula $P \cdot \sim P$ is a *wff* of R.S. but is not a tautology; therefore, by Metatheorem II, it is not a theorem of R.S. Hence R.S. contains a formula that is not provable as a theorem, so R.S. is consistent by the Post criterion.

We define the 'deductive completeness' of a logistic system for the propositional calculus as the converse of analyticity. An *analytic* system is one, all of whose theorems are tautologies. A propositional calculus will be called *deductively complete* when all tautologies are provable as theorems within it. The deductive completeness of R.S. is more difficult to establish than the first two Metatheorems and requires that we first develop some theorems of the system. The proof, then, will be postponed to Section 8.6, following the development of R.S. itself in Section 8.5, after we prove the independence of the axioms of R.S. in the next section.

8.4 Independence of the Axioms

A set of axioms is said to be independent if each of them is independent of the others, that is, if none of them can be derived as theorems from the others. To prove the independence of each axiom, it suffices to find a characteristic of *wffs* of the system such that:

1. The axiom to be proved independent lacks that characteristic.
2. All other axioms have that characteristic.
3. The characteristic in question is hereditary with respect to the rules of inference of the system.

Roughly the same sense of 'hereditary' is used here that is explained in Appendix A: a characteristic is *hereditary with respect to a set of rules of inference* if and only if whenever it belongs to one or more formulas, it also belongs to every formula deduced from these formulas by means of those rules of inference. The characteristic of being a tautology, on the normal or intended interpretation, is hereditary in the sense defined. But it cannot be used to prove the independence of any axiom, because *all* axioms of R.S. have this characteristic.

To prove the independence of the axioms of R.S., we must use a model containing more than two elements, a model similar to that used in Appendix A to prove the incompleteness of the original nineteen Rules of Inference. For R.S., a three-element model will suffice, but for other truth-functional axiom systems, models containing *more* than three elements may be required.

To prove the independence of Ax. 1 of R.S., we introduce a three-element model in terms of which the symbols and *wffs* of R.S. can be interpreted. The three elements are the 'values' **0**, **1**, and **2**, which play roles analogous to those of the truth values **T** and **F**. Every propositional symbol has one of these three values assigned to it. Every *wff* that is not a propositional symbol will have one of the three values **0**, **1**, **2** in accordance with the following tables:

P	$\sim P$
0	2
1	1
2	0

P	Q	$P \cdot Q$
0	0	0
0	1	1
0	2	2
1	0	1
1	1	2
1	2	2
2	0	2
2	1	2
2	2	2

The defined symbol '$\supset$', whose definition is given by

$$P \supset Q = \text{df} \sim (P \cdot \sim Q)$$

has, pursuant to that definition, the table

P	Q	$P \supset Q$
0	0	0
0	1	1
0	2	2
1	0	0
1	1	0
1	2	1
2	0	0
2	1	0
2	2	0

The characteristic of *wffs* that we utilize here is that of having the value **0**, regardless of the values assigned to its component propositional symbols. This characteristic is easily seen to belong to Ax. 2 and Ax. 3 of R.S. by the following tables (analogous to the table explained on page 30).

$(P$	$\cdot$	$Q)$	$\supset$	P
0	0	0	0	0
0	1	1	0	0
0	2	2	0	0
1	1	0	0	1
1	2	1	0	1
1	2	2	0	1
2	2	0	0	2
2	2	1	0	2
2	2	2	0	2

$(P$	$\supset$	$Q)$	$\supset$	$[\sim$	$(Q$	$\cdot$	$R)$	$\supset$	$\sim$	$(R$	$\cdot$	$P)]$
0	0	0	0	2	0	0	0	0	2	0	0	0
0	0	0	0	1	0	1	1	0	1	1	1	0
0	0	0	0	0	0	2	2	0	0	2	2	0
0	1	1	0	1	1	1	0	1	2	0	0	0
0	1	1	0	0	1	2	1	1	1	1	1	0
0	1	1	0	0	1	2	2	0	0	2	2	0
0	2	2	0	0	2	2	0	2	2	0	0	0
0	2	2	0	0	2	2	1	1	1	1	1	0
0	2	2	0	0	2	2	2	0	0	2	2	0
1	0	0	0	2	0	0	0	0	1	0	1	1
1	0	0	0	1	0	1	1	0	0	1	2	1
1	0	0	0	0	0	2	2	0	0	2	2	1
1	0	1	0	1	1	1	0	0	1	0	1	1
1	0	1	0	0	1	2	1	0	0	1	2	1
1	0	1	0	0	1	2	2	0	0	2	2	1
1	1	2	0	0	2	2	0	1	1	0	1	1
1	1	2	0	0	2	2	1	0	0	1	2	1
1	1	2	0	0	2	2	2	0	0	2	2	1
2	0	0	0	2	0	0	0	0	0	0	2	2
2	0	0	0	1	0	1	1	0	0	1	2	2
2	0	0	0	0	0	2	2	0	0	2	2	2
2	0	1	0	1	1	1	0	0	0	0	2	2
2	0	1	0	0	1	2	1	0	0	1	2	2
2	0	1	0	0	1	2	2	0	0	2	2	2
2	0	2	0	0	2	2	0	0	0	0	2	2
2	0	2	0	0	2	2	1	0	0	1	2	2
2	0	2	0	0	2	2	2	0	0	2	2	2

The characteristic is easily seen to be hereditary with respect to R 1 of R.S. by consulting the table given for '⊃'. In the only row in which both P and $P \supset Q$ have the value **0**, Q also has the value **0**. Hence if the characteristic belongs to one or more *wffs*, it also belongs to every *wff* deduced from them by R 1.

Finally, it is readily seen that the characteristic in question does *not* belong to Ax. 1. When P is assigned the value **1**, $P \supset (P \cdot P)$ has the value **1** rather than **0**, for $\mathbf{1} \supset (\mathbf{1} \cdot \mathbf{1})$ is $\mathbf{1} \supset \mathbf{2}$, which is **1**. Hence Ax. 1 is independent.

Three questions naturally arise in attempting to prove the independence of an axiom in a propositional calculus. First, how does one decide how big a model to use? Second, how does one decide what values (elements of the model) to assign in making a table for the primitive symbols of the system? Third, how does one decide what element (or elements) of the model to designate as the (hereditary) characteristic? There is no effective or mechanical answer to these questions. Methodology here is still a matter of trial and error. For simplicity and economy, one should try using a three-element model. If it works, good. If not, one should try a four-element model, and so on. The answers to the second and third questions are interrelated. To prove the independence of Axiom 1, we wanted it to take on an undesignated value in at least one case, where the other axioms take on only the designated value (or values) in every case, and the designated value is hereditary with respect to R 1. This need led us to assign, for at least one case, a different value for $P \cdot P$ than for P (by assigning a different value (**2**) to $P \cdot Q$ for at least one case in which both P and Q take on the same value (**1**)). Then to make Axiom 1, $P \supset (P \cdot P)$, take on an undesignated value in at least one case, we specified that the conditional with antecedent (value) **1** and consequent (value) **2** take on an undesignated value, in this case **1**. To show the independence of Axiom 2, $(P \cdot Q) \supset P$, that is, to make it take on an undesignated value in at least one case, we make its antecedent $P \cdot Q$ take on a designated value in at least one case in which its consequent P takes on an undesignated value. Several such assignments are possible, of course, but any chosen must be consistent with other constraints, such as making the other axioms take on only designated values, and making the designated value hereditary with respect to R 1.

To prove the independence of Ax. 2 of R.S., we use the same three-element model and the same table for '$\sim P$'. The difference lies in the table for '$P \cdot Q$', which follows, along with the derivative table for '$P \supset Q$'.

P	Q	$P \cdot Q$	$P \supset Q$
0	**0**	**0**	**0**
0	**1**	**0**	**2**
0	**2**	**2**	**2**
1	**0**	**0**	**0**
1	**1**	**0**	**2**
1	**2**	**2**	**2**

P	Q	$P \cdot Q$	$P \supset Q$
2	0	2	0
2	1	2	0
2	2	2	0

The characteristic of *wffs* that we utilize here is (again) that of having the value **0**, regardless of the values assigned to its component propositional symbols. This characteristic is easily seen to belong to Ax. 1 and Ax. 3 of R.S. by the following tables.

P	$\supset$	$(P$	$\cdot$	$P)$
0	0	0	0	0
1	0	1	0	1
2	0	2	2	2

$(P$	$\supset$	$Q)$	$\supset$	$[\sim$	$(Q$	$\cdot$	$R)$	$\supset$	$\sim$	$(R$	$\cdot$	$P)]$
0	0	0	0	2	0	0	0	0	2	0	0	0
0	0	0	0	2	0	0	1	0	2	1	0	0
0	0	0	0	0	0	2	2	0	0	2	2	0
0	2	1	0	2	1	0	0	0	2	0	0	0
0	2	1	0	2	1	0	1	0	2	1	0	0
0	2	1	0	0	1	2	2	0	0	2	2	0
0	2	2	0	0	2	2	0	2	2	0	0	0
0	2	2	0	0	2	2	1	2	2	1	0	0
0	2	2	0	0	2	2	2	0	0	2	2	0
1	0	0	0	2	0	0	0	0	2	0	0	1
1	0	0	0	2	0	0	1	0	2	1	0	1
1	0	0	0	0	0	2	2	0	0	2	2	1
1	2	1	0	2	1	0	0	0	2	0	0	1
1	2	1	0	2	1	0	1	0	2	1	0	1
1	2	1	0	0	1	2	2	0	0	2	2	1
1	2	2	0	0	2	2	0	2	2	0	0	1
1	2	2	0	0	2	2	1	2	2	1	0	1
1	2	2	0	0	2	2	2	0	0	2	2	1
2	0	0	0	2	0	0	0	0	0	0	2	2
2	0	0	0	2	0	0	1	0	0	1	2	2
2	0	0	0	0	0	2	2	0	0	2	2	2
2	0	1	0	2	1	0	0	0	0	0	2	2
2	0	1	0	2	1	0	1	0	0	1	2	2
2	0	1	0	0	1	2	2	0	0	2	2	2
2	0	2	0	0	2	2	0	0	0	0	2	2
2	0	2	0	0	2	2	1	0	0	1	2	2
2	0	2	0	0	2	2	2	0	0	2	2	2

The characteristic is easily seen to be hereditary with respect to R 1 of R.S. by consulting the table for '$\supset$'. In the only row in which both P and $P \supset Q$ have the value **0**, Q also has the value **0**. Hence if the characteristic belongs to one or more *wffs* it also belongs to every *wff* deduced from them by R 1.

Finally, it is readily seen that the characteristic in question does not belong to Ax. 2. When P and Q are both assigned the value **1**, $(P \cdot Q) \supset P$ has the value **2** rather than **0** for $(\mathbf{1} \cdot \mathbf{1}) \supset \mathbf{1}$ is $\mathbf{0} \supset \mathbf{1}$ which is **2**. Hence Ax. 2 is independent.

To prove the independence of Ax. 3 of R.S., we use the same three-element model and the same table for '$\sim P$'. The difference lies in the table for '$P \cdot Q$', which follows, along with the derivative table for '$P \supset Q$'.

P	Q	$P \cdot Q$	$P \supset Q$
0	0	0	0
0	1	1	1
0	2	2	2
1	0	2	0
1	1	2	0
1	2	2	0
2	0	2	0
2	1	2	0
2	2	2	0

The characteristic of *wffs* that we utilize here is (yet again) that of having the value **0**, regardless of the values assigned to its component propositional symbols. This characteristic is easily seen to belong to Ax. 1 and Ax. 2 of R.S. by the following tables.

P	$\supset$	$(P$	$\cdot$	$P)$
0	0	0	0	0
1	0	1	2	1
2	0	2	2	2

$(P$	$\cdot$	$Q)$	$\supset$	P
0	0	0	0	0
0	1	1	0	0
0	2	2	0	0
1	2	0	0	1
1	2	1	0	1
1	2	2	0	1
2	2	0	0	2
2	2	1	0	2
2	2	2	0	2

The characteristic is easily seen to be hereditary with respect to R 1 of R.S. by consulting the table for '$\supset$'. In the only row in which P and $P \supset Q$ have the value **0**, Q also has the value **0**. Hence if the characteristic belongs to one or more *wffs*, it also belongs to every *wff* deduced from them by R 1.

Finally, it is readily seen that the characteristic in question does *not* belong to Ax. 3. When P and Q are both assigned the value **1** and R is assigned the value **0**, $(P \supset Q) \supset [\sim(Q \cdot R) \supset \sim(R \cdot P)]$ has the value **1** rather than **0** for $(\mathbf{1} \supset \mathbf{1}) \supset [\sim(\mathbf{1} \cdot \mathbf{0}) \supset \sim(\mathbf{0} \cdot \mathbf{1})]$ reduces successively to $\mathbf{0} \supset [\sim\mathbf{2} \supset \sim\mathbf{1}]$, to $\mathbf{0} \supset [\mathbf{0} \supset \mathbf{1}]$, to $\mathbf{0} \supset \mathbf{1}$, and finally to **1**. Hence Ax. 3 is independent.

EXERCISES

For each of the following sets of axioms for propositional calculi, prove the independence of each axiom:

*1. Church's system P_N has as primitive operators $\sim$ and v, with $P \supset Q$ defined as $\sim P \vee Q$. Its Rule is: From P and $P \supset Q$ to infer Q. Its four axioms are

Axiom 1. $(P \vee P) \supset P$
Axiom 2. $P \supset (P \vee Q)$
Axiom 3. $[P \vee (Q \vee R)] \supset [Q \vee (P \vee R)]$
Axiom 4. $(Q \supset R) \supset [(P \vee Q) \supset (P \vee R)]$

2. The Götlind-Rasiowa system P_G has the same primitive operators, definition of $\supset$, and Rule as P_N above. Its three axioms are

Axiom 1. $(P \vee P) \supset P$
Axiom 2. $P \supset (P \vee Q)$
Axiom 3. $(Q \supset R) \supset [(P \vee Q) \supset (R \vee P)]$

3. Frege's system F.S. has as primitive operators $\sim$ and $\supset$, with $P \vee Q$ defined as $\sim P \supset Q$. Its Rule is: From P and $P \supset Q$ to infer Q. Its five axioms (after deleting Frege's original Axiom 3, $[P \supset (Q \supset R)] \supset [Q \supset (P \supset R)]$, which was not independent) are

Axiom 1. $P \supset (Q \supset P)$
Axiom 2. $[P \supset (Q \supset R)] \supset [(P \supset Q) \supset (P \supset R)]$
Axiom 3. $(P \supset Q) \supset (\sim Q \supset \sim P)$
Axiom 4. $\sim\sim P \supset P$
Axiom 5. $P \supset \sim\sim P$

4. Łukasiewicz's system L.S. has the same primitive operators, definition of v, and Rule as F.S. above. Its three axioms are

Axiom 1. $P \supset (Q \supset P)$
Axiom 2. $[P \supset (Q \supset R)] \supset [(P \supset Q) \supset (P \supset R)]$
Axiom 3. $(\sim P \supset \sim Q) \supset (Q \supset P)$

8.5 Development of the Calculus

In the actual development of derived rules and theorems of our object language, we regard all of its symbols as completely uninterpreted, since we are motivated by the desire for rigor in its development. *When interpreted normally,* the formulas $\sim RP$ and $P \sim R$ are logically equivalent. But they cannot be so regarded as *wffs* of R.S., *in its development,* and the *wff* $\sim RP \equiv P \sim R$ cannot be accepted as a theorem *until* it has been formally derived from the axioms of the system.

It is convenient to begin with the demonstration of a derived rule of inference for R.S., which will validate infinitely many arguments in it. We state it as

DR 1. $P \supset Q, Q \supset R \vdash \sim(\sim RP)$

Its demonstration requires a sequence of just five well formed formulas, the third of which we shall write twice, once in unabbreviated and once in abbreviated form:

$$
\begin{aligned}
&S_1\text{: } (P \supset Q) \supset [\sim(Q \sim R) \supset \sim(\sim RP)] \\
&S_2\text{: } P \supset Q \\
&S_3\text{: } \sim(Q \sim R) \supset \sim(\sim RP) \\
&S_3'\text{: } (Q \supset R) \supset \sim(\sim RP) \\
&S_4\text{: } Q \supset R \\
&S_5\text{: } \sim(\sim RP)
\end{aligned}
$$

That this sequence of S's is a demonstraction is easily verified. The first line, S_1, is Axiom 3 of R.S. It is true that there is an apparent difference between S_1 and our syntactical formulation of Axiom 3:

$$
\begin{aligned}
S_1\text{: } &(P \supset Q) \supset [\sim(Q \sim R) \supset \sim(\sim RP)] \\
\text{Ax. 3: } &(P \supset Q) \supset [\sim(QR) \supset \sim(RP)]
\end{aligned}
$$

because S_1 contains $\sim R$ wherever Axiom 3 contains R. The point is that we are here *talking about* well formed formulas *of* R.S. Both S_1 and Axiom 3 denote infinitely many *wffs* of R.S., and every *wff* of R.S. which is denoted by S_1 *is also denoted by Axiom* 3 (though not conversely). We can put the matter another way. Our first derived rule (DR 1) validates infinitely many arguments formulable within R.S., for example:

$$
\begin{array}{lcl}
\sim((\mathbf{A})\cdot(\sim(\mathbf{B}))) & & \sim((\mathbf{B})\cdot(\sim(\mathbf{C}))) \\
\sim((\mathbf{B})\cdot(\sim(\mathbf{C}))) & \text{and} & \sim((\mathbf{C})\cdot(\sim(\mathbf{D}))) \\
\sim((\sim(\mathbf{C}))\cdot(\mathbf{A})) & & \sim((\sim(\mathbf{D}))\cdot(\mathbf{B}))
\end{array}
$$

as well as

$$
\begin{array}{lcl}
\sim((\mathbf{A})\cdot(\sim(\mathbf{A}))) & & \sim(((\mathbf{A})\cdot(\mathbf{B}))\cdot(\sim(\sim(\mathbf{C})))) \\
\sim((\mathbf{A})\cdot(\sim(\mathbf{A}))) & \text{and} & \sim((\sim(\mathbf{C}))\cdot(\sim((\mathbf{D}_1)\cdot(\mathbf{D}_2)))) \\
\sim((\sim(\mathbf{A}))\cdot(\mathbf{A})) & & \sim((\sim((\mathbf{D}_1)\cdot(\mathbf{D}_2)))\cdot((\mathbf{A})\cdot(\mathbf{B})))
\end{array}
$$

The sequence of S's in the given demonstration denotes infinitely many sequences of *wffs* of R.S., in fact, one for each of the different arguments whose validity is being demonstrated. The first line in the demonstration *in* R.S. for the first of the four examples given is the *wff* of R.S. denoted by S_1 when 'P', 'Q', and 'R' are taken to denote '**A**', '**B**', and '**C**', respectively. But this is identically the same *wff* of R.S. denoted by our syntactical formulation of Axiom 3 when 'P', 'Q', and 'R' in it are taken to denote '**A**', '**B**', and '$\boldsymbol{\sim}(\boldsymbol{C})$', respectively. Hence the first *wff* of the demonstration sequence is one of the infinitely many axioms of R.S. supplied by Axiom 3. The situation is the same with respect to the demonstration of the validity of every other argument *in* R.S. that is validated by DR 1.

It is readily seen that the other lines of the sequence conform to the requirements laid down in our definition of demonstration. The two premisses of the argument occur as S_2 and S_4, while S_3 follows from S_1 and S_2 by R 1; and S_5 follows from $S_3(S_3')$ and S_4 by R 1. It is helpful to write in the 'justification' for each line in a demonstration: 'Ax. 3' to the right of S_1, 'premiss' to the right of S_2, and so on. These labels are not part of the demonstration, but are helpful to both the writer and the reader.

At this point the question naturally arises as to whether derived rules of inference can be appealed to in deriving theorems from the axioms of the system. It is most conveniently discussed in connection with an actual example. Let us take as our first theorem of R.S. the formula

TH. 1. $\vdash \sim(\sim PP)$

This *formula* follows directly from Axioms 1 and 2 by means of our first derived rule, DR 1. A sequence of formulas which has already been demonstrated to be a valid argument (by our demonstration of DR 1) is

S_1: $P \supset PP$	Ax. 1
S_2: $PP \supset P$	Ax. 2
S_3: $\sim(\sim PP)$	DR 1

That S_2 is Ax. 2 should be clear from the previous discussion. Every *wff* of R.S. denoted by our syntactical formulation of Ax. 2, '$PQ \supset P$', is an axiom, and every *wff* of R.S. denoted by the syntactical expression '$PP \supset P$' is (also) denoted by '$PQ \supset P$', so S_2 denotes infinitely many of the axioms of R.S. that are denoted by our syntactical formulation of Ax. 2. And that S_3 really follows by DR 1 from S_1 and S_2 is seen by observing that DR 1 validates any argument of the form

$$P \supset Q$$
$$Q \supset R$$
$$\sim(\sim RP)$$

no matter what *wffs* of R.S. the formulas P, Q, and R may be. Thus DR 1 includes the case in which 'P' and 'R' denote identically the same formulas, while 'Q' and 'PP' also denote the same *wffs*.

Although the sequence S_1, S_2, S_3 may be regarded as a '*proof*' of Theorem 1, it does not constitute a *demonstration*, for by definition, a *demonstration* involves the use of R 1 exclusively. But if we have a 'proof' in which a derived rule is used, as DR 1 is used in the given sequence, the *demonstration of the derived rule* can be *inserted* into the sequence to produce a demonstration proper. Thus in place of S_3 in the given sequence, the demonstration of DR 1 can be substituted to produce

S_1: $P \supset PP$	Ax. 1
S_2: $PP \supset P$	Ax. 2
S_3: $(P \supset PP) \supset [\sim(PP{\sim}P) \supset \sim(\sim PP)]$	Ax. 3
S_4: $\sim(PP{\sim}P) \supset \sim(\sim PP)$	R 1
S_4': $(PP \supset P) \supset \sim(\sim PP)$	
S_5: $\sim(\sim PP)$	R 1

This sequence *is* a demonstration because S_1, S_2, and S_3 are axioms; S_4 follows by R 1 from S_1 and S_3; and S_5 follows by R 1 from S_2 and S_4. Moreover, the present demonstration results from the earlier 'proof' by certain changes that are indicated in the proof itself. The original sequence, which we call a 'proof', is not a demonstration, but a *description* of a demonstration that can be given. A proof may be regarded as a prescription or recipe for the construction of a demonstration.

The situation is analogous to that which occurs when later theorems of a deductive system are derived, not directly from the axioms, but from earlier, already established theorems. Here, again, an example will aid the discussion. We shall take as our second theorem the formula

TH. 2. $\vdash \sim\sim P \supset P$

that follows directly from Theorem 1 by definition. Its proof may be written as:

S_1: $\sim(\sim\sim P{\sim}P)$	Th. 1
S_1': $\sim\sim P \supset P$	df.

This sequence is clearly *not a demonstration*, but it *is* a *proof*, for it tells us exactly how to construct a demonstration of Th. 2. In place of S_1 we need only write our demonstration of Th. 1—or rather the demonstration of that version of it which is relevant to the desired conclusion, Th. 2. The general statement of Th. 1 is

$$\vdash \sim(\sim PP)$$

which denotes every *wff* that is of this form, no matter which *wff* the syntactical variable 'P' denotes. Every *wff* denoted by '$\sim(\sim\sim P{\sim}P)$' is of that form, and is therefore included among the infinitely many provable formulas of R.S.

that are labeled Theorem 1. The demonstration described by the indicated proof of Th. 2 can be written as follows:

S_1:	$\sim P \supset \sim P \sim P$	Ax. 1
S_2:	$\sim P \sim P \supset \sim P$	Ax. 2
S_3:	$(\sim P \supset \sim P \sim P) \supset [\sim(\sim P \sim P \sim\sim P) \supset \sim(\sim\sim P \sim P)]$	Ax. 3
S_4:	$\sim(\sim P \sim P \sim\sim P) \supset \sim(\sim\sim P \sim P)$	R 1
S_4':	$(\sim P \sim P \supset \sim P) \supset \sim(\sim\sim P \sim P)$	df.
S_5:	$\sim(\sim\sim P \sim P)$	R 1
S_5':	$\sim\sim P \supset P$	df.

Compared with demonstrations, proofs are shorter and are therefore easier to write out. Since any proof can be made into a demonstration by replacing any line that is a previously established theorem by the demonstration of that theorem, and by replacing any line that results from the use of a derived rule by the demonstration of that rule, proofs can be regarded as shorthand notations for demonstrations. However, proofs are *different*, and should not be *confused* with demonstrations.

Before proceeding with the development of additional theorems and derived rules for R.S., it should be remarked that Th. 2 can be equally well expressed as $\vdash \sim P \vee P$, which is a version of the principle of the Excluded Middle. For, by our definition of the symbol 'v', '$\sim P \vee P$' is an abbreviation of '$\sim(\sim\sim P \sim P)$', which has (Th. 2) '$\sim\sim P \supset P$' as an alternative abbreviation. In the latter form it constitutes part of the principle of Double Negation.

Some additional theorems of R.S., together with their proofs (*not* demonstrations) are these:

TH. 3. $\vdash \sim(QR) \supset (R \supset \sim Q)$

Proof:	$\vdash \sim\sim Q \supset Q$	Th. 2
	$\vdash (\sim\sim Q \supset Q) \supset [\sim(QR) \supset \sim(R \sim\sim Q)]$	Ax. 3
	$\vdash \sim(QR) \supset \sim(R \sim\sim Q)$	R 1
	$\vdash \sim(QR) \supset (R \supset \sim Q)$	df.

TH. 4. $\vdash R \supset \sim\sim R$

Proof:	$\vdash \sim(\sim RR) \supset (R \supset \sim\sim R)$	Th. 3
	$\vdash \sim(\sim RR)$	Th. 1
	$\vdash R \supset \sim\sim R$	R 1

TH. 5. $\vdash (Q \supset P) \supset (\sim P \supset \sim Q)$

Proof:	$\vdash \sim(Q \sim P) \supset (\sim P \supset \sim Q)$	Th. 3
	$\vdash (Q \supset P) \supset (\sim P \supset \sim Q)$	df.

It will be observed that Th. 5 is part of the principle of Transposition, and that Theorems 2 and 4 are parts of the principle of Double Negation. But although both $P \supset \sim\sim P$ and $\sim\sim P \supset P$ have been proved to be theorems, *the* principle of Double Negation, $P \equiv \sim\sim P$, which abbreviates $(P \supset \sim\sim P) \cdot (\sim\sim P \supset P)$, has not (yet) been proved to be a theorem. It would follow from Theorems 2 and 4 by the principle of Conjunction, $P, Q \vdash P \cdot Q$, but the latter has not (yet) been established as a valid principle of inference or derived rule for R.S. These remarks are intended to throw additional light on the meaning of the '$\vdash$' symbol. Writing '$\vdash P$' asserts that there is a sequence of *wffs* ending with P, which is a demonstration. Writing '$\vdash P$ and $\vdash Q$' asserts that there are *two* sequences of *wffs*, both demonstrations, one ending with P, the other ending with Q. But writing '$\vdash P \cdot Q$' asserts that there is a *single* sequence of *wffs*, which is a demonstration, and ends with $P \cdot Q$. This assertion, although different, follows from the preceding by the principle of Conjunction, which will be established as DR 14.

The next derived rule is proved as follows:

DR 2. $\sim P \supset \sim Q \vdash Q \supset P$

Proof:	$(\sim P \supset \sim Q) \supset [\sim(\sim QQ) \supset \sim(Q \sim P)]$	Ax. 3
	$\sim P \supset \sim Q$	premiss
	$\sim(\sim QQ) \supset \sim(Q \sim P)$	R 1
	$\sim(\sim QQ)$	Th. 1
	$\sim(Q \sim P)$	R 1
	$Q \supset P$	df.

Although Theorem 5, $\vdash (Q \supset P) \supset (\sim P \supset \sim Q)$, is part of the principle of Transposition, DR 2, $\sim P \supset \sim Q \vdash Q \supset P$, is not. That principle asserts that $(Q \supset P) \equiv (\sim P \supset \sim Q)$, which is our abbreviation for $[(Q \supset P) \supset (\sim P \supset \sim Q)] \cdot [(\sim P \supset \sim Q) \supset (Q \supset P)]$. The left-hand conjunct is Theorem 5, but the right-hand conjunct is *not* DR 2. There is an important difference between

$$\sim P \supset \sim Q \vdash Q \supset P \qquad \text{and} \qquad \vdash (P \supset \sim Q) \supset (Q \supset P)$$

The first of these asserts that there is a sequence of *wffs*, each of which is either $\sim P \supset \sim Q$ or an axiom or follows from two preceding *wffs* by R 1 and whose last *wff* is $Q \supset P$. The second, however, asserts that there is a sequence of *wffs*, each of which is an axiom or follows from two preceding *wffs* by R 1 and whose last *wff* is $(\sim P \supset \sim Q) \supset (Q \supset P)$. (The second has not yet been established.) Of course there is some connection between them as there is between any two statements such as $P \vdash Q$ and $\vdash P \supset Q$. Given the latter, the former is easily established. To the sequence of *wffs* $S_1, S_2, \ldots, S_k$ (where S_k is $P \supset Q$), which constitutes a demonstration for $\vdash P \supset Q$, we need only add P as S_{k+1} and derive Q as S_{k+2}, for it follows from S_k and S_{k+1} by R 1. But although $\vdash P \supset Q$ follows from $P \vdash Q$, the proof that is does is less simple. It will be established as

Metatheorem III (the 'Deduction Theorem'); but until it has been proved, it cannot be assumed to hold for R.S.

Some additional derived rules, together with their proofs, are these:

DR 3. $P \supset Q \vdash RP \supset QR$

Proof:	$(P \supset Q) \supset [\sim(QR) \supset \sim(\mathrm{RP})]$	Ax. 3
	$P \supset Q$	premiss
	$\sim(QR) \supset \sim(RP)$	R 1
	$RP \supset QR$	DR 2

DR 4. $P \supset Q,\ R \supset S \vdash \sim[\sim(QS)(PR)]$

Proof:	$P \supset Q$	premiss
	$SP \supset QS$	DR 3
	$R \supset S$	premiss
	$PR \supset SP$	DR 3
	$\sim[\sim(QS)(PR)]$	DR 1

DR 5. $P \supset Q,\ Q \supset R,\ R \supset S \vdash P \supset S$

Proof:	$R \supset S$	premiss
	$(R \supset S) \supset (\sim S \supset \sim R)$	Th. 5
	$\sim S \supset \sim R$	R 1
	$(\sim S \supset \sim R) \supset [\sim(\sim RP) \supset \sim(P\sim S)]$	Ax. 3
	$\sim(\sim RP) \supset \sim(P\sim S)$	R 1
	$P \supset Q$	premiss
	$Q \supset R$	premiss
	$\sim(\sim RP)$	DR 1
	$\sim(P\sim S)$	R 1
	$P \supset S$	df.

The last derived rule mentioned may be thought of as a 'generalized' Hypothetical Syllogism. In developing R.S., it is convenient to establish DR 5 before proving the familiar Hypothetical Syllogism $P \supset Q,\ Q \supset R \vdash P \supset R$. The latter will be established as DR 6. For ease in proving the next derived rule, it is useful to prove three additional theorems first. Their proofs are left as exercises for the reader:

*TH. 6. $\vdash (R\sim\sim P) \supset (PR)$

TH. 7. $\vdash P \supset P$

TH. 8. $\vdash RP \supset PR$

It is convenient to state and prove a corollary of Theorem 7:

TH. 7, COR. $\vdash P \vee \sim P$

Proof: $\sim P \supset \sim P$ — Th. 7

$\sim(\sim P \sim\sim P)$ — df.

$P \vee \sim P$ — df.

It will be instructive for the reader to construct a *demonstration* (not merely a proof) for Theorem 7. It is so obvious a tautology as to seem trivial, and yet its demonstration in R.S. is not short.

Together, DR 5 and Theorem 7 provide an easy proof of the validity of the Hypothetical Syllogism

*DR 6. $P \supset Q, Q \supset R \vdash P \supset R$

Some additional theorems and derived rules are useful in proving the next Metatheorem.

Th. 9. $\vdash \sim(PR) \supset \sim(RP)$

DR 7. $P \supset Q, R \supset S \vdash PR \supset QS$

It is convenient to record two corollaries of DR 7:

DR 7, Cor. 1. $P \supset Q \vdash PR \supset QR$
DR 7, Cor. 2. $R \supset S \vdash PR \supset PS$
DR 8. $P \supset Q, P \supset R \vdash P \supset QR$

*Th. 10. $\vdash (PQ)R \supset P(QR)$

A useful corollary of Th. 10 is the other half of the principle of Association for '•':

Th. 10, Cor. $\vdash P(QR) \supset (PQ)R$

DR 9. $P \supset R, Q \supset S \vdash (P \vee Q) \supset (R \vee S)$
*DR 10. $P \supset R, Q \supset R \vdash (P \vee Q) \supset R$

Th. 11. $\vdash (P \vee Q) \supset (Q \vee P)$

Th. 12. $\vdash (P \vee Q) \vee R \supset P \vee (Q \vee R)$

A useful corollary of Th. 12 is the other half of the principle of Association for 'v':

Th. 12, Cor. $\vdash P \vee (Q \vee R) \supset (P \vee Q) \vee R$

Th. 13. $\vdash [P \supset (Q \supset R)] \supset [PQ \supset R]$

*Th. 14. $\vdash [PQ \supset R] \supset [P \supset (Q \supset R)]$

These last two theorems are the two halves of the principle of Exportation, but before the principle itself can be derived from them, the principle of Conjunction must be established (as DR 14).

DR 11. $P \supset Q, P \supset (Q \supset R) \vdash P \supset R$

TH. 15. $\vdash P \supset (Q \supset PQ)$

TH. 16. $\vdash P \supset (Q \supset P)$

Having established these derived rules and theorems, we are now able to prove the Deduction Theorem for R.S. as

METATHEOREM III. *If* $P_1, P_2, \ldots, P_{n-1}, P_n \vdash Q$ *then* $P_1, P_2, \ldots, P_{n-1} \vdash P_n \supset Q$.

Proof: We assume that $P_1, P_2, \ldots, P_{n-1}, P_n \vdash Q$, that is, that there is a demonstration or sequence of *wffs* $S_1, S_2, \ldots, S_s$, such that each S_i is either an axiom, or a $P_i (i = 1, 2, \ldots, n)$, or follows from two previous S's by R 1, and S_s is Q. Now consider the sequence of *wffs* $P_n \supset S_1, P_n \supset S_2, \ldots, P_n \supset S_s$. If we can 'fill in' *wffs* before each $P_n \supset S_i$ in such a way that the resulting total sequence is a demonstration from $P_1, P_2, \ldots, P_{n-1}$, so each line of the resulting total sequence is either an axiom or a P_i $(i = 1, 2, \ldots, n-1)$, or follows from two previous lines by R 1, then since $P_n \supset S_s$ is $P_n \supset Q$, we shall have a demonstration that $P_1, P_2, \ldots, P_{n-1} \vdash P_n \supset Q$. That we *can* 'fill in' to get the desired demonstration, is proved by weak induction on the number of formulas $P_n \supset S_i$ involved.

(α) In case $i = 1$, we have only the formula $P_n \supset S_1$ to consider. By assumption, S_1 is either an axiom or a P_i $(i = 1, 2, \ldots, n)$.

CASE 1. S_1 is an axiom. Here we fill in with the demonstration of Theorem 16, $\vdash S_1 \supset (P_n \supset S_1)$, and S_1 itself. From the last two formulas we derive $P_n \supset S_1$ by R 1. So the total sequence of *wffs* up to and including $P_n \supset S_1$ is a demonstration that $\vdash P_n \supset S_1$ and hence that $P_1, P_2, \ldots, P_{n-1} \vdash P_n \supset S_1$.

CASE 2. S_1 is a P_i $(i = 1, 2, \ldots, n-1)$. Here we fill in with the demonstration of Theorem 16, $\vdash S_1 \supset (P_n \supset S_1)$, and S_1, itself. From the last two formulas we derive $P_n \supset S_1$ by R 1, so the total sequence of *wffs* up to and including $P_n \supset S_1$ is a demonstration that $S_1 \vdash P_n \supset S_1$. Since S_1 is a P_i $(i = 1, 2, \ldots, n-1)$, we have a demonstration that $P_1, P_2, \ldots, P_{n-1} \vdash P_n \supset S_1$.

CASE 3. S_1 is P_n. Here we fill in with the demonstration of Theorem 7, $\vdash P_n \supset P_n$, that is, $\vdash P_n \supset S_1$. Thus the total sequence of *wffs* us to and including $P_n \supset S_1$ is a demonstration that $\vdash P_n \supset S_1$, and hence that $P_1, P_2, \ldots, P_{n-1} \vdash P_n \supset S_1$.

(β) Now suppose that we have properly filled in all the lines up to and including $P_n \supset S_{k-1}$, so we have a sequence of *wffs*, which is a demonstration that $P_1, P_2, \ldots, P_{n-1} \vdash P_n \supset S_{k-1}$. Under this assumption we show how to fill in to include $P_n \supset S_k$ in the sequence, which will then be a demonstration that $P_1, P_2, \ldots, P_{n-1} \vdash P_n \supset S_k$. By assumption, S_k is either an axiom, or a P_i ($i = 1, 2, \ldots, n$), or resulted in the original demonstration from the application of R 1 to two previous S's, say S_i and S_j ($i, j < k$).

CASE 1. S_k is an axiom. Insert the demonstration of Theorem 16, $\vdash S_k \supset (P_n \supset S_k)$, and S_k itself, and derive $P_n \supset S_k$ by R 1. The entire sequence will then be a demonstration that $P_1\ P_2, \ldots, P_{n-1} \vdash P_n \supset S_k$.

CASE 2. S_k is P_i ($i = 1, 2, \ldots, n - 1$). Insert the demonstration of Theorem 16, $\vdash S_k \supset (P_n \supset S_k)$, and S_k itself, and derive $P_n \supset S_k$ by R 1. The entire sequence will then be a demonstration that $P_1, P_2, \ldots, P_{n-1} \vdash P_n \supset S_k$.

CASE 3. S_k is P_n. Insert the demonstration of Theorem 7, $\vdash P_n \supset P_n$, that is, $\vdash P_n \supset S_k$, and the entire sequence will be a demonstration that $P_1, P_2, \ldots, P_{n-1} \vdash P_n \supset S_k$.

CASE 4. S_k resulted (in the original demonstration that $P_1, P_2, \ldots, P_n \vdash Q$), from the application of R 1 to two earlier S's, say S_i and S_j, where $i, j < k$ and S_i is of the form $S_j \supset S_k$. By the assumption of the β-case, we have already filled in up to and including both $P_n \supset S_j$ and $P_n \supset (S_j \supset S_k)$. By DR 11, stated as

$$P \supset Q, P \supset (Q \supset R) \vdash P \supset R$$

we have

$$P_n \supset S_j, P_n \supset (S_j \supset S_k) \vdash P_n \supset S_k$$

Insert the demonstration of this derived rule, whose last line is $P_n \supset S_k$, and the entire sequence will then be a demonstration that $P_1, P_2, \ldots, P_{n-1} \vdash P_n \supset S_k$.

Now by weak induction, we conclude that we can fill in for *any* number of lines $P_n \supset S_i$ in such a way that the resulting sequence will be a demonstration that $P_1, P_2, \ldots, P_{n-1} \vdash P_n \supset S_i$. We can, therefore, do it for the demonstration of $P_1, P_2, \ldots, P_{n-1}, P_n \vdash Q$ no matter how many lines $S_1, S_2, \ldots, S_s$ it contains. And since S_s is Q, we can construct a demonstration that $P_1, P_2, \ldots, P_{n-1} \vdash P_n \supset Q$. This concludes our proof of the Deduction Theorem.

An immediate consequence is

MT III, COROLLARY: *If* $P \vdash Q$ *then* $\vdash P \supset Q$.

An equally obvious corollary is the more general conclusion that: The Deduction Theorem holds for *any* propositional calculus that has only the rule *Modus Ponens* and that contains demonstrations for $P \supset P$, $P \supset (Q \supset P)$, and $(P \supset Q) \supset \{[P \supset (Q \supset R)] \supset (P \supset R)\}$.

The way in which Metatheorem III (abbreviated 'D.T.') may be used in proofs is indicated in the following proof of DR 6. First, we demonstrate the relatively trivial DR 6′: $P \supset Q$, $Q \supset R$, $P \vdash R$ by the following steps.

$P \supset Q$	premiss
P	premiss
Q	R 1
$Q \supset R$	premiss
R	R 1

Then we can prove DR 6 by simply applying the D.T. once, the proof reading

$P \supset Q, Q \supset R, P \vdash R$	DR 6′
$P \supset Q, Q \supset R \vdash P \supset R$	D.T.

Since the D.T. gives us an effective method of constructing a new demonstration for DR 6 on the basis of the old one for DR 6′, the preceding two step proof is a perfectly adequate prescription for the desired demonstration. It should not be thought that effort has been 'wasted' in constructing more difficult proofs of our earlier theorems, for they had to be established first in order to prove the Deduction theorem itself.

Some additional theorems and derived rules of R.S. are

Th. 17. $\vdash P \supset (Q \vee P)$

*Th. 17, Cor. $\vdash P \supset (P \vee Q)$

Th. 18. $\vdash (P \vee Q)R \supset (PR \vee QR)$

This theorem constitutes a part of the principle of Distribution—the distribution of '·' with respect to 'v'.

DR 12. $P \supset \sim Q \vdash P \supset \sim(QR)$
DR 13. $P \supset \sim R \vdash P \supset \sim(QR)$
*DR 14. $P, Q \vdash PQ$

Here, finally, we have the principle of Conjunction. It permits us to establish the principle of Double Negation as our next theorem, which follows directly from Th. 2 and Th. 4 by DR 14.

Th. 19. $\vdash P \equiv \sim\sim P$

Theorem 19, by itself, however, does not permit us to replace $\sim\sim P$ by P wherever it may occur in the interior of a larger *wff*. That is, where we are able to demonstrate a *wff* $\vdash (\text{---}\sim\sim P\ldots)$, the mere equivalence of $\sim\sim P$ and P does not permit us simply to infer that $\vdash (\text{---}P\ldots)$. The inference would be valid, but we must *prove* that it is valid within R.S. The legitimacy of any such replacement is asserted by our next Metatheorem. Before stating and proving it, we shall find it convenient to establish the following:

DR 15. $P \equiv Q \vdash \sim P \equiv \sim Q$
DR 16. $P \equiv Q,\ R \equiv S \vdash PR \equiv QS$
DR 16, Cor. $P \equiv Q,\ R \equiv S \vdash P \vee R \equiv Q \vee S$

Proofs will be left as exercises for the reader.

Now we are ready to prove the Rule of Replacement for R.S.

Metatheorem IV (*Rule of Replacement*). *Let $P_1, P_2, \ldots, P_n$ be any wffs, let Q be any wff that does not occur in any P_i, and let S be any wff that contains no components other than Q and P_i $(1 \le i \le n)$. Where S* is a wff that results from replacing any number of occurrences of Q in S by R, then* $Q \equiv R \vdash S \equiv S^*$.

Proof: Strong induction is used on the number of symbols in S, counting each occurrence of $\cdot$, $\sim$, Q, or any P_i as a single symbol.

(α) $n = 1$. In this case S is either Q alone or a P_i alone.

Case 1. S is Q and S^* is R. It is obvious that $Q \equiv R \vdash Q \equiv R$, which can be written $Q \equiv R \vdash S \equiv S^*$.

Case 2. S is Q and S^* is Q also. Since $\vdash Q \equiv Q$ by Th. 7 and DR 14, $\vdash S \equiv S^*$, whence $Q \equiv R \vdash S \equiv S^*$.

Case 3. S is a P_i. In this case S^* also is the same P_i. Since $\vdash P_i \equiv P_i$ by Th. 7 and DR 14, $\vdash S \equiv S^*$ whence $Q \equiv R \vdash S \equiv S^*$.

(β) Here the Metatheorem is assumed to be true for any S containing $< n$ symbols. Consider any S containing exactly n symbols $(n > 1)$. It is clear that S must be either $\sim S_1$ or $S_1 \cdot S_2$.

Case 1. S is $\sim S_1$. Because S contains n symbols, S_1 contains $< n$ symbols, so by the β-case assumption, $Q \equiv R \vdash S_1 \equiv S_1^*$, where S_1^* is a *wff* that results from replacing any number of occurrences of Q in S_1 by R. But $S_1 \equiv S_1^* \vdash \sim(S_1) \equiv \sim(S_1^*)$ by DR 15, and since any $\sim(S_1^*)$ is obviously the same as a $(\sim S_1)^*$ or S^*, we have $Q \equiv R \vdash S \equiv S^*$.

CASE 2. S is $S_1 \cdot S_2$. Here S_1 and S_2 each contains $< n$ symbols, so by the β-case assumption, $Q \equiv R \vdash S_1 \equiv S_1^*$ and $Q \equiv R \vdash S_2 \equiv S_2^*$. But by DR 16, $S_1 \equiv S_1^*,\ S_2 \equiv S_2^* \vdash S_1 \cdot S_2 \equiv S_1^* \cdot S_2^*$ and since any S^* is an $S_1^* \cdot S_2^*$, $Q \equiv R \vdash S \equiv S^*$.

Hence, by strong induction, we infer that regardless of the number of symbols in S, $Q \equiv R \vdash S \equiv S^*$.

MT IV, COROLLARY: *If Q, R, S, and S* are as in Metatheorem IV, then* $Q \equiv R,\ S \vdash S^*$.

The proof of the corollary is obvious.

Of the list of Rules of Inference used in proving the validity of arguments back in Chapter 3, the first nine were elementary valid argument forms proper, and the last ten were equivalences whose mutual replaceability was assumed. Of the first nine, the first, *Modus Ponens*, is the primitive rule R 1 of R.S. The third, the Hypothetical Syllogism, has already been established as DR 6; and the eighth, the principle of Conjunction, has been proved as DR 14. The remaining six rules can easily be proved as derived rules of R.S. They may be listed as

DR 17.	$P \supset Q,\ \sim Q \vdash \sim P$	(*Modus Tollens*)
DR 18.	$P \vee Q,\ \sim P \vdash Q$	(Disjunctive Syllogism)
DR 19.	$PQ \vdash P$	(Simplification)
DR 19, COR.	$PQ \vdash Q$	
*DR 20.	$(P \supset Q)(R \supset S),\ P \vee R \vdash Q \vee S$	(Constructive Dilemma)
DR 21.	$(P \supset Q)(R \supset S),\ \sim Q \vee \sim S \vdash \sim P \vee \sim R$	(Destructive Dilemma)
DR 22.	$P \vdash P \vee Q$	(Addition)

The various equivalences that made up the last ten of the nineteen Rules of Inference can easily be established. The principle of Double Negation has already been proved as Theorem 19. The various principles of Commutation and Association are obtained by simply applying DR 14 to already established theorems and corollaries.

TH. 20. $\vdash P \vee Q \equiv Q \vee P$ (Commutation of 'v')

TH. 21. $\vdash PQ \equiv QP$ (Commutation of '•')

By Theorem 21, we have $\vdash (P \equiv Q) \equiv (Q \equiv P)$, which can be used, without any special mention, in proving the following theorems.

TH. 22. $\vdash [P \vee (Q \vee R)] \equiv [(P \vee Q) \vee R]$ (Association of 'v')

TH. 23. $\vdash P(QR) \equiv (PQ)R$ (Association of '•')

The principle of Transposition is obtainable by DR 14 from Theorem 5 and the result of applying the Deduction Theorem to DR 2.

TH. 24. $\vdash (P \supset Q) \equiv (\sim Q \supset \sim P)$ (Transposition)

The proof of the principle of Exportation is even more obvious.

TH. 25. $\vdash [(PQ) \supset R] \equiv [P \supset (Q \supset R)]$ (Exportation)

The proofs of the final group of theorems of the R.S. will be left as exercises for the reader.

TH. 26. $\vdash P \equiv PP$ (Tautology)

TH. 26, COR. $\vdash P \equiv P \vee P$ (Tautology)

TH. 27. $\vdash \sim(PQ) \equiv (\sim P \vee \sim Q)$ (De Morgan's Theorem)

TH. 28. $\vdash \sim(P \vee Q) \equiv (\sim P \sim Q)$ (De Morgan's Theorem)

TH. 29. $\vdash (P \supset Q) \equiv (\sim P \vee Q)$ (Material Implication)

*TH. 30. $\vdash P(Q \vee R) \equiv PQ \vee PR$ (Distribution of '·' over 'v')

TH. 30, COR. $\vdash (P \vee Q)R \equiv PR \vee QR$

TH. 31. $\vdash (P \equiv Q) \equiv [PQ \vee \sim P \sim Q]$ (Material Equivalence)

TH. 32. $\vdash P \vee QR \equiv (P \vee Q)(P \vee R)$ (Distribution of 'v' over '·')

With the establishment of this final group of theorems, R.S. has been shown to contain all of the logical principles appealed to in validating extended arguments in Chapter 3. Containing also, as it does, the Deduction Theorem and the principle of Double Negation, it is adequate also for the methods of Conditional Proof and Indirect Proof that were discussed in Chapter 3. It still remains to be proved that the system is *deductively complete*, which will be established in the next section.

8.6 Deductive Completeness

In Theorems 22, 23, and 30 we already have the Association of 'v' and '·', and the Distribution of '·' with respect to 'v'. But as stated, these properties have been established only for cases involving exactly three *wffs*. In proving the deductive completeness of R.S., it is convenient to make use of more general Association and Distribution principles. These will be established as our next three Metatheorems. The first of them will establish the general Association and Commutation of the conjunction symbol, '·', stating that no matter in what order or grouping any number of *wffs* are conjoined, the resulting *wff* will be equivalent to the result of conjoining them in any other order or grouping. We may state this formally as follows:

METATHEOREM V. *Let $P_1, P_2, \ldots, P_n$ be any wffs and let Q and R be any wffs constructed out of them by means of '$\cdot$'. If each P_i $(1 \leq i \leq n)$ occurs exactly once in each of the wffs Q and R, then $\vdash Q \equiv R$.*

Proof: We use strong induction on the number of 'factors' (i.e., conjuncts) P_i in Q and R.

(α) Where $n = 1$, Q and R are identically the same *wff* P_1, so $\vdash Q \equiv R$ by Th. 7 and DR 14.

(β) Here we assume the Metatheorem to be true for every $k < n$ factors $P_1, P_2, \ldots, P_k$. Now consider Q and R, each constructed out of $n > 1$ factors $P_1, P_2, \ldots, P_n$. Q is $S \cdot T$ and R is $X \cdot Y$.

Each of the *wffs* S and T contains at least one of the factors $P_1, P_2, \ldots, P_n$. We can assume that P_1 is a factor of S, because if not we can apply Th. 21 and relabel to obtain $\vdash Q \equiv S \cdot T$ where S now does contain P_1 as a factor.

Because T contains at least one of $P_2, P_3, \ldots, P_n$ as a factor, S contains $< n$ of the factors $P_1, P_2, \ldots, P_n$. Hence either S is P_1 and $\vdash Q \equiv P_1 \cdot T$, or by the β-case assumption $\vdash S \equiv P_1 \cdot S'$, where S' is a *wff* that contains all the factors of S except P_1. In the latter case, by MT IV, Cor. we have

$$\vdash Q \equiv (P_1 \cdot S') \cdot T$$

and by Th. 23 and MT IV, Cor.

$$\vdash Q \equiv P_1 \cdot (S' \cdot T)$$

In either case, there is a *wff*, call it T', such that

$$\vdash Q \equiv P_1 \cdot T'$$

By the same reasoning, we can show that there is a *wff*, call it Y', such that

$$\vdash R \equiv P_1 \cdot Y'$$

Each of the *wffs* T' and Y' contains the $n - 1$ factors $P_2, P_3, \ldots, P_n$, so by the β-case assumption

$$\vdash T' \equiv Y'$$

By Th. 7 and DR 14 we have $\vdash P_1 \equiv P_1$, so by DR 16 we have

$$\vdash P_1 \cdot T' \equiv P_1 \cdot Y'$$

and by MT IV, Cor.

$$\vdash Q \equiv R$$

Metatheorem V now follows by strong induction.

The next principle concerns the general Association and Commutation of the disjunction symbol 'v'. No matter in what order or grouping any *wffs* are connected by 'v', the resulting disjunction or 'logical sum' will be equivalent to the result of connecting them by 'v' in any other order or grouping. We may state this formally as

METATHEOREM VI. *Let $P_1, P_2, \ldots, P_n$ be any wffs and let Q and R be any two wffs constructed out of them by means of 'v'. If each P_i $(1 \leq i \leq n)$ occurs exactly once in each of the wffs Q and R, then $\vdash Q \equiv R$.*

The proof will be left as an exercise for the reader.

MT VI, COR. *If Q and R are as in* Metatheorem VI, *then* $Q \vdash R$. The proof of the corollary is obvious.

Finally, we wish to establish a generalized statement of the Distribution of conjunction with respect to disjunction, that is, of '·' over 'v'. This is expressed as

METATHEOREM VII. *If Q is the logical sum of $P_1, P_2, \ldots, P_n$, that is $(P_1 \text{ v } P_2 \text{ v } \ldots \text{ v } P_n)$ with association to the left given as a convention, and S is the logical sum of $P_1R, P_2R, \ldots, P_nR$, then $\vdash QR \equiv S$.*

Proof: We use weak induction on the number of 'summands' (i.e. disjuncts) $P_1, P_2, \ldots, P_n$.

(α) Where $n = 1$, Q is P_1, QR is P_1R, and S is P_1R also. By Th. 7 and DR 14, $\vdash P_1R \equiv P_1R$ which is $\vdash QR \equiv S$.

(β) Assume the Metatheorem true for k summands $P_1, P_2, \ldots, P_k$. Now let Q be the logical sum or disjunction of $P_1, P_2, \ldots, P_k, P_{k+1}$, and let S be the logical sum of $P_1R, P_2R, \ldots, P_kR, P_{k+1}R$. Now we argue that

$\vdash (P_1 \text{ v } P_2 \text{ v } \ldots \text{ v } P_k)R \equiv P_1R \text{ v } P_2R \text{ v } \ldots \text{ v } P_kR$	by the β-case assumption
$\vdash P_{k+1}R \equiv P_{k+1}R$	Th. 7 and DR 14
$\vdash (P_1 \text{ v } P_2 \text{ v } \ldots \text{ v } P_k)R \text{ v } P_{k+1}R \equiv (P_1R \text{ v } P_2R \text{ v } \ldots \text{ v } P_kR) \text{ v } P_{k+1}R$	DR 16, Cor.

$\vdash [(P_1 \vee P_2 \vee \ldots \vee P_k) \vee P_{k+1}]R \equiv$
$(P_1 \vee P_2 \vee \ldots \vee P_k)R \vee P_{k+1}R$ Th. 30, Cor.
$\vdash [(P_1 \vee P_2 \vee \ldots \vee P_k) \vee P_{k+1}]R \equiv$
$(P_1R \vee P_2R \vee \ldots \vee P_kR) \vee P_{k+1}R$ MT IV, Cor.

By our convention of *association to the left*, the preceding line can also be written as

$$\vdash (P_1 \vee P_2 \vee \ldots \vee P_k \vee P_{k+1})R \equiv P_1R \vee P_2R \vee \ldots \vee P_kR \vee P_{k+1}R$$

which is $\vdash QR \equiv S$ where Q contains $k + 1$ summands. Metatheorem VII now follows by weak induction.

To prove that R.S. is deductively complete, we show that all tautologies are demonstrable as theorems in the system. Since all tautologies are expressible as *wffs* of R.S. (by MT I), the deductive completeness of R.S. is expressed as: If S is a *wff* that is a tautology, then $\vdash$ S. A criterion for deciding whether or not any *wff* is a tautology is supplied by the method of truth tables. Any *wff* S has a truth table with as many initial columns in it as there are distinct propositional symbols in S. Where there are n of them, say $P_1, P_2, \ldots, P_n$, the tautology S will have this truth table:

P_1	P_2	...	P_n	S
T	T	...	T	T
T	T	...	F	T
.	.	...	.	.
.	.	...	.	.
.	.	...	.	.
F	F	...	T	T
F	F	...	F	T

Any such truth table has 2^n rows, each of which represents a different assignment of **T**'s and **F**'s to the P_i's, one row for every possible assignment. That only **T**'s appear in the column under S indicates that every possible assignment of **T**'s and **F**'s to the P_i's must assign a **T** to S.

To show that $\vdash$ S for every such S, we establish the following:

1. Each row of its truth table, that is, each assignment of truth values to the P_i's, can be represented by a *wff* of R.S., the first row by Q_1, the second by $Q_2, \ldots$, and the last or $2^n th$ by Q_{2^n};
2. If $Q_1, Q_2, \ldots, Q_{2^n}$ are the 2^n *wffs* representing all possible assignments of **T**'s and **F**'s to the P_i's, then $\vdash (Q_1 \vee Q_2 \vee \ldots \vee Q_{2^n})$; and
3. If a particular assignment of **T**'s and **F**'s to the P_i's assigns a **T** to S, then where Q_j represents that particular assignment, we have $\vdash Q_j \supset$ S.

That these will suffice to prove that $\vdash S$ is easily seen. Where the truth table for S has all **T**'s in the column under S, then $\vdash Q_1 \supset S$, $\vdash Q_2 \supset S$, ..., $\vdash Q_{2^n} \supset S$. From these, by $2^n - 1$ uses of DR 10, we have $\vdash (Q_1 \vee Q_2 \vee \ldots \vee Q_{2^n}) \supset S$. And once we have established that $\vdash (Q_1 \vee Q_2 \vee \ldots \vee Q_{2^n})$, we obtain $\vdash S$ by R 1.

Now we proceed to attack the problem in detail. First, we must show that each possible assignment of **T**'s and **F**'s to the P_i's of a set $(P_1, P_2, \ldots, P_n)$ can be represented by a *wff*. It is convenient to lay down the following:

Definition. A *wff* is said to *represent* a particular assignment of truth values to propositional symbols $P_1, P_2, \ldots, P_n$ if and only if (on the normal interpretation of our operator symbols) that truth value assignment is the only one which makes that *wff* true.

Where **T**'s are assigned to every P_i, this assignment is represented by the conjunction $P_1 \cdot P_2 \cdot \ldots \cdot P_n$, which we denote by '$Q_1$'. Where **T**'s are assigned to every P_i except P_n, which is assigned an **F**, the assignment is represented by the conjunction $P_1 \cdot P_2 \cdot \ldots \cdot P_{n-1} \cdot \sim P_n$, which we denote by '$Q_2$', Similarly, we use this procedure for every other possible assignment corresponding to every row of the truth table, ending with the conjunction $\sim P_1 \cdot \sim P_2 \cdot \ldots \cdot \sim P_n$, which we denote by '$Q_{2^n}$'. In this way any row of any truth table can be represented by a *wff* of R.S., which establishes the *first* result mentioned in the preceding paragraph.

Next, we turn our attention to the *second* result, expressed as

METATHEOREM VIII. *If $Q_1, Q_2, \ldots, Q_{2^n}$ represent all possible distinct assignments of truth values to the n distinct propositional symbols $P_1, P_2, \ldots, P_n$, then $\vdash (Q_1 \vee Q_2 \vee \ldots \vee Q_{2^n})$.*

Proof: We use weak induction on the number of P_i's.

(α) Where $n = 1$, $2^n = 2$, and Q_1 is P_1 and Q_2 is $\sim P_1$. Here we have $\vdash P_1 \vee \sim P_1$ by Th. 7, Cor. which is $\vdash (Q_1 \vee Q_2)$.

(β) Assume that the Metatheorem is true for $P_1, P_2, \ldots, P_k$. Now consider the set $P_1, P_2, \ldots, P_k, P_{k+1}$. Where $Q_1, Q_2, \ldots, Q_{2^k}$ represent all possible distinct assignments of truth values to $P_1, P_2, \ldots, P_k$, we have $\vdash (Q_1 \vee Q_2 \vee \ldots \vee Q_{2^k})$ by the β-case assumption.

Then we continue the argument as follows:

$\vdash P_{k+1} \vee \sim P_{k+1}$	Th. 7, Cor.
$\vdash (Q_1 \vee Q_2 \vee \ldots \vee Q_{2^k})(P_{k+1} \vee \sim P_{k+1})$	DR 14
$\vdash (Q_1 \vee Q_2 \vee \ldots \vee Q_{2^k})(P_{k+1} \vee \sim P_{k+1}) \equiv$ $(Q_1 \vee Q_2 \vee \ldots \vee Q_{2^k})P_{k+1} \vee (Q_1 \vee Q_2 \vee \ldots \vee Q_{2^k}) \sim P_{k+1}$	Th. 30
$\vdash (Q_1 \vee Q_2 \vee \ldots \vee Q_{2^k})P_{k+1} \vee (Q_1 \vee Q_2 \vee \ldots \vee Q_{2^k}) \sim P_{k+1}$	MT IV, Cor.

$\vdash (Q_1 \vee Q_2 \vee \ldots \vee Q_{2^k})P_{k+1} \equiv (Q_1P_{k+1} \vee Q_2P_{k+1} \vee \ldots \vee Q_{2^k}P_{k+1})$ MT VII

$\vdash (Q_1 \vee Q_2 \vee \ldots \vee Q_{2^k}) {\sim} P_{k+1} \equiv$
$(Q_1 {\sim} P_{k+1} \vee Q_2 {\sim} P_{k+1} \vee \ldots \vee Q_{2^k} {\sim} P_{k+1})$ MT VII

$\vdash (Q_1P_{k+1} \vee Q_2P_{k+1} \vee \ldots \vee Q_{2^k}P_{k+1}) \vee (Q_1 {\sim} P_{k+1} \vee$
$Q_2 {\sim} P_{k+1} \vee \ldots \vee Q_{2^k} {\sim} P_{k+1})$ MT IV, Cor.

$\vdash Q_1P_{k+1} \vee Q_1 {\sim} P_{k+1} \vee Q_2P_{k+1} \vee Q_2 {\sim} P_{k+1} \vee \ldots \vee$
$Q_{2^k}P_{k+1} \vee Q_{2^k} {\sim} P_{k+1}$ MT VI, Cor.

The preceding expression contains $2^k + 2^k = 2 \cdot 2^k = 2^{k+1}$ distinct summands, each of which represents a different assignment of truth values to $P_1, P_2, \ldots, P_{k+1}$. Each Q_iP_{k+1} and each $Q_i {\sim} P_{k+1}$ is a different Q'_i, where the 2^{k+1} Q'_i's represent all possible distinct assignments of truth values to $P_1, P_2, \ldots, P_{k+1}$. Hence $\vdash (Q'_1 \vee Q'_2 \vee \ldots \vee Q'_{2^{k+1}})$. Metatheorem VIII now follows by weak induction.

We next prove that if the truth value assignment represented by Q_j assigns a **T** to S, then $\vdash Q_j \supset S$. We shall prove this by establishing a slightly more general result, which includes also the case in which the truth value assignment assigns an **F** to S instead. This is stated and established as our next metatheorem.

METATHEOREM IX. *Let Q_j represent any possible assignment of truth values to the n propositional symbols $P_1, P_2, \ldots, P_n$; and let S be any wff which has no propositional components other than P_i $(1 \leq i \leq n)$. If the truth value assignment represented by Q_j assigns a* **T** *to S, then $\vdash Q_j \supset S$; and if the truth value assignment represented by Q_j assigns an* **F** *to S, then $\vdash Q_j \supset {\sim}S$.*

Proof: We use strong induction on the number of symbols in S, counting each occurrence of $\cdot$, of $\sim$, and of any P_i as a single symbol.

(α) Where $n = 1$, S must be a single symbol, and since it is a *wff*, it must be a P_i $(1 \leq i \leq n)$.

CASE 1. Q_j assigns a **T** to S, that is, to P_i. Hence P_i rather than ${\sim}P_i$ must be a factor of Q_j. By MT V, $\vdash Q_j \equiv P_iR$, where R is a conjunction of all the factors of Q_j except P_i. Now we argue

$\vdash P_iR \supset P_i$ Ax. 2
$\vdash Q_j \supset P_i$ MT IV, Cor.

which is $\vdash Q_j \supset S$.

CASE 2. Q_j assigns an **F** to S, that is, to P_i. Hence ${\sim}P_i$ must be a factor of Q_j. By MT V, $\vdash Q_j \equiv {\sim}P_iR$, where R is a conjunction of all the factors of Q_j

except $\sim P_i$. Now we argue

$$\vdash \sim P_i R \supset \sim P_i \qquad \text{Ax. 2}$$
$$\vdash Q_j \supset \sim P_i \qquad \text{MT IV, Cor.}$$

which is $\vdash Q_j \supset \sim S$.

(β) Assume that the Metatheorem is true for any S containing any number $k < n$ symbols. Now consider any S containing n (>1) symbols. The *wff* S is either $S_1 \cdot S_2$ or $\sim S_3$.

CASE 1. S is $S_1 \cdot S_2$.
Subcase A: Q_j assigns a **T** to S. Here Q_j must assign a **T** to S_1 and a **T** to S_2. Since S_1, S_2 contain fewer than n symbols each, we argue as follows:

$$\vdash Q_j \supset S_1 \qquad \text{by the } \beta\text{-case assumption}$$
$$\vdash Q_j \supset S_2 \qquad \text{by the } \beta\text{-case assumption}$$
$$\vdash Q_j \supset S_1 \cdot S_2 \qquad \text{DR 8}$$

which is $\vdash Q_j \supset S$.
Subcase B: Q_j assigns an **F** to S. Here Q_j must assign an **F** to S_1 or an **F** to S_2. If to S_1, then $\vdash Q_j \supset \sim S_1$ by the β-case assumption, and hence by DR 12 $\vdash Q_j \supset \sim(S_1 \cdot S_2)$, which is $\vdash Q_j \supset \sim S$. If to S_2 then $\vdash Q_j \supset \sim S_2$ by the β-case assumption, and hence by DR 13 $\vdash Q_j \supset \sim(S_1 \cdot S_2)$, which is $\vdash Q_j \supset \sim S$.

CASE 2. S is $\sim S_3$.
Subcase A: Q_j assigns a **T** to S. Here Q_j must assign an **F** to S_3. Hence by the β-case assumption, $\vdash Q_j \supset \sim S_3$ which is $\vdash Q_j \supset S$.
Subcase B: Q_j assigns an **F** to S. Here Q_j must assign a **T** to S_3. Hence by the β-case assumption, $\vdash Q_j \supset S_3$. But $\vdash S_3 \supset \sim\sim S_3$ by Th. 4, so by DR 6 we have $\vdash Q_j \supset \sim\sim S_3$, which is $\vdash Q_j \supset \sim S$.

Metatheorem IX now follows by strong induction.

The deductive completeness of the system now follows easily, and may be proved as

METATHEOREM X. *R.S. is deductively complete (that is, if* S *is a tautology, then* $\vdash$ S).

Proof: If S is a tautology, then every possible assignment of truth values to its components $P_1, P_2, \ldots, P_n$ must assign a **T** to S. Hence by MT IX:

$$\vdash Q_1 \supset S$$
$$\vdash Q_2 \supset S$$
$$\cdots\cdots\cdots$$
$$\vdash Q_{2^n} \supset S$$

where $Q_1, Q_2, \ldots, Q_{2^n}$ represent all possible assignments of truth values to P_1, $P_2, \ldots, P_n$. Now by $2^n - 1$ uses of DR 10, we have

$$\vdash (Q_1 \vee Q_2 \vee \ldots \vee Q_{2^n}) \supset S$$

and by MT VIII,

$$\vdash (Q_1 \vee Q_2 \vee \ldots \vee Q_{2^n})$$

From these, we derive $\vdash S$ by R 1, which completes the proof of Metatheorem X.

The *decision problem* for any deductive system is the problem of stating an effective criterion for deciding whether or not any statement or well formed formula is a theorem of the system. In view of the analyticity and deductive completeness of R.S. (Metatheorems II and X), the method of truth tables constitutes a solution to the decision problem. Truth tables enable us to decide effectively whether or not any *wff* is a tautology. By MT II, *only* tautologies are theorems, and by MT X, *all* tautologies are theorems. Hence truth tables enable us to decide effectively whether or not any *wff* is a theorem. Moreover, the proofs up to and including that for MT X do not merely assure us that for any tautologous *wff* there exists a demonstration—they prescribe effectively a method of actually constructing its demonstration. The demonstration constructed by following the directions contained in the proof of deductive completeness will usually be longer than one discovered through the exercise of ingenuity and inventiveness. That is to be admitted. But it is significant and important that through the use of the directions contained in the proofs up to and including that of MT X, a demonstration within the logistic system can be written out for any tautology—*without any need for ingenuity or inventiveness*. Our effective solution of the decision problem for the system guarantees that this can be done.

It is clear from the foregoing that any argument whose validity can be established by the use of truth tables can be proved valid in R.S. In Chapter 3, the claim was made that any such argument could be proved valid by using the list of nineteen Rules of Inference augmented by the principles of Conditional Proof and Indirect Proof. We are now in a position to substantiate that claim, which is equivalent to the assertion that the method of deduction set forth in Chapter 3 is deductively complete. We can do so by showing that every argument which can be proved valid in R.S. can also be proved valid by the methods of Chapter 3.

Consider any argument $P_1, \ldots, P_n \therefore Q$ that can be proved valid in R.S. To say that it can be proved valid in R.S. is to say that there is a demonstration in R.S. for the derived rule of inference $P_1, \ldots, P_n \vdash Q$. By n uses of the Deduction Theorem, Exportation, and the Rule of Replacement, we have $\vdash P \supset Q$ where P is a conjunction of $P_1, \ldots, P_n$. By the analyticity of R.S., $P \supset Q$ is a truth table tautology, so $P \cdot {\sim}Q$ is a contradiction. Now there is a formal proof of validity for the argument

$$P_1, \ldots, P_n, \sim Q \quad \therefore P \cdot \sim Q \tag{1}$$

using the methods of Chapter 3 (by iterated uses of the principle of Conjunction). Hence there is a formal proof using the methods of Chapter 3 for the argument.

$$P_1, \ldots, P_n, \sim Q \quad \therefore N \tag{2}$$

where N is a disjunctive normal form[9] of the formula $P \cdot \sim Q$. This is so because the equivalences included among the Rules of Inference in Chapter 3 are sufficient to permit deriving the disjunctive normal form of any line in a formal proof.

Because $P \cdot \sim Q$ is a contradiction, N is a disjunction in which every disjunct contains a contradiction as a conjunct. Hence by repeated uses of variants of the formal proof of validity for

$$q \vee [(p \cdot \sim p) \cdot r] \quad \therefore q$$

the formal proof of validity for (2) can be extended to a formal proof of validity for

$$P_1, \ldots, P_n, \sim Q \quad \therefore N_1 \tag{3}$$

where N_1 is a single disjunct of N. If N_1 is not itself an explicit contradiction, it is a conjunction containing a contradiction as a conjunct. Hence by Com. and Simp. (and possibly Assoc.) the formal proof of validity for (3) can be extended to provide a derivation of an explicit contradiction from the set of premisses $P_1, \ldots, P_n, \sim Q$. And this derivation constitutes an Indirect Proof of validity for the original argument $P_1, \ldots, P_n \therefore Q$. Hence any argument that can be proved valid in R.S. can be proved valid by the methods of Chapter 3. This proof suffices to show that the method of deduction set forth in Chapter 3 is also a deductively complete system of logic.[10]

[9] See Appendix B, pages 337–343.

[10] The preceding proof is adapted from John Thomas Canty's 'Completeness of Copi's Method of Deduction,' *Notre Dame Journal of Formal Logic,* vol. 4 (1963), pp. 142–144. See also M. C. Bradley, 'Copi's Method of Deduction Again,' ibid., vol. 12 (1971), pp. 454–458.

9

Alternative Systems and Notations

9.1 Alternative Systems of Logic

There are three different senses in which the phrase 'alternative systems of logic' can be understood. These parallel the three senses of the phrase 'alternative systems of geometry', and can most easily be explained by analogy with the latter. We may speak of Euclidean plane geometry and Euclidean solid geometry as 'alternative systems' in the sense that the first can be studied independently of the second, and they are certainly different in that the second is *more inclusive* than the first. Analogously, we may speak of a Propositional Calculus and a Function Calculus as 'alternative systems of logic' in that the first can be studied independently of the second, and that the second is *more inclusive* than the first—where a Function Calculus contains all the tautologies and rules of the Propositional Calculus *plus* Quantification Axioms, Rules, and Theorems. It is not *this* sense of an alternative system with which we shall be concerned in the present chapter.

A second sense is that in which we can speak of Euclidean geometry and Riemannian (or Lobachevskian) geometry as alternative systems. These are alternative in the sense that, although they may possess some theorems in common, each contains some theorems not included in the other. Parallel to this situation in geometry, there are alternative systems of logic exhibiting the same sort of differences. An ordinary 'two-valued' system of logic, whose formulas—on interpretation—are either *true* or *false*, can be contrasted with 'three-valued' or 'many-valued' systems of logic whose formulas are supposed to take—on interpretation—either three or $n > 3$ different 'truth values'. Alternative systems of logic in this sense have been extensively developed, first by Jan Łukasiewicz in Poland and independently by E. L. Post in this country, more recently by J. B. Rosser and A. R. Turquette.[1] A study of these is beyond

[1] See J. Łukasiewicz, 'O logice trojwartosciowej', *Ruch Filozoficzny* (Lwow), vol. 5 (1920), pp. 169–171.

E. L. Post, 'Introduction to a General Theory of Elementary Propositions', *American Journal of Mathematics*, vol. 43 (1921), pp. 163–185.

J. B. Rosser, 'On the Many-Valued Logics', *American Journal of Physics*, vol. 9 (1941), pp. 207–212.

J. B. Rosser and A. R. Turquette, 'Axiom Schemes for M-Valued Propositional Calculi', *Journal* the scope of this book. It is not *this* sense of alternative systems with which we shall be concerned in the present chapter.

The third sense in which one can speak of alternative systems of geometry is that in which different axiomatic bases are assumed, but identically the same theorems are derivable from each of them. Thus many different axiom sets for Euclidean geometry have been devised, all of which yield the same theorems. In alternative systems of this sort, different terms may be taken as primitive or undefined, and different formulas are assumed as axioms or postulates. What may be a primitive term in one system may be defined by means of other primitives in the other, and what is assumed as an axiom in one may be derived as a theorem from the axioms of the other—where those axioms correspond to theorems of the first. It is *this* sense of alternative systems that will be discussed in the present chapter.

The logical truths whose systematization is under consideration are truth functional tautologies. Any system wholly adequate to their expression and development must be *functionally complete, analytic,* and *deductively complete,* in the senses in which R.S. was proved to possess these properties in Metatheorems I, II, and X of the preceding chapter. Any such system will be called a *Model System of Logic,* and any other axiom system for logic will be a genuine or acceptable alternative to R.S. if and only if it is a Model System. There are many different Model Systems, either different in that they start with different primitive or undefined terms, or different in that they assume different formulas as axioms. They are all equivalent, however, first in being able to express—on their normal interpretations—all truth functions, second in including *all* tautologies as theorems, and third in having *only* tautologies as theorems. One such alternative system will be proved to be a *Model System* in the following section.

9.2 The Hilbert-Ackermann System

The Hilbert-Ackermann system for the propositional calculus was acknowledged by D. Hilbert and W. Ackermann to be "due in essence to Whitehead and Russell (*Principia Mathematica,* 1st ed.)". It has as its primitive symbols infinitely many single letters with and without subscripts:

$$\begin{array}{lllll} A & A_1 & A_2 & A_3 & \ldots \\ B & B_1 & B_2 & B_3 & \ldots \\ C & C_1 & C_2 & C_3 & \ldots \\ D & D_1 & D_2 & D_3 & \ldots \end{array}$$

---

of Symbolic Logic, vol. 10 (1945), pp. 61–82, and *Many-Valued Logics,* North-Holland Publishing Company, Amsterdam, 1952.

See also: Alan Ross Anderson et al., 'Proceedings of a Colloquium on Modal and Many-Valued Logics', *Acta Philosophica Fennica,* Fasc. 16 (1963); and A. A. Zinov'ev, *Philosophical Problems of Many-Valued Logic,* G. Küng and D. D. Comey, eds. and trs., D. Reidel Publishing Company, Dordrecht-Holland, 1963.

which on their normal interpretations express noncompound propositions, and has in addition to parentheses, the two operator symbols '$\sim$' and 'v' (designated in our metalanguage by '$\sim$' and 'v'), which have as their normal interpretations the operations of negation and weak (or inclusive) disjunction. We continue to use the symbols 'P', 'Q', 'R', 'S', . . . , with and without subscripts, in our metalanguage, to denote well formed formulas of H.A. (the Hilbert–Ackermann system of logic). The notion of a *wff* of H.A. is defined recursively as follows:

RECURSIVE RULE FOR *wffs* IN H.A.

1. Any single letter of H.A. is a *wff*.
2. If P is a *wff*, then $\sim(P)$ is a *wff*.
3. If P and Q are *wffs*, then $(P) \vee (Q)$ is a *wff*.

(No formula of H.A. will be regarded as being a *wff* unless its being so follows from this definition.)

The symbols '$\supset$', '$\cdot$', '$\equiv$', are defined *syntactically* for our metalanguage by the following:

$$
\begin{aligned}
P \supset Q &= \text{df } \sim P \vee Q \\
P \cdot Q &= \text{df } \sim(\sim P \vee \sim Q) \\
PQ &= \text{df } P \cdot Q \\
P \equiv Q &= \text{df } (P \supset Q)(Q \supset P)
\end{aligned}
$$

We shall continue to use the same conventions regarding parentheses that were adopted in the preceding chapter.

Four (patterns of) axioms or postulates are assumed in H.A.

P 1. $(P \vee P) \supset P$
P 2. $P \supset (P \vee Q)$
P 3. $(P \vee Q) \supset (Q \vee P)$
P 4. $(P \supset Q) \supset [(R \vee P) \supset (R \vee Q)]$

Each of these syntactical expressions denotes infinitely many *wffs* of our object language H.A., just as in the metalogical development of R.S.

Finally, a single rule of inference is assumed, which we may state as

R′ 1. From P and $P \supset Q$ to infer Q.

It should be realized that R′ 1 is different from R 1, because R′ 1 legitimizes arguments within H.A. of the form

$$\begin{array}{l} P \\ \sim P \vee Q \\ Q \end{array}$$

whereas R 1 legitimizes arguments within R.S. of the form

$$\begin{array}{l} P \\ \sim(P \sim Q) \\ Q \end{array}$$

and these are clearly different. We may make the contrast more vivid by writing them as

R 1. From P and $\sim(P\sim Q)$ to infer Q.
R′ 1. From P and $\sim P \vee Q$ to infer Q.

By MT IV, Cor., and Th. 29 of R.S., any *wff* in R.S. that follows by R 1 from two other *wffs* must also follow by R′ 1 and conversely. But this statement cannot be assumed true of H.A. until it is *proved.*

A 'demonstration in H.A.' of the validity of an argument having premisses $P_1, P_2, \ldots, P_n$ and conclusion Q is defined to be a sequence of *wffs* $S_1, S_2, \ldots, S_t$ (of H.A.), each of which is either a postulate P 1, P 2, P 3, or P 4 or a P_i $(1 \leq i \leq n)$ or follows from two preceding S's by R′ 1, and such that S_t is Q. That there is such a demonstration in H.A. is written

$$P_1, P_2, \ldots, P_n \vdash_{\text{HA}} Q$$

Similarly, that the formula P is a theorem of H.A. is written

$$\vdash_{\text{HA}} P$$

which asserts that there is a sequence of *wffs* $S_1, S_2, \ldots, S_t$ (of H.A.) each of which is either a postulate P 1, P 2, P 3, or P 4 or follows from two preceding S's by R′ 1, and such that S_t is P.

The functional completeness of H.A. is easily established. (It was Exercise 1 on page 226 in the preceding chapter.) A proof of the analyticity of H.A. is easily given by using truth tables to show that any postulate P 1, P 2, P 3, P 4 is a tautology, and then proving that any *wff* which follows from tautologies by repeated applications of R′ 1 must be tautologous also.

The independence of the H.A. postulates is established by the following models.

To prove that Postulate 1 is independent, we use the three-element model

{**0**, **1**, **2**}, of which **0** is the designated element. *Wffs* are assigned values in accordance with the tables:

P	$\sim P$
0	2
1	1
2	0

P	Q	$P \vee Q$	$P \supset Q$
0	0	0	0
0	1	0	1
0	2	0	2
1	0	0	0
1	1	0	0
1	2	1	1
2	0	0	0
2	1	1	0
2	2	2	0

To prove Postulate 2 independent, we use the three-element model {**0**, **1**, **2**} of which **0** is designated, with the tables:

P	$\sim P$
0	1
1	0
2	2

P	Q	$P \vee Q$	$P \supset Q$
0	0	0	0
0	1	0	1
0	2	0	1
1	0	0	0
1	1	1	0
1	2	1	0
2	0	0	0
2	1	1	1
2	2	1	1

To prove Postulate 3 independent, we use {**0**, **1**, **2**} with **0** designated and tables:

P	$\sim P$
0	2
1	0
2	1

P	Q	$P \vee Q$	$P \supset Q$
0	0	0	0
0	1	0	2
0	2	0	2
1	0	0	0
1	1	1	0
1	2	0	0
2	0	0	0
2	1	2	1
2	2	2	0

To prove Postulate 4 independent, we use the four-element model {**0**, 1, 2, 3} with **0** designated and tables:

P	$\sim P$
0	1
1	**0**
2	3
3	**0**

P	Q	$P \vee Q$	$P \supset Q$
0	**0**	**0**	**0**
0	1	**0**	1
0	2	**0**	2
0	3	**0**	3
1	**0**	**0**	**0**
1	1	1	**0**
1	2	2	**0**
1	3	3	**0**
2	**0**	**0**	**0**
2	1	2	3
2	2	2	**0**
2	3	**0**	3
3	**0**	**0**	**0**
3	1	3	**0**
3	2	**0**	**0**
3	3	3	**0**

The deductive completeness of R.S. was proved in Chapter 7 by showing that all tautologies are derivable by its rule R 1 from its three axioms. One might think that, to prove the deductive completeness of H.A., it would suffice to derive the three axioms of R.S. as theorems of H.A., and the rule R 1 of R.S. as a derived rule of H.A. For would that not show all tautologies to be derivable, via those three theorems and one derived rule, from the four postulates of H.A. by its rule R′ 1? This question is not just rhetorical. In fact, the answer to it is negative. The trouble is that the systems H.A. and R.S. have different primitive symbols. Despite the fact that both R.S. and H.A. *are* deductively complete, not all tautologies can be derived by the R.S. rule R 1 from the three R.S. axioms *when these are regarded as constructed or formulated in the primitive base of H.A. rather than that of* R.S., that is, with $\sim$ and v as undefined symbols rather than $\sim$ and $\cdot$. This can be shown by proving the tautology $P \vee \sim P$ to be independent of the H.A. formulations of the three axioms of R.S.: $\sim[P \cdot \sim(P \cdot P)]$, $\sim[(P \cdot Q) \cdot \sim P]$, and $\sim\{\sim(P \cdot \sim Q) \cdot \sim\sim[\sim(Q \cdot R) \cdot \sim\sim(R \cdot P)]\}$, in the sense of not being derivable from them by the H.A. formulation of the R.S. Rule: $P, \sim(P \cdot \sim Q) \vdash_{\text{HA}} Q$.

To prove that a specified tautology cannot be proved in a given axiom system, one proceeds in the same way as in proving an axiom of a system to be independent of the other axioms. One regards the specified tautology as if it were an axiom to be proved independent, and attempts to show that it does not follow by the given system's rule from the (other) axioms of the system, just as described on pages 231 ff.

To prove independence (nonderivability) of the tautology $P \vee \sim P$, we use the six element model {**0**, **1**, **2**, **3**, **4**, **5**}, with the tables for ~ and v (and the derivative table for ·), given below.

P	Q	$P \vee Q$	$\sim P$	$\sim Q$	$\sim P \vee \sim Q$	$P \cdot Q$ $\sim(\sim P \vee \sim Q)$
0	0	0	5	5	5	0
0	1	0	5	5	5	0
0	2	3	5	4	5	0
0	3	3	5	1	0	5
0	4	0	5	0	0	5
0	5	0	5	0	0	5
1	0	0	5	5	5	0
1	1	0	5	5	5	0
1	2	3	5	4	5	0
1	3	3	5	1	0	5
1	4	0	5	0	0	5
1	5	0	5	0	0	5
2	0	3	4	5	5	0
2	1	3	4	5	5	0
2	2	3	4	4	5	0
2	3	3	4	1	0	5
2	4	3	4	0	0	5
2	5	3	4	0	0	5
3	0	3	1	5	0	5
3	1	3	1	5	0	5
3	2	3	1	4	0	5
3	3	3	1	1	0	5
3	4	3	1	0	0	5
3	5	3	1	0	0	5
4	0	0	0	5	0	5
4	1	0	0	5	0	5
4	2	3	0	4	0	5
4	3	3	0	1	0	5
4	4	5	0	0	0	5
4	5	5	0	0	0	5
5	0	0	0	5	0	5
5	1	0	0	5	0	5
5	2	3	0	4	0	5
5	3	3	0	1	0	5
5	4	5	0	0	0	5
5	5	5	0	0	0	5

In this model the three elements **0**, **1**, **2** are designated. The characteristic of taking only designated values is hereditary with respect to the rule: From *P* and

$\sim(P \cdot \sim Q)$ to infer Q: and the three H.A. formulations of the R.S. axioms take only designated values. But for the value 2 for P, we have $P \vee \sim P = 2 \vee \sim 2 = 2 \vee 4 = 3$, which is *not* a designated value.[2]

To prove the deductive completeness of H.A., we first establish somc theorems, derived rules, and metatheorems for it.

THEOREM 1. $\vdash_{\text{HA}} (Q \supset R) \supset [(P \supset Q) \supset (P \supset R)]$

Demonstration:	1. $(Q \supset R) \supset [(\sim P \vee Q) \supset (\sim P \vee R)]$	P 4
	2. $(Q \supset R) \supset [(P \supset Q) \supset (P \supset R)]$	df.

DR 1. $P \supset Q, Q \supset R \vdash_{\text{HA}} P \supset R$

Proof:	1. $(Q \supset R) \supset [(P \supset Q) \supset (P \supset R)]$	Th. 1
	2. $Q \supset R$	premiss
	3. $(P \supset Q) \supset (P \supset R)$	R′ 1
	4. $P \supset Q$	premiss
	5. $P \supset R$	R′ 1

THEOREM 2. $\vdash_{\text{HA}} P \supset (Q \vee P)$

Proof:	1. $P \supset (P \vee Q)$	P 2
	2. $(P \vee Q) \supset (Q \vee P)$	P 3
	3. $P \supset (Q \vee P)$	DR 1

DR 2. $Q \supset R \vdash_{\text{HA}} (P \vee Q) \supset (P \vee R)$

Demonstration:	1. $(Q \supset R) \supset [(P \vee Q) \supset (P \vee R)]$	P 4
	2. $Q \supset R$	premiss
	3. $(P \vee Q) \supset (P \vee R)$	R′ 1

THEOREM 3. $\vdash_{\text{HA}} P \supset P$

Proof:	1. $P \supset (P \vee P)$	P 2
	2. $(P \vee P) \supset P$	P 1
	3. $P \supset P$	DR 1

DR 3. $P \vee Q \vdash_{\text{HA}} Q \vee P$

Demonstration:	1. $(P \vee Q) \supset (Q \vee P)$	P 3
	2. $P \vee Q$	premiss
	3. $Q \vee P$	R′ 1

[2] See Henry Hiz, 'A Warning About Translating Axioms', *American Mathematical Monthly*, vol. 65 (1958), pp. 613 ff.; Thomas W. Scharle, 'Are Definitions Eliminable in Formal Systems?' (Abstract), *The Journal of Symbolic Logic*, vol. 35 (1970), pp. 182 ff.; and Alonzo Church, *Introduction to Mathematical Logic*, Princeton University Press, Princeton, N.J., 1956, pp. 125–128. On this matter I have benefited from correspondence with Professor Jean Porte and from discussion with Professor Anjan Shukla.

THEOREM 4. $\vdash_{HA} P \vee \sim P$

Proof: 1.	$P \supset P$	Th. 3
2.	$\sim P \vee P$	df.
3.	$P \vee \sim P$	DR 3

THEOREM 5. $\vdash_{HA} P \supset \sim\sim P$

Proof: 1.	$\sim P \vee \sim\sim P$	Th. 4
2.	$P \supset \sim\sim P$	df.

THEOREM 6. $\vdash_{HA} \sim\sim P \supset P$

Proof: 1.	$\sim P \supset \sim\sim\sim P$	Th. 5
2.	$(P \vee \sim P) \supset (P \vee \sim\sim\sim P)$	DR 2
3.	$P \vee \sim P$	Th. 4
4.	$P \vee \sim\sim\sim P$	R′ 1 (2, 3)
5.	$\sim\sim\sim P \vee P$	DR 3
6.	$\sim\sim P \supset P$	df.

THEOREM 7. $\vdash_{HA} [P \vee (Q \vee R)] \supset [Q \vee (P \vee R)]$

Proof: 1.	$R \supset (P \vee R)$	Th. 2
2.	$(Q \vee R) \supset [Q \vee (P \vee R)]$	DR 2
3.	$[P \vee (Q \vee R)] \supset \{P \vee [Q \vee (P \vee R)]\}$	DR 2
4.	$\{P \vee [Q \vee (P \vee R)]\} \supset \{[Q \vee (P \vee R)] \vee P\}$	P 3
5.	$[P \vee (Q \vee R)] \supset \{[Q \vee (P \vee R)] \vee P\}$	DR 1 (3, 4)
6.	$P \supset (P \vee R)$	P 2
7.	$(P \vee R) \supset [Q \vee (P \vee R)]$	Th. 2
8.	$P \supset [Q \vee (P \vee R)]$	DR 1 (6, 7)
9.	$\{[Q \vee (P \vee R)] \vee P\} \supset$ $\{[Q \vee (P \vee R)] \vee [Q \vee (P \vee R)]\}$	DR 2 (8)
10.	$\{[Q \vee (P \vee R)] \vee [Q \vee (P \vee R)]\} \supset$ $[Q \vee (P \vee R)]$	P 1
11.	$\{[Q \vee (P \vee R)] \vee P\} \supset [Q \vee (P \vee R)]$	DR 1 (9, 10)
12.	$[P \vee (Q \vee R)] \supset [Q \vee (P \vee R)]$	DR 1 (5, 11)

THEOREM 8. $\vdash_{HA} [P \vee (Q \vee R)] \supset [(P \vee Q) \vee R]$

Proof: 1.	$(Q \vee R) \supset (R \vee Q)$	P 3
2.	$[P \vee (Q \vee R)] \supset [P \vee (R \vee Q)]$	DR 2
3.	$[P \vee (R \vee Q)] \supset [R \vee (P \vee Q)]$	Th. 7
4.	$[P \vee (Q \vee R)] \supset [R \vee (P \vee Q)]$	DR 1 (2, 3)
5.	$[R \vee (P \vee Q)] \supset [(P \vee Q) \vee R]$	P 3
6.	$[P \vee (Q \vee R)] \supset [(P \vee Q) \vee R]$	DR 1 (4, 5)

THEOREM 9. $\vdash_{HA} [(P \vee Q) \vee R] \supset [P \vee (Q \vee R)]$

Proof:	1. $[(P \vee Q) \vee R] \supset [R \vee (P \vee Q)]$	P 3
	2. $[R \vee (P \vee Q)] \supset [P \vee (R \vee Q)]$	Th. 7
	3. $[(P \vee Q) \vee R] \supset [P \vee (R \vee Q)]$	DR 1
	4. $[P \vee (R \vee Q)] \supset [(P \vee R) \vee Q]$	Th. 8
	5. $[(P \vee Q) \vee R] \supset [(P \vee R) \vee Q]$	DR 1 (3, 4)
	6. $[(P \vee R) \vee Q] \supset [Q \vee (P \vee R)]$	P 3
	7. $[(P \vee Q) \vee R] \supset [Q \vee (P \vee R)]$	DR 1 (5, 6)
	8. $[Q \vee (P \vee R)] \supset [P \vee (Q \vee R)]$	Th. 7
	9. $[(P \vee Q) \vee R] \supset [P \vee (Q \vee R)]$	DR 1 (7, 8)

THEOREM 10. $\vdash_{HA} P \supset [Q \supset (PQ)]$

Proof:	1. $(\sim P \vee \sim Q) \vee \sim(\sim P \vee \sim Q)$	Th. 4
	2. $[(\sim P \vee \sim Q) \vee \sim(\sim P \vee \sim Q)] \supset \{\sim P \vee [\sim Q \vee \sim(\sim P \vee \sim Q)]\}$	Th. 9
	3. $\sim P \vee [\sim Q \vee \sim(\sim P \vee \sim Q)]$	R′ 1
	4. $P \supset [Q \supset (PQ)]$	df.

DR 4. $P, Q \vdash_{HA} PQ$

Proof:	1. $P \supset [Q \supset (PQ)]$	Th. 10
	2. P	premiss
	3. $Q \supset (PQ)$	R′ 1
	4. Q	premiss
	5. PQ	R′ 1

The next three theorems follow by DR 4 from Theorems 5 and 6, Theorems 8 and 9, and Theorem 3, respectively:

THEOREM 11. $\vdash_{HA} P \equiv \sim\sim P$

THEOREM 12. $\vdash_{HA} [P \vee (Q \vee R)] \equiv [(P \vee Q) \vee R]$

THEOREM 13. $\vdash_{HA} P \equiv P$

DR 5. $P \supset Q \vdash_{HA} \sim Q \supset \sim P$

Proof:	1. $P \supset Q$	premiss
	2. $Q \supset \sim\sim Q$	Th. 5
	3. $P \supset \sim\sim Q$	DR 1
	4. $\sim P \vee \sim\sim Q$	df.
	5. $\sim\sim Q \vee \sim P$	DR 3
	6. $\sim Q \supset \sim P$	df.

THEOREM 14. $\vdash_{HA} (PQ) \supset P$

Proof:	1. $\sim P \supset (\sim P \vee \sim Q)$	P2
	2. $\sim(\sim P \vee \sim Q) \supset \sim\sim P$	DR 5
	3. $\sim\sim P \supset P$	Th. 6
	4. $\sim(\sim P \vee \sim Q) \supset P$	DR 1
	5. $(PQ) \supset P$	df.

THEOREM 15. $\vdash_{HA} (PQ) \supset Q$

Proof:	1. $(\sim Q \vee \sim P) \supset (\sim P \vee \sim Q)$	P 3
	2. $\sim(\sim P \vee \sim Q) \supset \sim(\sim Q \vee \sim P)$	DR 5
	3. $(PQ) \supset (QP)$	df.
	4. $(QP) \supset Q$	Th. 14
	5. $(PQ) \supset Q$	DR 1

The next two derived rules follow from the two preceding theorems by R′ 1.

DR 6. $PQ \vdash_{HA} P$

DR 7. $PQ \vdash_{HA} Q$

DR 8. $P \equiv Q \vdash_{HA} \sim P \equiv \sim Q$

Proof:	1. $P \equiv Q$	premiss
	2. $(P \supset Q)(Q \supset P)$	df.
	3. $P \supset Q$	DR 6
	4. $\sim Q \supset \sim P$	DR 5
	5. $Q \supset P$	DR 7 (2)
	6. $\sim P \supset \sim Q$	DR 5
	7. $(\sim P \supset \sim Q)(\sim Q \supset \sim P)$	DR 4
	8. $\sim P \equiv \sim Q$	df.

DR 9. $P \equiv Q, Q \equiv R \vdash_{HA} P \equiv R$

Proof:	1. $P \equiv Q$	premiss
	2. $(P \supset Q)(Q \supset P)$	df.
	3. $P \supset Q$	DR 6
	4. $Q \supset P$	DR 7
	5. $Q \equiv R$	premiss
	6. $(Q \supset R)(R \supset Q)$	df.
	7. $Q \supset R$	DR 6
	8. $R \supset Q$	DR 7
	9. $P \supset R$	DR 1 (3, 7)
	10. $R \supset P$	DR 1 (8, 4)
	11. $(P \supset R)(R \supset P)$	DR 4
	12. $P \equiv R$	df.

THEOREM 16. $\vdash_{HA} P \equiv (P \vee P)$

Proof:	1. $P \supset (P \vee P)$	P 2
	2. $(P \vee P) \supset P$	P 1
	3. $[P \supset (P \vee P)][(P \vee P) \supset P]$	DR 4
	4. $P \equiv (P \vee P)$	df.

THEOREM 17. $\vdash_{HA} P \equiv (PP)$

Proof:	1. $\sim P \equiv (\sim P \vee \sim P)$	Th. 16
	2. $\sim\sim P \equiv \sim(\sim P \vee \sim P)$	DR 8
	3. $P \equiv \sim\sim P$	Th. 11
	4. $P \equiv \sim(\sim P \vee \sim P)$	DR 9
	5. $P \equiv (PP)$	df.

DR 10. $Q \supset R \vdash_{HA} (P \vee Q) \supset (R \vee P)$

Proof:	1. $Q \supset R$	premiss
	2. $(P \vee Q) \supset (P \vee R)$	DR 2
	3. $(P \vee R) \supset (R \vee P)$	P 3
	4. $(P \vee Q) \supset (R \vee P)$	DR 1

DR 11. $P \supset Q, R \supset S \vdash_{HA} (P \vee R) \supset (Q \vee S)$

Proof:	1. $R \supset S$	premiss
	2. $(P \vee R) \supset (S \vee P)$	DR 10
	3. $P \supset Q$	premiss
	4. $(S \vee P) \supset (Q \vee S)$	DR 10
	5. $(P \vee R) \supset (Q \vee S)$	DR 1 (2, 4)

DR 12. $P \equiv Q, R \equiv S \vdash_{HA} (P \vee R) \equiv (Q \vee S)$

Proof:	1. $P \equiv Q$	premiss
	2. $(P \supset Q)(Q \supset P)$	df.
	3. $P \supset Q$	DR 6
	4. $Q \supset P$	DR 7
	5. $R \equiv S$	premiss
	6. $(R \supset S)(S \supset R)$	df.
	7. $R \supset S$	DR 6
	8. $S \supset R$	DR 7
	9. $(P \vee R) \supset (Q \vee S)$	DR 11 (3, 7)
	10. $(Q \vee S) \supset (P \vee R)$	DR 11 (4, 8)
	11. $[(P \vee R) \supset (Q \vee S)][(Q \vee S) \supset (P \vee R)]$	DR 4
	12. $(P \vee R) \equiv (Q \vee S)$	df.

At this point it will be useful to prove the Rule of Replacement as our first metatheorem.

METATHEOREM I. (*Rule of Replacement*). *Let* $P_1, P_2, \ldots, P_n$ *be any wffs, let* Q *be any wff that does not occur in any* P_i*, and let* S *be any wff which contains no components other than* Q *and* P_i $(1 \leq i \leq n)$. *If* S* *is any wff which results from replacing any number of occurrences of* Q *in* S *by* R*, then* $Q \equiv R \vdash_{\text{HA}} S \equiv S^*$.

Proof: Strong induction on the number of symbols in S, counting each occurrence of v, $\sim$, Q, or any P_i as a single symbol.

(α) $n = 1$. Here S is either Q alone or a P_i alone.

CASE 1. S is Q and S* is R. It is obvious that $Q \equiv R \vdash_{\text{HA}} Q \equiv R$, which can be written $Q \equiv R \vdash_{\text{HA}} S \equiv S^*$.

CASE 2. S is Q and S* is Q also. Here $\vdash_{\text{HA}} Q \equiv Q$ (Th. 13), that is, $\vdash_{\text{HA}} S \equiv S^*$, whence $Q \equiv R \vdash_{\text{HA}} S \equiv S^*$.

CASE 3. S is a P_i. Here S* is P_i, also. Here $\vdash_{\text{HA}} P_i \equiv P_i$ (Th. 13), that is, $\vdash_{\text{HA}} S \equiv S^*$, whence $Q \equiv R \vdash_{\text{HA}} S \equiv S^*$.

(β) Here the Metatheorem is assumed to be true for any S containing fewer than n symbols. Now consider any S containing exactly n (>1) symbols. S is either $\sim S_1$ or $S_1 \text{ v } S_2$.

CASE 1. S is $\sim S_1$. Here S_1 contains fewer than n symbols, so by the β-case assumption, $Q \equiv R \vdash_{\text{HA}} S_1 \equiv S_1^*$. But $S_1 \equiv S_1^* \vdash_{\text{HA}} \sim S_1 \equiv \sim S_1^*$ by DR 8. It is obvious that $\sim(S_1^*) = (\sim S_1)^*$, so $\sim S_1^*$ is S*, whence $Q \equiv R \vdash_{\text{HA}} S \equiv S^*$.

CASE 2. S is $S_1 \text{ v } S_2$. Here S_1 and S_2 each contains fewer than n symbols, so by the β-case assumption, $Q \equiv R \vdash_{\text{HA}} S_1 \equiv S_1^*$ and $Q \equiv R \vdash_{\text{HA}} S_2 \equiv S_2^*$. Now by DR 12: $S_1 \equiv S_1^*, S_2 \equiv S_2^* \vdash_{\text{HA}} (S_1 \text{ v } S_2) \equiv (S_1^* \text{ v } S_2^*)$, so $Q \equiv R \vdash_{\text{HA}} (S_1 \text{ v } S_2) \equiv (S_1^* \text{ v } S_2^*)$. Since any $S_1^* \text{ v } S_2^*$ is an S*, we have $Q \equiv R \vdash_{\text{HA}} S \equiv S^*$.

Hence by strong induction we infer that regardless of the number of symbols in S, $Q \equiv R \vdash_{\text{HA}} S \equiv S^*$.

MT I, COROLLARY: *If* Q, R, S, *and* S* *are as in* Metatheorem I, *then* $Q \equiv R, S \vdash_{\text{HA}} S^*$.

The proof of this corollary is obvious.

A few more theorems and derived rules will move us closer to a proof of deductive completeness for H.A.

THEOREM 18. $\vdash_{HA} (P \vee Q) \equiv (Q \vee P)$

Proof:	1. $(P \vee Q) \supset (Q \vee P)$	P 3
	2. $(Q \vee P) \supset (P \vee Q)$	P 3
	3. $[(P \vee Q) \supset (Q \vee P)][(Q \vee P) \supset (P \vee Q)]$	DR 4
	4. $(P \vee Q) \equiv (Q \vee P)$	df.

DR 13. $P \supset Q, P \supset R \vdash_{HA} P \supset (QR)$

Proof:	1. $P \supset Q$	premiss
	2. $\sim Q \supset \sim P$	DR 5
	3. $P \supset R$	premiss
	4. $\sim R \supset \sim P$	DR 5
	5. $(\sim Q \vee \sim R) \supset (\sim P \vee \sim P)$	DR 11
	6. $\sim(\sim P \vee \sim P) \supset \sim(\sim Q \vee \sim R)$	DR 5
	7. $(PP) \supset (QR)$	df.
	8. $P \supset (QR)$	MT I, Cor., Th. 17

THEOREM 19. $\vdash_{HA} [P \vee (QR)] \supset [(P \vee Q)(P \vee R)]$

Proof:	1. $(QR) \supset Q$	Th. 14
	2. $[P \vee (QR)] \supset (P \vee Q)$	DR 2
	3. $(QR) \supset R$	Th. 15
	4. $[P \vee (QR)] \supset (P \vee R)$	DR 2
	5. $[P \vee (QR)] \supset [(P \vee Q)(P \vee R)]$	DR 13

DR 14. $P \supset (Q \supset R) \vdash_{HA} Q \supset (P \supset R)$

Proof:	1. $P \supset (Q \supset R)$	premiss
	2. $\sim P \vee (\sim Q \vee R)$	df.
	3. $[\sim P \vee (\sim Q \vee R)] \supset [\sim Q \vee (\sim P \vee R)]$	Th. 7
	4. $\sim Q \vee (\sim P \vee R)$	R′ 1
	5. $Q \supset (P \supset R)$	df.

DR 15. $P \supset (Q \supset R) \vdash_{HA} (PQ) \supset R$

Proof:	1. $P \supset (Q \supset R)$	premiss
	2. $\sim P \vee (\sim Q \vee R)$	df.
	3. $(\sim P \vee \sim Q) \vee R$	MT I, Cor., Th. 12
	4. $\sim\sim(\sim P \vee \sim Q) \vee R$	MT I, Cor., Th. 11
	5. $(PQ) \supset R$	df.

THEOREM 20. $\vdash_{\text{HA}} [(P \vee Q)(P \vee R)] \supset [P \vee (QR)]$

Proof:	1. $Q \supset [R \supset (QR)]$	Th. 10
	2. $[R \supset (QR)] \supset \{(P \vee R) \supset [P \vee (QR)]\}$	P 4
	3. $Q \supset \{(P \vee R) \supset [P \vee (QR)]\}$	DR 1
	4. $(P \vee R) \supset \{Q \supset [P \vee (QR)]\}$	DR 14
	5. $\{Q \supset [P \vee (QR)]\} \supset$ $\{(P \vee Q) \supset \{P \vee [P \vee (QR)]\}\}$	P 4
	6. $(P \vee R) \supset \{(P \vee Q) \supset \{P \vee [P \vee (QR)]\}\}$	DR 1 (4, 5)
	7. $(P \vee R) \supset \{(P \vee Q) \supset [(P \vee P) \vee (QR)]\}$	MT I, Cor., Th. 12
	8. $(P \vee R) \supset \{(P \vee Q) \supset [P \vee (QR)]\}$	MT I, Cor., Th. 16
	9. $(P \vee Q) \supset \{(P \vee R) \supset [P \vee (QR)]\}$	DR 14
	10. $[(P \vee Q)(P \vee R)] \supset [P \vee (QR)]$	DR 15

THEOREM 21. $\vdash_{\text{HA}} [P \vee (QR)] \equiv [(P \vee Q)(P \vee R)]$

Proof: Th. 21 follows from Th. 19 and Th. 20 by DR 4.

THEOREM 22. $\vdash_{\text{HA}} \sim(PQ) \equiv (\sim P \vee \sim Q)$

Proof:	1. $\sim(PQ) \equiv \sim(PQ)$	Th. 13
	2. $\sim(PQ) \equiv \sim\sim(\sim P \vee \sim Q)$	df.
	3. $\sim(PQ) \equiv (\sim P \vee \sim Q)$	MT I, Cor., Th. 11

THEOREM 23. $\vdash_{\text{HA}} \sim(P \vee Q) \equiv (\sim P \sim Q)$

Proof:	1. $\sim(P \vee Q) \equiv \sim(P \vee Q)$	Th. 13
	2. $\sim(P \vee Q) \equiv \sim(\sim\sim P \vee \sim\sim Q)$	MT I, Cor., Th. 11
	3. $\sim(P \vee Q) \equiv (\sim P \sim Q)$	df.

DR 16. $P \vdash_{\text{HA}} P \vee Q$

Demonstration:	1. $P \supset (P \vee Q)$	P 2
	2. P	premiss
	3. $P \vee Q$	R′ 1

At this point it is useful to state and prove

METATHEOREM II (*Generalized Association and Commutation of* v). *Let* $P_1, P_2, \ldots, P_n$ *be any wffs and let* Q *and* R *be any two wffs constructed out of them by means of* v. *If each* P_i $(1 \leq i \leq n)$ *occurs exactly once in each of the wffs* Q *and* R *then* $\vdash_{\text{HA}} Q \equiv R$.

Proof: Strong induction on the number of disjuncts P_i in Q and in R. (α) $n = 1$. Here Q and R are identically the same *wff* P_1, so $\vdash_{\text{HA}} Q \equiv R$ by Th. 13.

(β) Here the Metatheorem is assumed to be true for any $k < n$ disjuncts $P_1, P_2, \ldots, P_k$. Now consider Q and R each constructed out of exactly n (>1) disjuncts $P_1, P_2, \ldots, P_n$. Q is S v T and R is X v Y.

Each of the *wffs* S and T contains at least one of the *wffs* P_i ($1 \leq i \leq n$). We can assume that P_1 is a disjunct of S, because if not, we can use Th. 18 and MT I, Cor. to obtain $\vdash_{HA} Q \equiv (S \text{ v } T)$, where S now does contain P_1 as a disjunct.

Because T contains at least one of $P_2, P_3, \ldots, P_n$ as a disjunct, S contains fewer than n of the disjuncts P_i. Hence either S is P_1 and $\vdash_{HA} Q \equiv (P_1 \text{ v } T)$, or by the β-case assumption $\vdash_{HA} S \equiv (P_1 \text{ v } S')$, where S$'$ is a *wff* that contains all the disjuncts of S except P_1. In the latter case, by MT I, Cor. we have

$$\vdash_{HA} Q \equiv [(P_1 \text{ v } S') \text{ v } T]$$

and by Th. 12 and MT I, Cor.

$$\vdash_{HA} Q \equiv [P_1 \text{ v } (S' \text{ v } T)]$$

In either case there is a *wff*, call it T', such that

$$\vdash_{HA} Q \equiv (P_1 \text{ v } T')$$

By the same reasoning we can show that there is a *wff*, call it Y', such that

$$\vdash_{HA} R \equiv (P_1 \text{ v } Y')$$

Each of the *wffs* T' and Y' contains the $n - 1$ disjuncts $P_2, P_3, \ldots, P_n$, so by the β-case assumption

$$\vdash_{HA} T' \equiv Y'.$$

By Th. 13 we have $\vdash_{HA} P_1 \equiv P_1$, so by DR 12 we have

$$\vdash_{HA} (P_1 \text{ v } T') \equiv (P_1 \text{ v } Y')$$

which by MT I, Cor. gives

$$\vdash_{HA} Q \equiv R.$$

Metatheorem II now follows by strong induction.

To prove the deductive completeness of H.A., we use a somewhat different method than that used in Chapter 7 to prove the deductive completeness of R.S. Here we make use of the notion of a Conjunctive Normal Form, as discussed in Appendix B. Our representation of a *wff* is in Conjunctive Normal Form (abbreviated as C.N.F.) if and only if: (1) it contains only propositional symbols, parentheses, and the symbols v, $\sim$, and $\cdot$; (2) the negation signs apply only to propositional symbols; and (3) no disjunct is a conjunction, that is, the symbol v is nowhere adjacent to a conjunction. In symbols, the *wff* S is in C.N.F. if and only if S is

$$S_1 \cdot S_2 \cdot \ldots \cdot S_s$$

(associated in any way at all), where each S_i is

$$T_{i_1} \vee T_{i_2} \vee \ldots \vee T_{i_n}$$

(associated in any way at all), where each T_i is either a propositional symbol or the negation of a propositional symbol.

Next we state and prove

METATHEOREM III. *Given any representation of a wff S there is a Conjunctive Normal Form formula* S_{CNF} *such that* $\vdash_{HA} S \equiv S_{CNF}$.

Proof: If S is itself in C.N.F., then $\vdash_{HA} S \equiv S_{CNF}$ by Th. 13. If S fails to be in C.N.F., it can only be through failing to satisfy conditions 1, 2, or 3 in the preceding paragraph. If S fails to satisfy condition 1, it contains defined symbols $\equiv$ or $\supset$. Here we replace every well formed part of S of the form $S_1 \equiv S_2$ by $(\sim S_1 \vee S_2)(\sim S_2 \vee S_1)$ and every well formed part of S of the form $S_1 \supset S_2$ by $\sim S_1 \vee S_2$. The result of these replacements is S′, where $\vdash_{HA} S \equiv S'$ (by df.) and S′ satisfies condition 1.

If S (or S′) fails to satisfy condition 2, we replace every w.f. part of the form $\sim(S_1 S_2)$ by $\sim S_1 \vee \sim S_2$ and every w.f. part of the form $\sim(S_1 \vee S_2)$ by $\sim S_1 \sim S_2$. After all such replacements have been made, we replace every w.f. part of the form $\sim\sim S_1$ by S_1. The result of these replacements is S″, where $\vdash_{HA} S' \equiv S''$ and hence $\vdash_{HA} S \equiv S''$, by MT I and Theorems 22, 23, and 11, and S″ satisfies conditions 1 and 2.

If S (or S′ or S″) fails to satisfy condition 3, it can only be because it contains w.f. parts of either the form $S_1 \vee (S_2 S_3)$ or the form $(S_2 S_3) \vee S_1$. Now we replace each w.f. part. $S_1 \vee (S_2 S_3)$ by $(S_1 \vee S_2)(S_1 \vee S_3)$, and each w.f. part $(S_2 S_3) \vee S_1$ first by $S_1 \vee (S_2 S_3)$ and then by $(S_1 \vee S_2)(S_1 \vee S_3)$. The result of these replacements satisfies all the conditions 1, 2, and 3 and is therefore an S_{CNF}. Moreover, $\vdash_{HA} S \equiv S_{CNF}$ by MT I and Theorems 18 and 19.

Hence for any *wff* S there is a *wff* S_{CNF} in C.N.F. such that $\vdash_{HA} S \equiv S_{CNF}$.

All H.A. equivalences are truth preserving, so if S is a tautology, then any S_{CNF} such that $\vdash_{HA} S \equiv S_{CNF}$ is a tautology also.

It is clear that if a *wff* S is a disjunction of *wffs* $T_1, T_2, \ldots, T_n$ where each T_j is either a propositional symbol or the negation of a propositional symbol, then if S is a tautology there must be a propositional symbol *P* such that both *P* and $\sim P$ are disjuncts in S. For if not, then suitable truth value assignments (*false* to each T_j which is a propositional symbol and *true* to each propositional symbol whose negation is a T_j) will make every disjunct of S false and hence S itself false. This is contrary to the assumption that S is a tautology and therefore true for any assignment of truth values to its component propositional symbols.

With these remarks in mind, we can state and prove

METATHEOREM IV. *If a wff in* C.N.F. S_{CNF} *is a tautology, then* $\vdash_{HA} S_{CNF}$.

Proof: Let S_{CNF} be a *wff* in C.N.F., which is a tautology. Then S_{CNF} must be a conjunction $S_1 S_2 \ldots S_s$ each of whose conjuncts S_i $(1 \leq i \leq s)$ is a disjunction of *wffs* $T_1, T_2, \ldots, T_{n_i}$ each of which is either a propositional symbol or the negation of a propositional symbol. Since S_{CNF} is a tautology, each of its conjuncts S_i is tautologous, because a conjunction is true if and only if all of its conjuncts are true. We have already remarked that a disjunction S_i of propositional symbols and negations of propositional symbols can be a tautology only if there is a propositional symbol P such that both P and $\sim P$ are disjuncts of S_i. But $\vdash_{HA} P \vee \sim P$ by Theorem 4, from which we derive $\vdash_{HA} (P \vee \sim P) \vee Q$ by DR 16, where in this case Q is a disjunction of all of the disjuncts of S_i other than P and $\sim P$. But by MT II, $\vdash_{HA} S_i \equiv [(P \vee \sim P) \vee Q]$, whence $\vdash_{HA} S_i$ by MT I, Cor. Hence every conjunct S_i of S_{CNF} is a theorem, that is: $\vdash_{HA} S_1$, $\vdash_{HA} S_2$, . . . , $\vdash_{HA} S_s$. Now by $s - 1$ uses of DR 4 we obtain $\vdash_{HA} S_1 S_2 \ldots S_s$ which is $\vdash_{HA} S_{CNF}$. This completes our proof of Metatheorem IV.

Now the deductive completeness of H.A. follows.

METATHEOREM V. *The system* H.A. *is deductively complete.*

Proof: By the functional completeness of H.A., any tautology is expressible in H.A., say by the *wff* S. By MT III there is a C.N.F. formula S_{CNF} such that $\vdash_{HA} S \equiv S_{CNF}$. The *wff* S_{CNF} is also tautologous, whence by MT IV, $\vdash_{HA} S_{CNF}$, and by MT I, Cor., $\vdash_{HA}$ S. So H.A. is deductively complete.

EXERCISES

Prove that each of the propositional calculi described in the exercises on page 237 is a model system of logic.

9.3 The Use of Dots as Brackets

It has been remarked that the language of symbolic logic requires punctuation if ambiguity is to be avoided. It shares this characteristic with natural languages as well as with other artificial languages like (ordinary) algebra. We have been using three kinds of punctuation marks in our logical language: parentheses, brackets, and braces. In the following discussion, for convenience, we shall use the word 'brackets' to refer to all of these.

Even moderately complicated formulas require many pairs of brackets, which make them difficult to read. Any use of paired punctuation marks involves redundancy. In the formula

$$(P \supset Q) \vee (P \supset \sim Q)$$

some punctuation is necessary to avoid ambiguity. Brackets are essential, but not *paired* brackets, for the outermost parentheses can be dropped without incurring any ambiguity, leaving

$$P \supset Q) \vee (P \supset \sim Q$$

The same punctuation effect can be obtained by replacing the remaining parentheses by dots, which would give

$$P \supset Q . \vee . P \supset \sim Q$$

There is no danger of confusing the punctuation dot with the conjunction dot because the punctuation dot can occur only adjacently to a connective symbol such as 'v', '⊃', or '≡', whereas the conjunction dot can never do so.

Dots are symmetrical in shape, in contrast to the asymmetry of brackets. Thus '(', '[', '{' are all concave on the right, which indicates that they group or 'operate toward' the right, while ')', ']', '}' are concave on and 'operate toward' the left. The symmetry of the dot notation is compensated for by introducing the convention that punctuation dots always operate *away* from the connective symbol to which they are adjacent.

There is another analogy between bracketing logical formulas and punctuating sentences of natural languages. The latter have punctuation marks of different degrees of strength, the stronger of which 'take precedence over' or 'extend over' the weaker. For example, a period is stronger than a semicolon, and a semicolon is stronger than a comma. Of the three kinds of brackets used in logical formulas, the unstated convention has been to regard braces as stronger than brackets, and brackets as stronger than parentheses. We have used parentheses to group symbols *within* brackets, but not conversely, and we have used brackets to group symbols *within* braces, but not conversely.

When this convention is adhered to, it permits additional redundant brackets to be dropped. Thus the formula

$$(P \supset Q) \supset [(R \vee P) \supset (R \vee Q)]$$

remains unambiguous when rewritten as

$$P \supset Q) \supset [R \vee P) \supset (R \vee Q$$

If we use a single dot in place of a parenthesis, and a double dot in place of a bracket, the preceding formula can be rewritten as

$$P \supset Q . \supset : R \vee P . \supset . R \vee Q$$

where the convention is that two dots bind more strongly or have greater scope than one dot. Using three dots as a punctuation mark that has greater scope than either one or two dots, the formula

$$[P \vee (Q \vee R)] \supset \{P \vee [Q \vee (P \vee R)]\}$$

can be rewritten as

$$P \vee . Q \vee R : \supset :. P \vee : Q \vee . P \vee R$$

In writing some formulas, we were forced to include one pair of braces within another, because we had only three kinds of brackets available. Now the use of dots permits the generation of as many different punctuation marks of different degrees of strength as may be desired, by the simple expedient of adding dots one at a time. The general convention here is that the scope of a group of *n* dots extends over that of any number of groups of less than *n* dots, and that the scope of a group of *n* dots extends to but not beyond the nearest group of *n* or more dots. The formula

$$\{(Q \vee R) \supset [Q \vee (P \vee R)]\} \supset \{[P \vee (Q \vee R)] \supset \{P \vee [Q \vee (P \vee R)]\}\}$$

can be written as

$$Q \vee R . \supset : Q \vee . P \vee R :. \supset :: P \vee . Q \vee R : \supset :. P \vee : Q \vee . P \vee R$$

For the sake of symmetry and greater ease of reading, one frequently adds dots which are not strictly necessary for the avoidance of ambiguity. Thus the formula

$$P \supset . Q \supset P$$

is frequently written

$$P . \supset . Q \supset P$$

The preceding lengthy formula may be more easily read when rewritten according to this convention as

$$Q \vee R : \supset : Q . \vee . P \vee R :: \supset :: P . \vee . Q \vee R :. \supset :. P : \vee : Q . \vee . P \vee R$$

Now what should be done in case two bracketed expressions are connected by the conjunction symbol, as in the formula

$$(P \supset Q) \cdot (Q \supset P)$$

Applying the technique already described, we should emerge with the awkward expression

$$P \supset Q \therefore Q \supset P$$

To avoid this awkwardness, it is customary to let the conjunction dot itself do the bracketing, writing the formula simply as

$$P \supset Q . Q \supset P$$

Here the single dot is to be thought of as operating in *both* directions, left and right. This convention is satisfactory, as can be seen by observing that the different formulas

1. $P \supset [(Q . Q) \supset P]$
2. $[P \supset (Q . Q)] \supset P$
3. $[(P \supset Q) . Q] \supset P$
4. $P \supset [Q . (Q \supset P)]$

can be written in distinguishable and unambiguous fashion as

1′. $P \supset :. Q . Q : \supset P$
2′. $P \supset : Q . Q :. \supset P$
3′. $P \supset Q . Q : \supset P$
4′. $P \supset : Q . Q \supset P$

The use of dots as brackets has the advantage of economy, and the further advantage of providing an infinite number of different punctuation marks of different degrees of strength, where the strength or scope of each may be determined by the simple expedient of counting its constituent dots.[3]

9.4 A Parenthesis-Free Notation

A logical notation that dispenses with brackets entirely has been devised by the Polish logician, J. Łukasiewicz, and has been extensively used by logicians of the Polish School. Corresponding to the four most commonly used operator symbols

$$\sim \quad \supset \quad \cdot \quad \vee$$

[3] A more extended discussion of this topic will be found in C. I. Lewis and C. H. Langford, *Symbolic Logic*, The Century Co., Appendix I, New York, 1932, pp. 486–489. For more technical discussions, the reader is referred to H. B. Curry, 'On the Use of Dots as Brackets in Logical Expressions', *The Journal of Symbolic Logic*, vol. 2 (1937), pp. 26–28 and A. M. Turing, 'The Use of Dots as Brackets in Church's System', ibid., vol. 7 (1942), pp. 146–156.

are the four symbols

$$N\ C\ K\ A$$

In writing their formulas, they use lower-case letters '*p*', '*q*', '*r*', '*s*', . . . instead of the capital letters '*P*', '*Q*', '*R*', '*S*', . . . Instead of the connective symbols '*C*', '*K*', '*A*' being written *between* the formulas they connect, they are placed directly to the left of the two formulas to be joined. Thus

$$\sim P \text{ is written } Np$$
$$P \supset Q \text{ is written } Cpq$$
$$P \cdot Q \text{ is written } Kpq$$
$$P \vee Q \text{ is written } Apq$$

That this notation is unambiguous can be seen by comparing the following formulas:

$$P \supset (Q \supset R) \text{ is written } CpCqr$$
$$(P \supset Q) \supset R \text{ is written } CCpqr$$
$$P \supset (Q \cdot R) \text{ is written } CpKqr$$
$$(P \supset Q) \cdot R \text{ is written } KCpqr$$
$$(P \vee Q) \supset (R \cdot S) \text{ is written } CApqKrs$$

The axioms of R.S. translate into the Polish notation as

Ax. 1. *CpKpp*
Ax. 2. *CKpqp*
Ax. 3. *CCpqCNKqrNKrp*

And the three postulates of Łukasiewicz's own system are written in his notation as

P 1. *CpCqp*
P 2. *CCpCqrCCpqCpr*
P 3. *CCNqNpCpq*

The Polish notation has the obvious advantage of dispensing with all special punctuation marks, for the *order* in which its symbols are written suffices to make any formula unambiguous.

EXERCISES

*1. Translate the postulates of H.A. into the Polish notation.
2. Translate the postulates of F.S. into the Polish notation.
3. Translate the postulates of P_N into the Polish notation.

9.5 The Stroke and Dagger Operators

Any of the following pairs of operators provide a functionally complete logic: $\sim$ and $\cdot$, $\sim$ and v, $\sim$ and $\supset$, or $\supset$ and $+$. We can construct a functionally complete system of logic containing just a single operator, and we can do it in either of two ways.

The first way is to adopt as the single primitive operator the so-called 'stroke function'. This operator symbol, called 'alternative denial' by Quine,[4] operates on or connects two formulas and is written '$P \mid Q$'. Its standard interpretation is to deny that both of the formulas P and Q are true, which is the same as affirming that at least one is false. It is defined by the truth table

P	Q	$P \mid Q$
T	T	F
T	F	T
F	T	T
F	F	T

The other operators, $\sim$, $\cdot$, v, and $\supset$ can all be defined in terms of the stroke function. That the following definitions preserve the standard interpretations of the symbols being defined is easily verified by the method of truth tables:

$$\begin{aligned} \sim P &= \text{df} \quad P \mid P \\ P.Q &= \text{df} \quad P \mid Q. \mid .P \mid Q \\ P \text{ v } Q &= \text{df} \quad P \mid P. \mid .Q \mid Q \\ P \supset Q &= \text{df} \quad P. \mid .Q \mid Q \end{aligned}$$

The other operator that suffices for a functionally complete logic is that of 'joint denial', symbolized by a dagger with its point down. Written '$P{\downarrow}Q$', its standard interpretation is to deny that either of the formulas P or Q is true, which is the same as affirming that they are both false. It is defined by the truth table

P	Q	$P{\downarrow}Q$
T	T	F
T	F	F
F	T	F
F	F	T

[4] See W. V. O. Quine, *Mathematical Logic*, Harvard University Press, Cambridge, Mass., 1947, pp. 48–49. The stroke function is frequently referred to as the 'Sheffer stroke function' after Professor H. M. Sheffer, although it was first used by C. S. Peirce.

The other operators can all be defined in terms of the dagger function alone. That the following definitions preserve the standard interpretations of the symbols that are being defined is easily verified by the method of truth tables:

$$\begin{aligned} \sim P &= \text{df} \quad P{\downarrow}P \\ P.Q &= \text{df} \quad P{\downarrow}P.{\downarrow}.Q{\downarrow}Q \\ P \vee Q &= \text{df} \quad P{\downarrow}Q.{\downarrow}.P{\downarrow}Q \\ P \supset Q &= \text{df} \quad P{\downarrow}P.{\downarrow}.Q{:}{\downarrow}{:}P{\downarrow}P.{\downarrow}.Q \end{aligned}$$

Thus we see that functionally complete systems of logic based on a single operator can be constructed in terms either of the stroke or of the dagger. It is interesting to note the parallel between the definition of the stroke function in terms of the dagger and the definition of the dagger function in terms of the stroke. These are

$$\begin{aligned} P{\downarrow}Q &= \text{df} \quad P|P.|.Q|Q{:}|{:}P|P.|.Q|Q \\ P|Q &= \text{df} \quad P{\downarrow}P.{\downarrow}.Q{\downarrow}Q{:}{\downarrow}{:}P{\downarrow}P.{\downarrow}.Q{\downarrow}Q \end{aligned}$$

as may easily be verified.

EXERCISES

1. Express Ax. 1 of R.S. in terms of the stroke function.
*2. Express Ax. 1 of R.S. in terms of the dagger function.
3. Express P 1 of H.A. in terms of the dagger function.
4. Express Ax. 3 of R.S. in terms of the stroke function.
5. Express Ax. 1 of L.S. in terms of the dagger function.
*6. Express Ax. 2 of R.S. in terms of the stroke function.
7. Express Ax. 5 of F.S. in terms of the stroke function.
*8. Express Ax. 5 of F.S. in terms of the dagger function.
9. Express Ax. 1 of F.S. in terms of the stroke function.
10. Express P 2 of H.A. in terms of the dagger function.

9.6 The Nicod System

Thus far in the text and exercises, a number of alternative Model Systems of Logic have been set forth. Each of them is based on two primitive operator symbols, and the number of their axioms or postulates ranges from three (for R.S.) to five (for F.S.). A system that is more economical both in primitive operators and postulates is attributable to J. G. P. Nicod.[5] The Nicod System can be set forth as follows.

[5] J. G. P. Nicod, 'A Reduction in the Number of the Primitive Propositions of Logic', *Proceedings of the Cambridge Philosophical Society*, vol. 19 (1916), pp. 32–40.

See also J. Jørgensen, *A Treatise of Formal Logic*, vol. 2, Russell and Russell Publishers, New York, 1962, pp. 149–172.

The primitive symbols are infinitely many propositional symbols P, Q, R, S, T, with and without subscripts, parentheses (or dots), and the single operator symbol '$|$'. The recursive rule for well formed formulas in N may be stated as:

1. Any single letter of N is a *wff*.
2. If P and Q are *wffs*, then $(P)|(Q)$ is a *wff*.

(No formula of N will be regarded as being a *wff* unless its being so follows from this definition.)

It should be noted that even the recursive rule for well formed formulas is simpler in the Nicod System, since it needs only two rather than three clauses. We have here the first fruit of using only one operator symbol.

The single axiom required (or rather, the single pattern for infinitely many axioms of the object logic) is stated in our metalanguage, in which we use dots for brackets, as

Ax. $P.|.Q|R:|::T.|.T|T:.|:.S|Q:|:P|S.|.P|S$

The single rule of inference that is required may be stated as

Rule. From P and $P.|.R|Q$ to infer Q.

The definitions of 'valid argument in N' and of 'theorem of N', that is, of

$$P_1, P_2, \ldots, P_n \vdash_{\overline{\mathrm{N}}} Q \quad \text{and} \quad \vdash_{\overline{\mathrm{N}}} P$$

are strictly analogous to those given for '$\vdash_{\overline{\mathrm{RS}}}$' and '$\vdash_{\overline{\mathrm{HA}}}$'.

Although more economical in the respects indicated, the Nicod System can scarcely be said to be *simpler* than the systems already mentioned. There is only one axiom for N, but it is more complicated than any axiom or postulate of any of the other systems. Not only is it longer, but it involves five distinct propositional symbols, 'P', 'Q', 'R', 'S', and 'T', whereas the entire set of axioms for any of the other systems can be stated in terms of just three distinct propositional symbols. The situation is similar with respect to the rules of inference of the several systems. The Nicod Rule is not *Modus Ponens*, but a more powerful

---

W. V. Quine, 'A Note on Nicod's Postulate', *Mind*, n.s. vol. 41 (1932), pp. 345–350.

B. A. Bernstein, 'Remark on Nicod's Reduction of *Principia Mathematica*', *The Journal of Symbolic Logic*, vol. 2 (1937), pp. 165–166.

Thomas W. Scharle, 'Axiomatization of Propositional Calculus with Sheffer Functors', *Notre Dame Journal of Formal Logic*, vol. 6 (1965), pp. 209–217.

Several alternative, one-axiom systems are listed in A. N. Prior, *Formal Logic*, Oxford University Press, England, 1962, p. 306. Two in particular are noteworthy because they use only four propositional variables, in contrast to Nicod's five. They are also different from Nicod's axiom in that they contain no proper part that is itself a theorem. These systems are attributable to Wajsberg (1931) and to Łukasiewicz (1931). I am indebted to Professor P. T. Geach of the University of Leeds for calling them to my attention.

instrument for deduction. *Modus Ponens,* which may be stated as

From P and $P.|.Q|Q$ to infer Q

is simply a special case of the Nicod Rule, in which 'R' and 'Q' denote identically the same well formed formula of the object language. Nicod's Rule as well as his axiom is more complicated than those required in less economical systems of logic.

That the Nicod System is functionally complete has already been indicated in our discussion of the stroke function in Section 9.5. The analyticity of N is simply enough shown by establishing via truth tables that the Axiom is tautologous, and that the Rule can lead only to tautologies from tautologies. The deductive completeness of N remains to be proved to complete the proof that the Nicod System is a Model Logic. When this is attempted, we find that for all the economy of its single axiom and the greater deductive strength of its Rule, it is very difficult indeed to derive theorems in the Nicod System. Nevertheless, it is interesting from the point of view of seeing how far one can go in the direction of reducing the number of postulates and still have a deductively complete system of logic.

We shall develop seventeen theorems of the Nicod System, and one derived rule. The last four theorems will be the four axioms of the Hilbert-Ackermann System, and the derived rule will be the Hilbert-Ackermann System's rule R′ 1. Of course, these must be stated in their unabbreviated form, and then expressed in the Nicod System's notation. The Hilbert-Ackermann axioms, expressed in terms of that system's primitive symbols $\sim$ and v are

1. $\sim(P \vee P) \vee P$
2. $\sim P \vee (P \vee Q)$
3. $\sim(P \vee Q) \vee (Q \vee P)$
4. $\sim(\sim P \vee Q) \vee [\sim(R \vee P) \vee (R \vee Q)]$

The Hilbert-Ackermann rule, similarly unabbreviated, is

From P and $\sim P \vee Q$ to infer Q

These, expressed in the Nicod System's notation, in which '$\sim P$' is written '$P|P$', and '$P \vee Q$' is written '$P|P.|.Q|Q$', are stated as:

1. $P|P.|.P|P:|:P|P.|.P|P:.|:.P|P.|.P|P:|:P|P.|.P|P::|::P|P.$
2. $P|P.|.P|P:.|:.P|P.|.Q|Q:|:P|P.|.Q|Q.$
3. $P|P.|.Q|Q:|:P|P.|.Q|Q:.|:.P|P.|.Q|Q:|:P|P.|.Q|Q::|::Q|Q.|.P|P:|:Q|Q.|.P|P.$
4. $P|P.|.P|P:|:Q|Q:.|:.P|P.|.P|P:|:Q|Q::|::P|P.|.P|P:|:Q|Q:.|:.P|P.|.P|P:|:Q|Q:::|:::R|R.|.P|P:|:R|R.|.P|P:.|:.R|R.|.P|P:|:R|R.|.P|P::|::R|R.|.Q|Q:|:R|R.|.Q::.|::.R|R.|.P|P:|:R|R.|.P|P:.|:.R|R.|.P|P:|:R|R.|.P|P::|::R|R.|.Q|Q:|:R|R.|.Q|Q.$

R′ 1. From P and $P|P.|.P|P:|:Q|Q$ to infer Q.

Our development of the Nicod System will be entirely in terms of the stroke function. Since the formulas of the proofs are of almost intolerable length, they will be *described* rather than written out. Our descriptions, however, will be sufficiently complete to permit the reader to write them out for himself.

THEOREM 1. $\vdash_N$ Q:. | :.T. | .T | T: | :S::. | ::.S: | :T. | .T | T:. | :.Q:: | ::S: | :T. | .T | T:. | :.Q.

Proof: Line 1 is the Nicod Axiom with T in place of P, of Q, and of R. Line 2 is the Nicod Axiom with T. | .T | T in place of P and of Q, and with S | T: | :T | S. | .T | S in place of R. Line 3 is the result of applying the Nicod Rule to lines 1 and 2. Line 4 is the Nicod Axiom with S: | :T. | .T | T in place of P, with T. | .T | T: | :S in place of Q and of R, and with Q in place of S. Line 5 is the result of applying the Nicod Rule to lines 3 and 4.

THEOREM 2. $\vdash_N$ T. | .T | T

Proof: Line 1 is Theorem 1 with T. | .T | T in place of Q, and with S | T: | :T | S. | .T | S in place of S. Line 2 is the Nicod Axiom with T. | .T | T:: | ::T. | .T | T:. | :.S | T: | :T | S. | .T | S in place of P, with S | T: | :T | S. | .T | S:. | :.T. | .T | T:: | ::T. | .T | T in place of Q and of R, and with S | T::: | :::S | T: | :T | S. | .T | S:. | :.T. | .T | T:: | ::S::. | ::.S | T: | :T | S. | .T | S:. | :.T. | .T | T:: | ::S:::. | :::.T. | .T | T in place of S. Line 3 is the result of applying the Nicod Rule to lines 1 and 2. Line 4 is the Nicod Axiom with S | T: | :T | S. | .T | S:. | :.T. | .T | T in place of P, with T in place of Q, and with T | T in place of R. Line 5 is Theorem 1 with S | T: | :T | S. | .T | S:. | :.T. | .T | T:: | ::T. | .T | T in place of Q and with S | T::: | :::S | T: | :T | S. | .T | S:. | :.T. | .T | T:: | ::S::. | ::.S | T: | :T | S. | .T | S:. | :.T. | .T | T:: | ::S in place of S. Line 6 is the result of applying the Nicod Rule to lines 4 and 5. Line 7 is the result of applying the Nicod Rule to lines 3 and 6. Line 8 is the Nicod Axiom with T in place of P, of Q, and of R. Line 9 is the result of applying the Nicod Rule to lines 7 and 8.

THEOREM 3. $\vdash_N$ S | P: | :P | S. | .P | S

Proof: Line 1 is the Nicod Axiom with P in place of Q and of R. Line 2 is Theorem 2 with P in place of T. Line 3 is the result of applying the Nicod Rule to lines 1 and 2.

THEOREM 4. $\vdash_N$ P | P. | .P

Proof: Line 1 is Theorem 3 with P | P in place of P and with P in place of S. Line 2 is Theorem 2 with P in place of T. Line 3 is the result of applying the Nicod Rule to lines 1 and 2.

THEOREM 5. $\vdash_N$ P | P: | :S | P. | .S | P

Proof: Line 1 is Theorem 3 with P | S in place of S and with S | P. | .S | P in place of P. Line 2 is Theorem 3 with P in place of S and S in place of P. Line 3 is the result of applying the Nicod Rule to lines 1 and 2. Line 4 is the Nicod

Axiom with S|P.|.S|P in place of P, with P in place of Q, with S in place of R, and with P|P in place of S. Line 5 is the result of applying the Nicod Rule to lines 3 and 4. Line 6 is Theorem 4. Line 7 is the result of applying the Nicod Rule to lines 5 and 6. Line 8 is Theorem 3 with S|P.|.S|P in place of S, and with P|P in place of P. Line 9 is the result of applying the Nicod Rule to lines 7 and 8.

THEOREM 6. $\vdash_N$ P.|.Q|R::|::S|Q:|:P|S.|.P|S:.|:.S|Q:|:P|S.|.P|S

Proof: Line 1 is the Nicod Axiom with S|Q:|:P|S.|.P|S:.|:.S|Q:|:P|S.|.P|S in place of P, with T.|.T|T:.|:.S|Q:|:P|S.|.P|S in place of Q and of R, and with P.|.Q|R in place of S. Line 2 is Theorem 5 with S|Q:|:P|S.|.P|S in place of P, and with T.|.T|T in place of S. Line 3 is the result of applying the Nicod Rule to lines 1 and 2. Line 4 is the Nicod Axiom. Line 5 is the result of applying the Nicod Rule to lines 3 and 4. Line 6 is Theorem 3 with S|Q:|:P|S.|.P|S:.|:.S|Q:|:P|S.|.P|S in place of S, and with P.|.Q|R in place of P. Line 7 is the result of applying the Nicod Rule to lines 5 and 6.

THEOREM 7. $\vdash_N$ Q|S.|.U:|:Q|S.|.U:.|:.S|Q.|.U

Proof: Line 1 is Theorem 6 with Q|S in place of P, with S|Q in place of Q and of R, and with U in place of S. Line 2 is Theorem 3 with Q in place of S and with S in place of P. Line 3 is the result of applying the Nicod Rule to lines 1 and 2. Line 4 is Theorem 3 with U.|.S|Q in place of S and with Q|S.|.U:|:Q|S.|.U in place of P. Line 5 is the result of applying the Nicod Rule to lines 3 and 4. Line 6 is Theorem 6 with S|Q.|.U in place of P, with U.|.S|Q in place of Q and of R, and with Q|S.|.U:|:Q|S.|.U in place of S. Line 7 is Theorem 3 with S|Q in place of S and with U in place of P. Line 8 is the result of applying the Nicod Rule to lines 6 and 7. Line 9 is the result of applying the Nicod Rule to lines 5 and 8. Line 10 is Theorem 3 with S|Q.|.U in place of S, and with Q|S.|.U:|:Q|S.|.U in place of P. Line 11 is the result of applying the Nicod Rule to lines 9 and 10.

THEOREM 8. $\vdash_N$ P.|.Q|R::|::Q|S:|:P|S.|.P|S:.|:.Q|S:|:P|S.|.P|S

Proof: Line 1 is Theorem 6 with P.|.Q|R in place of P, with S|Q:|:P|S.|.P|S in place of Q and of R, and with Q|S:|:P|S.|.P|S:.|:.Q|S:|:P|S.|.P|S in place of S. Line 2 is Theorem 6. Line 3 is the result of applying the Nicod Rule to lines 1 and 2. Line 4 is Theorem 7 with P|S.|.P|S in place of U. Line 5 is the result of applying the Nicod Rule to lines 3 and 4.

THEOREM 9. $\vdash_N$ S:.|:.P.|.S|S:|:P.|.S|S

Proof: Line 1 is Theorem 3 with S|S.|.P in place of S, and with P.|.S|S:|:P.|.S|S in place of P. Line 2 is Theorem 1 with S|S in place of S. Line 3 is the result of applying the Nicod Rule to lines 1 and 2. Line 4 is the Nicod Axiom with P.|.S|S:|:P.|.S|S in place of P, with S|S in place of Q, and with P in place of *R*. Line 5 is the result of applying the Nicod Rule to lines 3 and 4. Line 6 is Theorem 2 with S in place of *T*. Line 7 is the result of applying the Nicod Rule to lines 5 and 6. Line 8 is Theorem 3 with *P*.|.*S*|S:|:*P*.|.*S*|S in place of S, and with S in place of *P*. Line 9 is the result of applying the Nicod Rule to lines 7 and 8.

THEOREM 10. $\vdash_N$ *Q*|*Q*:|:*Q*|S.|.*Q*|S

Proof: Line 1 is Theorem 8 with *Q*|*Q* in place of *P*, with S|*Q* in place of *Q* and of *R*, and with *Q*|S.|.*Q*|S in place of S. Line 2 is Theorem 5 with *Q* in place of *P*. Line 3 is the result of applying the Nicod Rule to lines 1 and 2. Line 4 is Theorem 3 with *Q* in place of *P*. Line 5 is the result of applying the Nicod Rule to lines 3 and 4.

THEOREM 11. $\vdash_N$ *Q*:.|:.*Q*|*P*.|.*P*:|:*Q*|*P*.|.*P*

Proof: Line 1 is Theorem 8 with *Q* in place of *R* and with *P* in place of S. Line 2 is Theorem 9 with *Q* in place of S. Line 3 is Theorem 8 with *Q* in place of *P*, with *P*.|.*Q*|*Q* in place of *Q* and of *R*, and with *Q*|*P*:|: *P*|*P*.|.*P*|*P*:.|:.*Q*|*P*:|:*P*|*P*.|.*P*|*P* in place of S. Line 4 is the result of applying the Nicod Rule to lines 2 and 3. Line 5 is the result of applying the Nicod Rule to lines 1 and 4. Line 6 is Theorem 9 with *P*|*P*.|.*P* in place of S and with *Q* in place of *P*. Line 7 is Theorem 4. Line 8 is the result of applying the Nicod Rule to lines 6 and 7. Line 9 is Theorem 8 with *Q*|*P* in place of *P*, with *P*|*P* in place of *Q* and of *R*, and with *P* in place of S. Line 10 is Theorem 8 with *Q* in place of *P*, with *Q*|*P*:|:*P*|*P*.|.*P*|*P* in place of *Q* and of *R*, and with *P*|*P*.|.*P*:.|:.*Q*|*P*.|.*P*:|:*Q*|*P*.|.*P*::|::*P*|*P*.|.*P*:.|:.*Q*|*P*.|.*P*:|:*Q*|*P*.|.*P* in place of S. Line 11 is the result of applying the Nicod Rule to lines 5 and 10. Line 12 is the result of applying the Nicod Rule to lines 9 and 11. Line 13 is Theorem 8 with *Q* in place of *P*, with *P*|*P*.|.*P* in place of *Q* and of *R*, and with *Q*|*P*.|.*P*:|:*Q*|*P*.|.*P* in place of S. Line 14 is the result of applying the Nicod Rule to lines 8 and 13. Line 15 is Theorem 8 with *Q* in place of *P*, with *P*|*P*.|.*P*:.|:.*Q*|*P*.|.*P*:|:*Q*|*P*.|.*P* in place of *Q* and of *R*, and with *Q*:.|:.*Q*|*P*.|.*P*:|:*Q*|*P*.|.*P*::|::*Q*:.|:.*Q*|*P*.|.*P*:|:*Q*|*P*.|.*P* in place of S. Line 16 is the result of applying the Nicod Rule to lines 12 and 15. Line 17 is the result of applying the Nicod Rule to lines 14 and 16. Line 18 is Theorem 10 with *Q*|*P*.|.*P*:|:*Q*|*P*.|.*P* in place of S. Line 19 is Theorem 3 with *Q*|*Q* in place of S, and with *Q*:.|:.*Q*|*P*.|.*P*:|:*Q*|*P*.|.*P*::|::*Q*:.|:.*Q*|*P*.|.*P*:|:*Q*|*P*.|.*P* in place of *P*. Line 20 is the result of applying the Nicod Rule to lines 18 and 19. Line 21 is Theorem 8 with *Q*:.|:.*Q*|*P*.|.*P*:|:*Q*|*P*.|.*P*::|::*Q*:.|:.*Q*|*P*.|.*P*:|:*Q*|*P*.|.*P* in place of *P* and of S, and with *Q* in place of *R*. Line 22 is the result of applying the Nicod Rule to lines 20 and 21. Line 23 is the result of applying the Nicod Rule to lines 17 and 22. Line 24 is Theorem 4 with *Q*:.|:.*Q*|*P*.|.*P*:|:*Q*|*P*.|.*P*::|::*Q*:.|:.*Q*|*P*.|.*P*:|:*Q*|*P*.|.*P* in place of *P*. Line 25 is the result of applying the Nicod Rule to lines 23 and 24.

THEOREM 12. $\vdash_{\mathrm{N}} P:|:Q|R.|.Q|R::|::Q:|:P|R.|.P|R:.|:.Q:|:P|R.|.P|R$

Proof: Line 1 is Theorem 8 with Q in place of P, with $Q|R.|.R$ in place of Q and of R, and with $P|R.|.P|R$ in place of S. Line 2 is Theorem 11 with R in place of P. Line 3 is the result of applying the Nicod Rule to lines 1 and 2. Line 4 is Theorem 8 with $P:|:Q|R.|.Q|R$ in place of P, with $Q|R.|.R:|:P|R.|.P|R$ in place of Q and of R, and with $Q:|:P|R.|.P|R:.|:.Q:|:P|R.|.P|R$ in place of S. Line 5 is Theorem 8 with $Q|R$ in place of Q and of R, and with R in place of S. Line 6 is the result of applying the Nicod Rule to lines 4 and 5. Line 7 is the result of applying the Nicod Rule to lines 3 and 6.

THEOREM 13. $\vdash_{\mathrm{N}} P.|.Q|Q::.|::.R|R.|.P|P:.|:.R|R.|.Q|Q:|:R|R.|.Q|Q::|::R|R.|.P|P:.|:.R|R.|.Q|Q:|:R|R.|.Q|Q$

Proof: Line 1 is Theorem 12 with $R|R.|.P|P$ in place of P, with $P.|.Q|Q$ in place of Q, and with $R|R.|.Q|Q:|:R|R.|.Q|Q$ in place of R. Line 2 is Theorem 8 with $R|R$ in place of P, with P in place of Q and of R, and with $Q|Q$ in place of S. Line 3 is the result of applying the Nicod Rule to lines 1 and 2.

THEOREM 14. $\vdash_{\mathrm{N}} P|P.|.P|P:|:P|P.|.P|P:.|:.P|P.|.P|P:|:P|P.|.P|P::|::P|P$ (Ax. 1 of H.A.)

Proof: Line 1 is Theorem 4 with $P|P.|.P|P:|:P|P.|.P|P$ in place of P. Line 2 is Theorem 4 with $P|P$ in place of P. Line 3 is Theorem 8 with $P|P.|.P|P:|:P|P.|.P|P:.|:.P|P.|.P|P:|:P|P.|.P|P$ in place of P, with $P|P.|.P|P$ in place of Q and of R, and with $P|P$ in place of S. Line 4 is the result of applying the Nicod Rule to lines 1 and 3. Line 5 is the result of applying the Nicod Rule to lines 2 and 4.

THEOREM 15. $\vdash_{\mathrm{N}} P|P.|.P|P:.|:.P|P.|.Q|Q:|:P|P.|.Q|Q$ (Ax. 2 of H.A.)

Proof: Line 1 is Theorem 10 with $P|P$ in place of Q, and with $Q|Q$ in place of S.

THEOREM 16. $\vdash_{\mathrm{N}} P|P.|.Q|Q:|:P|P.|.Q|Q:.|:.P|P.|.Q|Q:|:P|P.|.Q|Q::|::Q|Q.|.P|P:|:Q|Q.|.P|P$ (Ax. 3 of H.A.)

Proof: Line 1 is Theorem 3 with $P|P$ in place of S, and with $Q|Q$ in place of P. Line 2 is Theorem 4 with $P|P.|.Q|Q:|:P|P.|.Q|Q$ in place of P. Line 3 is Theorem 8 with $P|P.|.Q|Q:|:P|P.|.Q|Q:.|:.P|P.|.Q|Q:|:P|P.|.Q|Q$ in place of P, with $P|P.|.Q|Q$ in place of Q and of R, and with $Q|Q.|.P|P:|:Q|Q.|.P|P$ in place of S. Line 4 is the result of applying the Nicod Rule to lines 2 and 3. Line 5 is the result of applying the Nicod Rule to lines 1 and 4.

THEOREM 17. $\vdash_N P|P.|.P|P:|:Q|Q:.|:.P|P.|.P|P:|:Q|Q::|::P|P.|.P|P:|:Q|Q:.|:.P|P.|.P|P:|:Q|Q:::|:::R|R.|.P|P:|:R|R.|.P|P:.|:.R|R.|.P|P:|:R|R.|.P|P::|::R|R.|.Q|Q:|:R|R.|.Q|Q::.|::.R|R.|.P|P:|:R|R.|.P|P:.|:.R|R.|.P|P:|:R|R.|.P|P::|::R|R.|.Q|Q:|:R|R.|.Q|Q$ (Ax. 4 of H.A.)

Proof: Line 1 is Theorem 4 with $P|P.|.P|P:|:Q|Q:.|:.P|P.|.P|P:|:Q|Q$ in place of P. Line 2 is Theorem 9 with P for S and $P|P$ for P. Line 3 is Theorem 8 with $P|P.|.P|P$ in place of Q and of R, and with $Q|Q$ in place of S. Line 4 is the result of applying the Nicod Rule to lines 2 and 3. Line 5 is Theorem 8 with $P|P.|.P|P:|:Q|Q:.|:.P|P.|.P|P:|:Q|Q::|::P|P.|.P|P:|:Q|Q:.|:.P|P.|.P|P:|:Q|Q$ in place of P, with $P|P.|.P|P:|:Q|Q$ in place of Q and of R, and with $P.|.Q|Q:|:P.|.Q|Q$ in place of S. Line 6 is the result of applying the Nicod Rule to lines 1 and 5. Line 7 is the result of applying the Nicod Rule to lines 4 and 6. Line 8 is Theorem 13. Line 9 is Theorem 4 with $R|R.|.P|P:|:R|R.|.P|P$ in place of P. Line 10 is Theorem 8 with $R|R.|.P|P:|:R|R.|.P|P:.|:.R|R.|.P|P:|:R|R.|.P|P$ in place of P, with $R|R.|.P|P$ in place of Q and of R, and with $R|R.|.Q|Q:|:R|R.|.Q|Q$ in place of S. Line 11 is the result of applying the Nicod Rule to lines 9 and 10. Line 12 is Theorem 8 with $P.|.Q|Q$ in place of P, with $R|R.|.P|P:.|:.R|R.|.Q|Q:|:R|R.|.Q|Q$ in place of Q and of R, and with $R|R.|.P|P:|:R|R.|.P|P:.|:.R|R.|.P|P:|:R|R.|.P|P::|::R|R.|.Q|Q:|:R|R.|.Q|Q::.|::.R|R.|.P|P:|:R|R.|.P|P:.|:.R|R.|.P|P:|:R|R.|.P|P::|::R|R.|.Q|Q:|:R|R.|.Q|Q$ in place of S. Line 13 is the result of applying the Nicod Rule to lines 8 and 12. Line 14 is the result of applying the Nicod Rule to lines 11 and 13. Line 15 is Theorem 8 with $P|P.|.P|P:|:Q|Q:.|:.P|P.|.P|P:|:Q|Q::|::P|P.|.P|P:|:Q|Q:.|:.P|P.|.P|P:|:Q|Q$ in place of P, with $P.|.Q|Q$ in place of Q and of R, and with $R|R.|.P|P:|:R|R.|.P|P:.|:.R|R.|.P|P:|:R|R.|.P|P::|::R|R.|.Q|Q:|:R|R.|.Q|Q::.|::.R|R.|.P|P:|:R|R.|.P|P:.|:.R|R.|.P|P:|:R|R.|.P|P::|::R|R.|.Q|Q:|:R|R.|.Q|Q$ in place of S. Line 16 is the result of applying the Nicod Rule to lines 7 and 15. Line 17 is the result of applying the Nicod Rule to lines 14 and 16.

DR 1. $P, P|P.|.P|P:|:Q|Q \vdash_N Q$ (R′ 1 of H.A.)

Proof: Line 1 is Theorem 9 with P in place of S, and with $P|P$ in place of P. Line 2 is the premiss P. Line 3 is the result of applying the Nicod Rule to lines 1 and 2. Line 4 is the premiss $P|P.|.P|P:|:Q|Q$. Line 5 is the result of applying the Nicod Rule to lines 3 and 4.

Deriving all tautologies from a single axiom (axiom form) by a single rule in terms of a single operator is thus seen to be possible. But it is a tedious business. As Dr. Samuel Johnson is said to have remarked in quite another connection, it is '. . . like a dog's walking on his hind legs. It is not done well; but you are surprised to find it done at all.'

10

A First-Order Function Calculus

10.1 The New Logistic System RS_1

In Chapter 4 and the first four sections of Chapter 5, we *used* logical principles governing the quantification of individual variables to prove the validity of arguments and to demonstrate logical truths. An axiomatic development of those principles is called a 'first-order function calculus', or, alternatively, a 'lower' or 'restricted' function calculus.[1] In this chapter we construct such a logistic system, develop some of its theorems, and prove that it has such (desirable) characteristics as consistency and completeness. Again our metalanguage will be ordinary English plus some elementary arithmetic, and some special symbols that will be introduced and defined as needed. Our object language or object logic is the new system RS_1, which will now be described.

The logistic system RS_1 contains infinitely many primitive symbols of the following categories.

1. Infinitely many capital letters from the first part of the alphabet, with and without subscripts:

$$A, B, C, A_1, B_1, C_1, A_2, B_2, C_2, \ldots$$

These are *propositional constants*, and on the system's intended interpretation will express noncompound propositions.

2. Infinitely many capital letters from the middle part of the alphabet, with and without subscripts:

$$P, Q, R, P_1, Q_1, R_1, P_2, Q_2, R_2, \ldots$$

These are *propositional variables*, and on the system's intended interpretation will be statement variables of the kind discussed in Section 2.3 of Chapter 2. Symbols of the first two categories are *propositional symbols*.

3. Infinitely many capital letters from the first part of the alphabet, with and without subscripts, having right-hand superscripts '1', '2', '3'. . . .

[1] Also a 'first-order', 'lower', or 'restricted' *predicate* calculus.

$$\begin{array}{l} A^1, B^1, C^1, A^1_1, B^1_1, C^1_1, A^1_2, B^1_2, C^1_2, \ldots \\ A^2, B^2, C^2, A^2_1, B^2_1, C^2_1, A^2_2, B^2_2, C^2_2, \ldots \\ A^3, B^3, C^3, A^3_1, B^3_1, C^3_1, A^3_2, B^3_2, C^3_2, \ldots \\ \ldots\ldots\ldots\ldots\ldots\ldots\ldots\ldots\ldots\ldots \end{array}$$

These are *predicate constants,* and on the system's intended interpretation each of them will designate a particular attribute or binary relation or ternary relation . . . or n-ary relation according as its right superscript is '1' or '2' or '3' or . . . or 'n'.

4. Infinitely many capital letters from the middle part of the alphabet, with and without subscripts, having right-hand superscripts '1', '2', '3', . . .

$$\begin{array}{l} P^1, Q^1, R^1, P^1_1, Q^1_1, R^1_1, P^1_2, Q^1_2, R^1_2, \ldots \\ P^2, Q^2, R^2, P^2_1, Q^2_1, R^2_1, P^2_2, Q^2_2, R^2_2, \ldots \\ P^3, Q^3, R^3, P^3_1, Q^3_1, R^3_1, P^3_2, Q^3_2, R^3_2, \ldots \\ \ldots\ldots\ldots\ldots\ldots\ldots\ldots\ldots\ldots\ldots \end{array}$$

These are *predicate variables,* and on the system's intended interpretation will be symbols for which names of particular attributes, binary relations, and so forth, can be substituted. Symbols of the third and fourth categories are *predicate symbols.*

5. Infinitely many lower-case letters from the first part of the alphabet, with and without subscripts:

$$a, b, c, a_1, b_1, c_1, a_2, b_2, c_2, \ldots$$

These are *individual constants,* and on the system's intended interpretation will be proper names of individuals.

6. Infinitely many lower-case letters from the latter part of the alphabet, with and without subscripts:

$$x, y, z, x_1, y_1, z_1, x_2, y_2, z_2, \ldots$$

These are *individual variables,* and on the system's intended interpretation will be individual variables of the kind discussed in Chapter 4. Symbols of the fifth and sixth categories are *individual symbols.*

7. Just four additional symbols complete the list of primitive symbols of RS_1; these are the tilde, the dot, and left- and right-hand parentheses:

$$\sim, \cdot, (,)$$

In addition to the primitive symbols we introduce some defined symbols into our object language RS_1. Before doing so, however, we must indicate the use of certain special symbols of our metalanguage. As in Chapter 7, we introduce the special symbols '$\sim$', '$\cdot$', '(', and ')' to denote the symbols '$\sim$', '$\cdot$', '(', and ')' of the object language, and we shall also use brackets and braces to denote the object language's parentheses when that is conducive to easier reading. Capital letters, with and without subscripts, will be used in the metalanguage as syntactical variables, that is, as symbols for which designations of any symbols or sequences of symbols of the object language can be substituted. Lower case letters, with and without subscripts, will be used as syntactical individual variables, that is, symbols for which designations of individual symbols of the object language can be substituted. Finally, we adopt the convention that the juxtaposition of two symbols of the object language will be denoted in the metalanguage by the juxtaposition of their names. Thus in any context where 'F' denotes '$\mathbf{A}^1$' and 'x' denotes '$\boldsymbol{x}$', '$F(x)$' will denote '$\mathbf{A}^1(\boldsymbol{x})$'. We shall also find it convenient to insert commas into any sequence of symbols of the metalanguage that designate individual symbols of the object language. Thus where 'F' denotes '$\boldsymbol{B}^3$' and 'x_1', 'x_2', and 'x_3' denote '$\boldsymbol{a}_1$', '$\boldsymbol{a}_2$', and '$\boldsymbol{a}_3$', respectively, we shall use '$F(x_1, x_2, x_3)$' to denote '$\boldsymbol{B}^3(\boldsymbol{a}_1\boldsymbol{a}_2\boldsymbol{a}_3)$'.

We introduce the symbols '$\mathbf{v}$', '$\boldsymbol{\supset}$', '$\boldsymbol{\equiv}$', and '$\boldsymbol{\exists}$' into the object language *by definition*, and denote them in the metalanguage by the symbols 'v', '$\supset$', '$\equiv$', and '$\exists$'. The new symbols of the object language are introduced as definitional abbreviations:

Df. $P \vee Q$ is defined to be an abbreviation of $\sim(\sim P \cdot \sim Q)$.
Df. $P \supset Q$ is defined to be an abbreviation of $\sim(P \cdot \sim Q)$.
Df. $P \equiv Q$ is defined to be an abbreviation of $(P \supset Q) \cdot (Q \supset P)$.
Df. $(\exists x)P$ is defined to be an abbreviation of $\sim(x)\sim P$.

We shall feel free to drop parentheses (in the metalanguage), retaining only those required to avoid ambiguity, or to make for greater ease of interpretation. We shall also, on occasion, write '$P \cdot Q$' as 'PQ'. Although we shall not always take advantage of the following convention by dropping unnecessary parentheses, we set up the following *order of precedence* among the symbols of our metalanguage, each symbol having precedence over any that lies in a column to its right:

$\equiv$	$\supset$	v	$\cdot$	$\sim$	
					(x) where not immediately preceded by a predicate symbol
					$(\exists x)$

This convention dictates that an expression such as

$$P \equiv (x)Q \vee \sim RS \supset (\exists x)T \cdot U$$

is understood to denote the same formulas of the object language that are denoted by

$$\{P\} \equiv \{\{[(x)Q] \vee [(\sim R)\cdot(S)]\} \supset \{[(\exists x)T]\cdot[U]\}\}$$

Our final convention is that of *association to the left*, which means that where the convention of *order of precedence* does not suffice to remove the ambiguity of an expression, its parts should be grouped by parentheses *to the left*. That is, when an expression contains two (or more) occurrences of the same connective, and their relative scopes within the expression are not otherwise indicated, the occurrence to the right shall be understood to have the wider (or widest) scope.

We define *formula of* RS_1 as any finite sequence of symbols of RS_1. Among these are included such sequences as

$$)\sim)($$
$$(\mathbf{x})(\mathbf{A}^1(\mathbf{x}))$$
$$((\exists \mathbf{y})(\mathbf{B}^2(\mathbf{xy}))) \supset (\mathbf{C}^1(\mathbf{a}))$$
$$(\exists \vee \sim \equiv$$
$$\mathbf{Q}^3(\mathbf{ab})$$

of which we shall want to include only the second and third as *well formed*, that is, as meaningful on the system's intended or normal interpretation.

Now we define *well formed formula of* RS_1 by the following recursive rules:

(α) 1. If F is a propositional symbol, then F is a *wff*.
2. If F is an n-adic predicate symbol and $x_1, x_2, \ldots, x_n$ are n (not necessarily distinct) individual symbols (where $n = 1, 2, 3, \ldots$), then $F(x_1, x_2, \ldots, x_n)$ is a *wff*.

(β) 1. If F is a *wff, then* $\sim(F)$ is a *wff*.
2. If F is a *wff* and G is a *wff*, then $(F)\cdot(G)$ is a *wff*.
3. If F is a *wff* and x is an individual variable, then $(x)(F)$ is a *wff*.

No formula of RS_1 is a *wff* unless its being so follows from these rules, or from these rules together with the definitions of the defined symbols that have been stated.

Now that we have an effective criterion for *wff*, we shall restrict our discussion in the remainder of this chapter to *wffs*, which alone interest us.

It is convenient at this point to introduce and define some additional special terms. (There is no loss of generality in phrasing our definitions in terms of undefined symbols only, for defined symbols are always eliminable.)

Df. If x is an individual variable, then (x), when not immediately preceded by a predicate symbol, is the *universal quantifier* of x (or on x).

Df. If Q is a *wff* that occurs in a *wff* P, then Q is a *well formed part* of P.

Df. An occurrence of a variable x in a *wff* P will be called a *bound* occurrence of x in P if it is in a well formed part of P of the form $(x)Q$.

Df. An occurrence of a variable x in a *wff* P will be called a *free occurrence* if it is not bound.

We assume infinitely many postulates for our object logic. Every *wff* of any of the five following patterns is a postulate:

P 1. $P \supset (P \cdot P)$
P 2. $(P \cdot Q) \supset P$
P 3. $(P \supset Q) \supset [\sim(Q \cdot R) \supset \sim(R \cdot P)]$
P 4. $(x)(P \supset Q) \supset [P \supset (x)Q]$, where x is any individual variable, P is any *wff* containing no free occurrences of x, and Q is any *wff*.
P 5. $(x)P \supset Q$, where x is any individual variable, y is any individual variable or constant, P is any *wff*, Q is the result of replacing each free occurrence of x in P by y, and if y is a variable, then it must occur free in Q at all places where x occurs free in P (that is, no bound occurrence of y in Q is the result of replacing a free occurrence of x in P by y).

The restrictions on P 4 and P 5 serve to prevent the inclusion of such manifest falsehoods (on the intended interpretation) as

$$(x)[(x = 1) \supset (x + x = 2)] \supset [(x = 1) \supset (x)(x + x = 2)]$$

this statement has a true antecedent and a false consequent, but it is *not* an instance of P 4, since x occurs free in $(x = 1)$ Also prevented is such a patent falsehood as

$$(x)[(\exists y)(y \neq x)] \supset [(\exists y)(y \neq y)]$$

which is *not* an instance of P 5, since there is a bound occurrence of the variable y in $(\exists y)(y \neq y)$ which is the result of replacing a free occurrence of the variable x in $(\exists y)(y \neq x)$ by y.

We assume *two* rules of inference for RS_1:

R 1. From P and $P \supset Q$ to infer Q.
R 2. From P to infer $(x)P$.

Next we define 'demonstration' for RS_1. Formally,

$$P_1, P_2, \ldots, P_n \vdash Q$$

asserts that there is a finite sequence of *wffs* $S_1, S_2, \ldots, S_t$ such that for every S_j $(1 \leq j \leq t)$ either:

(a) S_j is one of the Postulates P 1–P 5; or
(b) S_j is one of the premisses P_i $(1 \leq i \leq n)$; or

(c) S_j is the result of applying R 1 to two earlier S's of the sequence, say S_i and S_k where $i < j$ and $k < j$; or

(d) S_j is the result of applying R 2 to an earlier S_i of the sequence, so that S_j is $(x)S_i$ where $i < j$; and S_t is Q.

Informally, we regard $P_1, P_2, \ldots, P_n \vdash Q$ as asserting that Q is validly inferred from $P_1, P_2, \ldots, P_n$ in RS_1.

Next we define 'theorem of RS_1' as any *wff* Q such that $\vdash Q$. It should be noted that '$\vdash$' is a special symbol of our metalanguage, and does not occur in RS_1 itself.

We can now establish the *consistency* of RS_1. We begin by stating the following definition:

Df. Where F is any *wff* of RS_1, its *associated propositional formula* (abbreviated to 'a.p.f.' and symbolized F°) is the formula that results from F by first deleting all occurrences of quantifiers in F, together with all parentheses that were required by those quantifiers, and then replacing every well formed part of F of the form $P^n(x_1, x_2, \ldots, x_n)$ and every propositional symbol in F by the propositional constant $\boldsymbol{A}$.

As an example, where F is

$$((x_1)((x_2)(\boldsymbol{A}^2(x_1x_2)))) \supset (((x_1)(\boldsymbol{B}^1(x_1))) \cdot ((x_2)(\boldsymbol{B}^1(x_2))))$$

it's a.p.f. F° is

$$(\boldsymbol{A}) \supset ((\boldsymbol{A}) \cdot (\boldsymbol{A}))$$

Next, we state and prove the following:

LEMMA: If $\vdash F$, then F° is a (truth table) tautology.

Proof. 1. First we show that every postulate of RS_1 has only tautologies for its a.p.f.'s. An a.p.f. of any instance of P 1 is of the form $P^\circ \supset (P^\circ \cdot P^\circ)$, which is a tautology. Similarly, for P 2 and P 3. The a.p.f. of any instance of P 4 has the form $(P^\circ \supset Q^\circ) \supset (P^\circ \supset Q^\circ)$, which is a tautology. In P 5, since Q differs from P only in the individual symbols they contain, their a.p.f.'s are identical, so the a.p.f. of any instance of P 5 has the form $P^\circ \supset P^\circ$ which is a tautology.

2. Next, we show that any *wff* obtained by applying the rules of RS_1 to *wffs* that have tautologies for their a.p.f.'s has a tautology for its a.p.f. By R 2 we derive $(x)P$ from P. But $(x)P$ and P have identically the same a.p.f. Therefore, if the a.p.f. of P is a tautology, then the a.p.f. of $(x)P$ is also. By R 1 we derive Q from P and $P \supset Q$. So if P° and $P^\circ \supset Q^\circ$ are tautologies, then Q° is also.

Since all theorems of RS_1 follow by the rules from the postulates, all theorems of RS_1 have tautologies as their a.p.f.'s. This completes our proof of the lemma. The consistency of RS_1 will be stated as

METATHEOREM I. *RS_1 is consistent.*

Proof: (Here we use Post's criterion for consistency: a system is consistent if it contains a *wff* that is not a theorem.) The formula '$(\mathbf{A})\cdot(\sim(\mathbf{A}))$' is a *wff* of RS_1 that has an a.p.f. (itself) which is not a tautology. Hence, by our lemma, it is not a theorem of RS_1, so RS_1 is consistent. The consistency of RS_1 also follows from its analyticity, which is remarked (in passing) in Section 10.6.

10.2 Development of RS_1

In developing the present system, we shall borrow freely from the results obtained in Chapter 8, in which it was established that every tautology can be demonstrated as a theorem in the Propositional Calculus based on P 1, P 2, P 3, and R 1. This is noted in the following theorem and derived rule.

THEOREM O. *All tautologies are theorems, and if in any tautology T that contains the propositional symbols $P_1, P_2, \ldots, P_n$ we replace all occurrences of $P_1, P_2, \ldots, P_n$, respectively, by any wffs $F_1, F_2, \ldots, F_n$ of RS_1, the result is a theorem F of RS_1.*

Proof: By the completeness of the Propositional Calculus based on P 1, P 2, P 3, and R 1, there is a demonstration in RS_1 for every tautology. If in every step of that demonstration, we replace all occurrences of $P_1, P_2, \ldots, P_n$, by $F_1, F_2, \ldots, F_n$, we obtain a demonstration in RS_1 of F, since making the indicated substitutions in the postulates P 1, P 2, P 3 gives us simply other instances of those postulates.

DERIVED RULE O. All tautologically valid argument forms are demonstrable in RS_1, and if in any tautologically valid derived rule $S_1, S_2, \ldots, S_m \vdash T$ that contains the propositional symbols $P_1, P_2, \ldots, P_n$, we replace all occurrences of $P_1, P_2, \ldots, P_n$, respectively, by any *wffs* $F_1, F_2, \ldots, F_n$ of RS_1, the result is a demonstrably valid derived rule $G_1, G_2, \ldots, G_m \vdash H$ of RS_1.

Proof. Exactly parallels the proof of Theorem O.

In any *proof* in RS_1, any step that is justified by Th. O or DR O will be noted simply by Ⓟ (for 'Propositional Calculus').

At this point we proceed to establish the first few derived rules and theorems of RS_1.

DR 1. If P contains no free occurrence of x, then $P \supset Q \vdash P \supset (x)Q$.

Demonstration:		
	S_1: $P \supset Q$	premiss
	S_2: $(x)(P \supset Q)$	R 2
	S_3: $(x)(P \supset Q) \supset [P \supset (x)Q]$	P 4
	S_4: $P \supset (x)Q$	R 1

DR 2. $(x)[F(x) \supset G(x)], (x)F(x) \vdash (x)G(x)$

Demonstration:	S_1: $(x)[F(x) \supset G(x)]$	premiss
	S_2: $(x)[F(x) \supset G(x)] \supset [F(x) \supset G(x)]$	P 5
	S_3: $F(x) \supset G(x)$	R 1
	S_4: $(x)F(x)$	premiss
	S_5: $(x)F(x) \supset F(x)$	P 5
	S_6: $F(x)$	R 1
	S_7: $G(x)$	R 1
	S_8: $(x)G(x)$	R 2

DR 3. $(x)(P \cdot Q) \vdash (x)P \cdot (x)Q$

Proof:	$(x)(P \cdot Q)$	premiss
	$(x)(P \cdot Q) \supset P \cdot Q$	P 5
	$P \cdot Q$	R 1
	P	Ⓟ
	$(x)P$	R 2
	Q	Ⓟ
	$(x)Q$	R 2
	$(x)P \cdot (x)Q$	Ⓟ

DR 4. $(x)(P \equiv Q) \vdash (x)P \equiv (x)Q$

Proof:	$(x)(P \equiv Q)$	premiss
	$(x)(P \equiv Q) \supset (P \equiv Q)$	P 5
	$P \equiv Q$	R 1
	$P \supset Q$	Ⓟ
	$(x)P \supset P$	P 5
	$(x)P \supset Q$	Ⓟ
	$(x)P \supset (x)Q$	DR 1
	$Q \supset P$	Ⓟ
	$(x)Q \supset Q$	P 5
	$(x)Q \supset P$	Ⓟ
	$(x)Q \supset (x)P$	DR 1
	$[(x)P \supset (x)Q] \cdot [(x)Q \supset (x)P]$	Ⓟ
	$(x)P \equiv (x)Q$	df.

TH. 1. $\vdash (x)(P \cdot Q) \equiv (x)P \cdot (x)Q$

Proof: First we establish A:	$\vdash (x)(P \cdot Q) \supset (x)P \cdot (x)Q$	
	$\vdash (x)(P \cdot Q) \supset P \cdot Q$	P 5
	$\vdash P \cdot Q \supset P$	P 2
	$\vdash (x)(P \cdot Q) \supset P$	Ⓟ
	$\vdash (x)(P \cdot Q) \supset (x)P$	DR 1

By steps similar to the above we obtain:

$\vdash (x)(P \cdot Q) \supset (x)Q$

whence $\vdash (x)(P \cdot Q) \supset (x)P \cdot (x)Q$ Ⓟ

Next we establish B: $\vdash (x)P \cdot (x)Q \supset (x)(P \cdot Q)$

$\vdash (x)P \supset P$	P 5
$\vdash (x)Q \supset Q$	P 5
$\vdash (x)P \cdot (x)Q \supset P \cdot Q$	Ⓟ
$\vdash (x)P \cdot (x)Q \supset (x)(P \cdot Q)$	DR 1

Now from A and B,

$\vdash [(x)(P \cdot Q) \supset (x)P \cdot (x)Q] \cdot [(x)P \cdot (x)Q \supset (x)(P \cdot Q)]$	Ⓟ
$\vdash (x)(P \cdot Q) \equiv (x)P \cdot (x)Q$	df.

TH. 2. $\vdash (x)(P \supset Q) \supset [(x)P \supset (x)Q]$

Proof:

$\vdash [(x)(P \supset Q) \cdot (x)P] \supset (x)(P \supset Q)$	P 2
$\vdash [(x)(P \supset Q) \cdot (x)P] \supset (x)P$	Ⓟ
$\vdash (x)(P \supset Q) \supset (P \supset Q)$	P 5
$\vdash (x)P \supset P$	P 5
$\vdash [(x)(P \supset Q) \cdot (x)P] \supset [(P \supset Q) \cdot P]$	Ⓟ
$\vdash [(P \supset Q) \cdot P] \supset Q$	Ⓟ
$\vdash [(x)(P \supset Q) \cdot (x)P] \supset Q$	Ⓟ
$\vdash [(x)(P \supset Q) \cdot (x)P] \supset (x)Q$	DR 1
$\vdash (x)(P \supset Q) \supset [(x)P \supset (x)Q]$	Ⓟ

Next we state and prove the Deduction Theorem for RS_1, which corresponds to the strengthened method of Conditional Proof used in Section 3.8 and Chapters 4 and 5.

METATHEOREM II. *(Deduction Theorem—D.T.) If there is a demonstration that $P_1, P_2, \ldots, P_{n-1}, P_n \vdash Q$ in which no variable occurring free in P_n is quantified by R 2 in any step, then there is a demonstration that $P_1, P_2, \ldots, P_{n-1} \vdash P_n \supset Q$ in which exactly those variables are quantified by R 2 that were quantified by R 2 in the original demonstration.*[2]

Proof: We assume that there is a sequence of *wffs* $S_1, S_2, \ldots, S_t$ such that every S_j $(1 \leq j \leq t)$ is (a) one of the Postulates P 1–P 5, or (b) one of the P_i's $(1 \leq i \leq n)$, or (c) results from applying R 1 to two earlier S's of the sequence,

[2]The restriction that no variable that is free in P_n is quantified by *R* 2 parallels exactly the restriction on **UG** in Chapter 4, because P_n here has the role of an assumption within whose scope lie subsequent lines of the first demonstration.

or (d) results from applying R 2 to an earlier S; and S_t is Q. Now consider the sequence of *wffs:* $P_n \supset S_1, P_n \supset S_2, \ldots, P_n \supset S_t$. If we can 'fill in' *wffs* before each $P_n \supset S_j$ in such a way that the resulting total sequence is a demonstration from $P_1, P_2, \ldots, P_{n-1}$, so each line of the resulting total sequence is either (a) one of the Postulates P 1–P 5, or (b) one of the P_i's $(1 \leq i \leq n-1)$, or (c) results from applying R 1 to two earlier lines of the sequence, or (d) results from applying R 2 to an earlier line of the sequence, then since the last line $P_n \supset S_t$ is $P_n \supset Q$, we shall have a demonstration that $P_1, P_2, \ldots, P_{n-1} \vdash P_n \supset Q$.

The proof proceeds by weak induction on the number of lines (t) in the original demonstration.

(α) In case $t = 1$ we have only the formula $P_n \supset S_1$ to consider. We wish to show that $P_1, P_2, \ldots, P_{n-1} \vdash P_n \supset S_1$. By assumption, S_1 is either a Postulate P 1–P 5 or a P_i $(1 \leq i \leq n)$.

CASE 1. S_1 is a Postulate. Here we fill in with the demonstration of $S_1 \supset (P_n \supset S_1)$ and S_1 itself. From these we have $P_n \supset S_1$ by R 1, so the total sequence up to and including $P_n \supset S_1$ is a demonstration that $\vdash P_n \supset S_1$ and hence that $P_1, P_2, \ldots, P_{n-1} \vdash P_n \supset S_1$.

CASE 2. S_1 is a P_i $(1 \leq i < n)$. Here we fill in with the demonstration of $S_1 \supset (P_n \supset S_1)$ and S_1 itself, from which we have $P_n \supset S_1$ by R 1. Here the total sequence is a demonstration that $S_1 \vdash P_n \supset S_1$, and since S_1 is a P_i $(1 \leq i \leq n-1)$ it is a demonstration that $P_1, P_2, \ldots, P_{n-1} \vdash P_n \supset S_1$.

CASE 3. S_1 is P_n. Here we fill in with the demonstration of $S_1 \supset S_1$, which is a demonstration that $\vdash P_n \supset S_1$, and hence that $P_1, P_2, \ldots, P_{n-1} \vdash P_n \supset S_1$.

(β) Now suppose we have filled in for all lines $P_n \supset S_1, P_n \supset S_2, \ldots,$ up to and including $P_n \supset S_{k-1}$, so we have a sequence of *wffs*, which is a demonstration that $P_1, P_2, \ldots, P_{n-1} \vdash P_n \supset S_{k-1}$. Under this assumption, we show how to fill in to include $P_n \supset S_k$ in the sequence. This will then be a demonstration that $P_1, P_2, \ldots, P_{n-1} \vdash P_n \supset S_k$. By assumption, S_k is either a postulate, a P_i $(1 \leq i \leq n)$, resulted in the original demonstration from applying R 1 to two earlier S's, or resulted in the original demonstration from applying R 2 to an earlier S.

CASE 1. S_k is a Postulate. Fill in as in the α-case.

CASE 2. S_k is a P_i $(1 \leq i \leq n)$. Fill in as in the α-case.

CASE 3. S_k results from applying R 1 to two previous S's, say S_i and S_j where S_j is $S_i \supset S_k$. By the β-case assumption, since $i < k$, $j < k$ we already have $P_n \supset S_i$ and $P_n \supset S_j$, which is $P_n \supset (S_i \supset S_k)$. Here we insert the demonstration of $[P_n \supset (S_i \supset S_k)] \supset [(P_n \supset S_i) \supset (P_n \supset S_k)]$ (a tautology), and $P_n \supset S_k$ follows by two applications of R 1.

CASE 4. S_k results from applying R 2 to an earlier S, say S_j where $j < k$. By the β-case assumption we already have $P_n \supset S_j$. S_k is $(x)S_j$ where x, by our original assumption, is not free in P_n. Hence by R 2, $(x)(P_n \supset S_j)$, and by P 4, $(x)(P_n \supset S_j) \supset (P_n \supset (x)S_j)$. Now R 1 will give $P_n \supset (x)S_j$ which is $P_n \supset S_k$.

From α, β by weak induction we can thus fill in for *any* number of steps S_j of the original demonstration. Moreover, no variable is quantified in the 'filled in' sequence which was not quantified in the original sequence. This proves Metatheorem II (D.T.).

MT II. COROLLARY: *The D.T. as stated above holds for any system of logic which has only rules* R 1 *and* R 2 *and which contains demonstrations for*

$P \supset P$, $P \supset (Q \supset P)$, $[P \supset (Q \supset R)] \supset [(P \supset Q) \supset (P \supset R)]$, and
$(x)(P \supset Q) \supset (P \supset (x)Q)$ where no free x's occur in P.

Proof: Obvious.

We can illustrate the *use* of the D.T. by using it to prove DR 5:

DR 5. $(x)[F(x) \supset G(x)] \vdash (x)F(x) \supset (x)G(x)$

Proof: $(x)[F(x) \supset G(x)], (x)F(x) \vdash (x)G(x)$ DR 2
$(x)[F(x) \supset G(x)] \vdash (x)F(x) \supset (x)G(x)$ D.T.

Some additional theorems which can be established quite easily using the Deduction Theorem are

TH. 3. $\vdash (x)(P \equiv Q) \supset [(x)P \equiv (x)Q]$
*TH. 4. $\vdash (x)(P \supset Q) \supset [(\exists x)P \supset (\exists x)Q]$

Their proofs will be left as exercises for the reader. We now proceed to some theorems that state equivalences. Since $(\exists x)$ was introduced as an abbreviation for $\sim(x)\sim$, the next theorem,

TH. 5. $\vdash (\exists x)P \equiv \sim(x)\sim P$

follows immediately from $\vdash \sim(x)\sim P \equiv \sim(x)\sim P$ (Ⓟ) by definition.

The following theorems, however, require proofs that are simple but not quite so simple as that of Th. 5.

TH. 6. $\vdash (x)P \equiv \sim(\exists x)\sim P$

TH. 7. $\vdash \sim(x)P \equiv (\exists x)\sim P$

TH. 8. $\vdash \sim(\exists x)P \equiv (x)\sim P$

The proofs of these will also be left as exercises.

The next theorem provides for the permutation of universal quantifiers:

TH. 9. $\vdash (x)(y)P \equiv (y)(x)P$

Proof:	$\vdash (y)P \supset P$	P 5
	$\vdash (x)[(y)P \supset P]$	R 2
	$\vdash (x)(y)P \supset (x)P$	DR 5
	$\vdash (x)(y)P \supset (y)(x)P$	DR 1

We obtain $\vdash (y)(x)P \supset (x)(y)P$ in the same fashion, and then have

$\vdash (x)(y)P \equiv (y)(x)P$ by Ⓟ

Another theorem that can be proved quite simply is

TH. 10. $\vdash [(x)P \vee (x)Q] \supset (x)(P \vee Q)$

Proof:	$\vdash (x)P \supset P$	P 5
	$\vdash (x)Q \supset Q$	P 5
	$\vdash [(x)P \vee (x)Q] \supset (P \vee Q)$	Ⓟ
	$\vdash [(x)P \vee (x)Q] \supset (x)(P \vee Q)$	DR 1

Having derived a number of equivalences as theorems, it will be convenient to establish a rule of replacement which will permit the interchange of equivalent formulas in any context. This will be proved as the next Metatheorem.

METATHEOREM III (*Rule of Replacement*—R.R.). *Let $P_1, P_2, \ldots, P_n$, be any wffs, let A be any wff that does not occur in any P_i, and let W be any wff that contains no symbols other than the P_i's $(1 \leq i \leq n)$, A, $\cdot$, $\sim$, (x), and parentheses.*[3] *Let W^* be the result of replacing any number of occurrences of A in W by B. Then $A \equiv B \vdash W \equiv W^*$.*

Proof: We use strong induction on the number of symbols in *W*, counting each occurrence of P_i $(1 \leq i \leq n)$, *A*, *B*, $\sim$, $\cdot$, (x), as a single symbol.

(α) In case *W* contains a single symbol, *W* is either a P_i or *A*.

CASE 1. *W* is P_i. Here W^* is P_i, and since $\vdash P_i \equiv P_i$, we have $\vdash W \equiv W^*$, whence $A \equiv B \vdash W \equiv W^*$.

CASE 2. *W* is *A*. Here W^* is either *A* or *B*.

[3] The last restriction does not limit the generality of the Metatheorem, since all occurrences of the defined symbols v, $\supset$, $\equiv$, and $\exists$ can be replaced by undefined symbols.

Subcase A: W^* is A. Since $\vdash A \equiv A$, we have $\vdash W \equiv W^*$, whence $A \equiv B \vdash W \equiv W^*$.

Subcase B: W^* is B. Since $A \equiv B \vdash A \equiv B$, we have $A \equiv B \vdash W \equiv W^*$.

(β) Now suppose the Metatheorem true for any *wff* W that contains k or fewer symbols, and consider a *wff* W that contains $k + 1$ symbols. W must be $\sim L$, $(x)L$, or $M \cdot N$.

CASE 1. W is $\sim L$. L must contain just k symbols, so by the β-case assumption $A \equiv B \vdash L \equiv L^*$. But $(L \equiv L^*) \supset (\sim L \equiv \sim L^*)$ is provable, so $A \equiv B \vdash \sim L \equiv \sim L^*$, which is $A \equiv B \vdash W \equiv W^*$. Note that $(\sim L)^* = \sim(L^*)$.

CASE 2. W is $(x)L$. Again L contains just k symbols, so by the β-case assumption $A \equiv B \vdash L \equiv L^*$. Hence by R 2 we have $A \equiv B \vdash (x)(L \equiv L^*)$, and by DR 4 we have $(x)(L \equiv L^*) \vdash (x)L \equiv (x)L^*$, whence $A \equiv B \vdash W \equiv W^*$. Note that $(x)(L^*) = ((x)(L))^*$.

CASE 3. W is $M \cdot N$. Since M and N each contains fewer than k symbols, by the β-case assumption $A \equiv B \vdash M \equiv M^*$ and $A \equiv B \vdash N \equiv N^*$. Since $\vdash [(M \equiv M^*) \cdot (N \equiv N^*)] \supset [M \cdot N \equiv M^* \cdot N^*]$, and every $M^* \cdot N^*$ is a W^*, we have $A \equiv B \vdash W \equiv W^*$.

This completes the induction and finishes the proof.

MT III. COROLLARY: *If W and W^* are as in* MT III, *then* $A \equiv B, W \vdash W^*$. The proof is quite obvious.

10.3 Duality

We begin our discussion of duality with a quite complicated definition:

Df. Let W be any *wff* that contains no occurrences of $\supset$ or $\equiv$ (any *wff* can be made into such a W by rewriting every well formed part of the form $P \supset Q$ as $\sim P \vee Q$ and every well formed part of the form $P \equiv Q$ as $(\sim P \vee Q) \cdot (\sim Q \vee P)$). Where $P_1, \ldots, P_n$ are propositional symbols or composed of predicate symbols followed by the appropriate number of individual symbols, W will be constructed out of $P_1, P_2, \ldots, P_n, \cdot, \sim, \vee, (x), (\exists x)$ exclusively. Then the *dual* of W (written W^Δ) is formed by

replacing every occurrence of P_i in W by $\sim P_i$,[4]
replacing every occurrence of $\sim P_i$ in W by P_i,
replacing every occurrence of (x) in W by $(\exists x)$,
replacing every occurrence of $(\exists x)$ in W by (x),
replacing every occurrence of $\cdot$ in W by $\vee$,
replacing every occurrence of $\vee$ in W by $\cdot$.

[4] Except those occurrences of P_i in well formed parts of the form $\sim P_i$.

Examples (where P, Q, R, S are P_i's):

1. W: $P \cdot Q$
   W^Δ: $\sim P \vee \sim Q$
2. W: $(x)(P \vee Q)$
   W^Δ: $(\exists x)(\sim P \cdot \sim Q)$
3. W: $(y)(\exists z)[P \vee (\sim Q \vee R \cdot S)]$
   W^Δ: $(\exists y)(z)[\sim P \cdot Q \cdot (\sim R \vee \sim S)]$

There are several immediate consequences of our definition: First, where W is any *wff* that contains no part of the form $\sim\sim P_i$:

1. $W = W^{\Delta\Delta}$

Where W and U are any *wffs* whatever:

2. $(W \cdot U)^\Delta = W^\Delta \vee U^\Delta$
3. $(W \vee U)^\Delta = W^\Delta \cdot U^\Delta$
4. $((x)W)^\Delta = (\exists x)W^\Delta$
5. $((\exists x)W)^\Delta = (x)W^\Delta$
6. $(\sim W)^\Delta$ { a. If W is a P_i, then $(\sim W)^\Delta$ is P_i.
   b. If W contains at least two symbols, then $(\sim W)^\Delta = \sim(W^\Delta)$.

We can now establish a general duality result.

METATHEOREM IV. *(Duality Theorem) If W^Δ is the dual of W, then* $\vdash \sim W \equiv W^\Delta$.

Proof: We use strong induction on the structure of W (i.e., the number of symbols it contains, counting each P_i as a single symbol).

(α) If W contains just one symbol, W is a P_i. Here W^Δ is $\sim P_i$, and since $\vdash \sim P_i \equiv \sim P_i$ by Ⓟ, we have $\vdash \sim W \equiv W^\Delta$.

(β) Assume the Metatheorem true for any W containing k or fewer symbols. Now consider any W containing $k + 1$ symbols.

CASE 1. W is $\sim R$.

Subcase A: R contains more than one symbol. Then $(\sim R)^\Delta$ is $\sim(R^\Delta)$. By the β-case assumption, $\vdash \sim R \equiv R^\Delta$, hence by Ⓟ $\vdash \sim\sim R \equiv \sim(R^\Delta)$, or $\vdash \sim\sim R \equiv (\sim R)^\Delta$ which is $\vdash \sim W \equiv W^\Delta$.

Subcase B: R contains just one symbol, that is, R is a P_i. Then W is $\sim P_i$ and W^Δ is P_i. By Ⓟ $\vdash \sim\sim P_i \equiv P_i$, whence $\vdash \sim W \equiv W^\Delta$.

CASE 2. W is $(x)R$. By β we have $\vdash \sim R \equiv R^\Delta$. We also have, by Ⓟ and R.R., $\vdash \sim(x)R \equiv \sim(x)\sim\sim R$, so, by R.R., $\vdash \sim(x)R \equiv \sim(x)\sim R^\Delta$, which (by definition) is $\vdash \sim(x)R \equiv (\exists x)R^\Delta$ or $\vdash \sim W \equiv W^\Delta$.

CASE 3. W is $(\exists x)R$. By β, $\vdash \sim R \equiv R^{\Delta}$. We also have (Theorem 8) $\vdash \sim(\exists x)R \equiv (x)\sim R$. By R.R. we have $\vdash \sim(\exists x)R \equiv (x)R^{\Delta}$, which is $\vdash \sim W \equiv W^{\Delta}$.

CASE 4. W is $A \cdot B$. By β, $\vdash \sim A \equiv A^{\Delta}$ and $\vdash \sim B \equiv B^{\Delta}$. By Ⓟ, $\vdash \sim(A \cdot B) \equiv \sim A \vee \sim B$, so by R.R., $\vdash \sim(A \cdot B) \equiv A^{\Delta} \vee B^{\Delta}$, which is $\vdash \sim W \equiv W^{\Delta}$.

CASE 5. W is $A \vee B$. By β, $\vdash \sim A \equiv A^{\Delta}$ and $\vdash \sim B \equiv B^{\Delta}$. By Ⓟ, $\vdash \sim(A \vee B) \equiv \sim A \cdot \sim B$, so by R.R., $\vdash \sim(A \vee B) \equiv A^{\Delta} \cdot B^{\Delta}$, which is $\vdash \sim W \equiv W^{\Delta}$.

MT IV. COROLLARY: $\vdash (W \equiv U) \supset (W^{\Delta} \equiv U^{\Delta})$
The proof of the corollary is obvious.

Duality has many uses. The traditional 'Square of Opposition'[5] which displays ***I*** and ***O*** propositions as the negations of ***E*** and ***A*** propositions, respectively, is clearly a special case of the duality result established above:

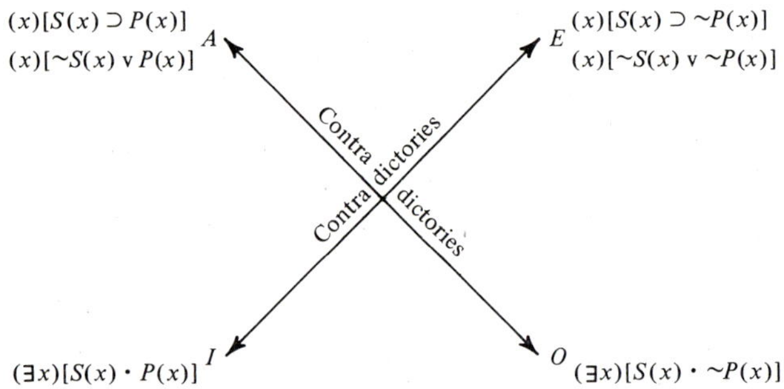

The familiar theorems of De Morgan are also special cases of the duality theorem. And the general duality theorem permits us to formulate negations with ease. For example, the proposition

There is a course that all students take.

may be symbolized as

$$(\exists x)\{C^1(x) \cdot (y)[S^1(y) \supset T^2(y, x)]\}$$

and may be rewritten without using the implication sign to become

$$(\exists x)\{C^1(x) \cdot (y)[\sim S^1(y) \vee T^2(y, x)]\}$$

[5]Discussed in Chapter 4, page 68.

The dual (and hence the negation) of this formula is

$$(x)\{\sim C^1(x) \vee (\exists y)[S^1(y)\cdot\sim T^2(y, x)]\}$$

which is most 'naturally' written using the implication sign as

$$(x)\{C^1(x) \supset (\exists y)[S^1(y)\cdot\sim T^2(y, x)]\}$$

and is the symbolic translation of the English sentence

For every course there is some student who does not take it.

Having established the Rule of Replacement and the Duality Theorem, some additional theorems are very easily proved.

TH. 11. $\vdash (\exists x)(P \vee Q) \equiv (\exists x)P \vee (\exists x)Q$

Proof: $\vdash (x)(\sim P\cdot\sim Q) \equiv (x)\sim P\cdot(x)\sim Q$ — Th. 1
$\vdash (\exists x)(P \vee Q) \equiv (\exists x)P \vee (\exists x)Q$ — Duality Theorem Cor. and R 1

The proofs of the next two theorems are equally simple, and may be left as exercises.

TH. 12. $\vdash (\exists x)(\exists y)P \equiv (\exists y)(\exists x)P$

*TH. 13. $\vdash (\exists x)(P\cdot Q) \supset (\exists x)P\cdot(\exists x)Q$

Now that we have the Rule of Replacement, it is desirable to augment our supply of equivalences. For the next ten theorems (Th. 14 through Th. 23) and the next derived rule (DR 6), we make the blanket assumption that *there are no free occurrences* of x in Q. We shall give proofs of some of these next results, and leave others as exercises for the reader:

TH. 14. $\vdash (x)Q \equiv Q$

Proof: $\vdash (x)Q \supset Q$ — P 5
$\vdash Q \supset Q$ — Ⓟ
$\vdash Q \supset (x)Q$ — DR 1
$\vdash (x)Q \equiv Q$ — Ⓟ

TH. 15. $\vdash (\exists x)Q \equiv Q$

TH. 16. $\vdash (x)(P\cdot Q) \equiv (x)P\cdot Q$

Proof: $\vdash (x)(P\cdot Q) \equiv (x)P\cdot(x)Q$ — Th. 1
$\vdash (x)Q \equiv Q$ — Th. 14
$\vdash (x)(P\cdot Q) \equiv (x)P\cdot Q$ — R.R.

TH. 17. $\vdash (\exists x)(P \vee Q) \equiv (\exists x)P \vee Q$

Proof: $\vdash (x)(\sim P \cdot \sim Q) \equiv (x) \sim P \cdot \sim Q$ — Th. 16
$\vdash (\exists x)(P \vee Q) \equiv (\exists x)P \vee Q$ — Duality Theorem Cor. and R 1

TH. 18. $\vdash (\exists x)(P \supset Q) \equiv (x)P \supset Q$

TH. 19. $\vdash (\exists x)(Q \supset P) \equiv Q \supset (\exists x)P$

TH. 20. $\vdash (x)(Q \supset P) \equiv Q \supset (x)P$

*TH. 21. $\vdash (x)(P \supset Q) \equiv (\exists x)P \supset Q$

TH. 22. $\vdash (x)(P \vee Q) \equiv (x)P \vee Q$

TH. 23. $\vdash (\exists x)(P \cdot Q) \equiv (\exists x)P \cdot Q$

*DR 6. $P \supset Q \vdash (\exists x)P \supset Q$

The next theorem requires no such restriction on free occurrences of variables in the formulas it contains:

TH. 24. $\vdash (\exists x)(y)P \supset (y)(\exists x)P$

Proof: $\vdash (x) \sim P \supset \sim P$	P 5
$\vdash \sim\sim P \supset \sim(x) \sim P$	Ⓟ
$\vdash P \supset \sim(x) \sim P$	Ⓟ
$\vdash P \supset (\exists x)P$	df.
$\vdash (y)P \supset P$	P 5
$\vdash (y)P \supset (\exists x)P$	Ⓟ
$\vdash (y)P \supset (y)(\exists x)P$	DR 1
$\vdash (\exists x)(y)P \supset (y)(\exists x)P$	DR 6

For the next two theorems we make the general assumptions that x and y are any two individual variables, and $F(x)$ and $F(y)$ are exactly alike except that x is free in $F(x)$ at all and only those places that y is free in $F(y)$:

TH. 25. $\vdash (x)F(x) \equiv (y)F(y)$

Proof: $\vdash (x)F(x) \supset F(y)$	P 5
$\vdash (x)F(x) \supset (y)F(y)$	DR 1

We obtain $\vdash (y)F(y) \supset (x)F(x)$ the same way, and then have
$\vdash (x)F(x) \equiv (y)F(y)$ — by Ⓟ

TH. 26. $\vdash (\exists x)F(x) \equiv (\exists y)F(y)$

Proof: $\vdash (x) \sim F(x) \equiv (y) \sim F(y)$	Th. 25
$\vdash \sim(x) \sim F(x) \equiv \sim(y) \sim F(y)$	Ⓟ
$\vdash (\exists x)F(x) \equiv (\exists y)F(y)$	df.

Theorems 25 and 26, together with the Rule of Replacement, permit the 'interchange of bound variables'—subject to the restrictions mentioned in the statement of the theorems.

We end the present group of theorems with

TH. 27. If P and Q are as in P 5, then $\vdash Q \supset (\exists x)P$

Proof:	$\vdash (x){\sim}P \supset {\sim}Q$	P 5
	$\vdash {\sim}{\sim}Q \supset {\sim}(x){\sim}P$	Ⓟ
	$\vdash Q \supset {\sim}(x){\sim}P$	Ⓟ
	$\vdash Q \supset (\exists x)P$	df.

10.4 RS_1 and the 'Natural Deduction' Techniques

We wish now to show that RS_1 *is* an axiomatization of the logic we *used* in validating inferences by our list of nineteen Rules of Inference, the strengthened method of Conditional Proof, and the Quantification Rules set forth in Chapter 4. That RS_1 contains all the nineteen Rules of Inference has been shown by our completeness proof for the system of logic based on P 1, P 2, P 3, and R 1, and our discussion in connection with Th. 0 and DR 0. In a sense, the Deduction Theorem for RS_1 corresponds to the strengthened method of Conditional Proof.

In the presence of R 1, P 5 corresponds to **UI**. The restrictions on **UI** are exactly the same as the restrictions on P 5. We may establish **UI** as a derived rule of RS_1:

DR 7. If P and Q are as in P 5 then $(x)P \vdash Q$.

Demonstration:	$(x)P$	premiss
	$(x)P \supset Q$	P 5
	Q	R 1

Again in the presence of R 1, **EG** follows from P 5 via Th. 27. The restrictions on **EG** are exactly the same as those on P 5. We may establish **EG** as a derived rule of RS_1:

DR 8. If P and Q are as in P 5, then $Q \vdash (\exists x)P$.

Proof:	Q	premiss
	$Q \supset (\exists x)P$	Th. 27
	$(\exists x)P$	R 1

The rule **UG** corresponds roughly to R 2, although it seems from one point of view weaker than R 2, and from another point of view stronger. The second restriction on **UG** prevents its use to infer $(\mu)(\Phi\mu)$ from $\Phi\nu$, where ν is a variable that occurs free in an assumption within whose scope $\Phi\nu$ lies. But there is no such restriction on R 2. That restriction on **UG**, however, serves only to limit its application in the context of a Conditional Proof. In RS_1 the same limitation is imposed by placing the restriction on the rule of Conditional Proof itself, in our statement of the Deduction Theorem—which does not permit going from $P \vdash Q$ to $\vdash P \supset Q$ if R 2 is used on a variable occurring free in *P*. Hence in the context of RS_1 the second restriction on **UG** applies to R 2, although it is stated for the Deduction Theorem rather than for R 2 itself. So **UG** is not really weaker (more restricted) than R 2.

The rule **UG** seems stronger than R 2 in that it permits a change of variable in its application, which R 2 does not. But in the presence of the rest of the system RS_1, **UG** is easily proved as a derived rule, combining the power of R 2 and Th. 25. The general convention governing $\Phi\mu$ and $\Phi\nu$ in Chapter 4, and the first restriction on **UG** (that ν does not occur free in $(\mu)(\Phi\mu)$), are together equivalent to the conditions stated for Th. 25, that $\Phi\mu$ and $\Phi\nu$ are exactly alike except that μ is free in $\Phi\mu$ at all and only those places that ν is free in $\Phi\nu$. We have $F(x) \vdash (x)F(x)$ directly by R 2, and hence by Th. 25 we have

DR 9. If $F(x)$ and $F(y)$ are exactly alike except that x is free in $F(x)$ at all and only those places that y is free in $F(y)$, then $F(x) \vdash (y)F(y)$.

We can do everything in RS_1 by DR 9 that we could do earlier by **UG**.

In Th. 0 and DR 0, and in DR 7, DR 8, DR 9, and the Deduction Theorem, the system RS_1 is seen to comprise all of the deductive strength available in our nineteen Rules of Inference, Quantification Rules **UI**, **EG**, **UG**, and the strengthened principle of Conditional Proof.

Instead of attempting to formulate a derived rule in RS_1 corresponding to **EI**, we prove a Metatheorem that legitimizes any use of **EI** in a proof subject to the restrictions involved in its statement.

METATHEOREM V. (The legitimacy of **EI**—perhaps better, the eliminability or avoidability of **EI**.)

If there is a sequence of wffs which would constitute a demonstration that $P_1, P_2, \ldots, P_n \vdash Q$ *except that some wffs of the sequence are derived from preceding ones by* **EI** *(here construed simply as the rule that if R can be derived in a given context from* $F(y)$*, then it can be derived in that context from* $(\exists x)F(x)$*) such that:*

1. *there are m uses of* **EI** *of which the ith use is to infer* R_i *from* $(\exists x_i)F_i(x_i)$ *on the basis of its having been derived from* $F_i(y_i)$*, where the free* x_i*'s in* $F_i(x_i)$ *correspond exactly to the free* y_i*'s in* $F_i(y_i)$*, and*
2. *no* y_i *occurs free either in Q, or in any premiss* P_j*, or in any wff of the sequence that precedes* $F_i(y_i)$*, or in the wff* R_i *derived from* $F_i(y_i)$*, and*
3. R 2 *is not used to infer* $(y_i)S$ *from any S;*

then there is a genuine demonstration that $P_1, P_2, \ldots, P_n \vdash Q$ *which makes no use of* **EI**. *And in the latter (genuine) demonstration there are no uses of* R 2 *to* *quantify variables free in* $P_1, P_2, \ldots, P_n$ *which were not quantified by* R 2 *in the original sequence or 'quasi-demonstration'.*

Proof: In any such quasi-demonstration every *wff* $F_i(y_i)$ is preceded by a *wff* $(\exists x_i)F_i(x_i)$ and followed by a *wff* R_i. Except for the m *formulas* $F_1(y_1)$, $F_2(y_2), \ldots, F_m(y_m)$ every *wff* of the indicated sequence is either an axiom or a P_j or follows from preceding *wffs* by R 1 or R 2. Suppose there are exactly t uses of R 2 of which the qth use is to infer $S_{k_q}(=(x_{j_q})S_{j_q})$ from S_{j_q}.

We define S'_{k_q} to be $(x_{j_1})\ldots(x_{j_{q-1}})(x_{j_{q+1}})\ldots(x_{j_t})S_{k_q}$, which is $(x_{j_1})\ldots(x_{j_{q-1}})$ $(x_{j_{q+1}})\ldots(x_{j_t})(x_{j_q})S_{j_q}$. Since $S'_{k_q} \vdash S_{k_q}$ by $t-1$ uses of P 5 and R 1, it is obvious from the existence of the original quasi-demonstration that there exists a genuine demonstration that

$$(1) \qquad P_1, P_2, \ldots, P_n, S'_{k_1}, S'_{k_2}, \ldots, S'_{k_t}, F_1(y_1), F_2(y_2), \ldots, F_m(y_m) \vdash Q$$

which involves no use of either **EI** or R 2. (Where $F_i(y_i)$ occurred without justification in the quasi-demonstration, here it is a premiss; and where S_{k_q} was inferred by R 2, here it is inferred from the premiss S'_{k_q} by $t-1$ uses of P 5 and R 1.)

Since no use is made of R 2 in the demonstration that (1), we can use the Deduction Theorem to obtain

$$(1') \quad P_1, P_2, \ldots, P_n, S'_{k_1}, S'_{k_2}, \ldots, S'_{k_t}, F_1(y_1), F_2(y_2), \ldots, F_{m-1}(y_{m-1}) \vdash F_m(y_m) \supset Q$$

Now by successive uses of R 2, Theorems 21 and 26, and the Rule of Replacement, we obtain from those same premisses first

$$\vdash (y_m)[F_m(y_m) \supset Q]$$

then

$$\vdash (\exists y_m)F_m(y_m) \supset Q$$

and finally

$$\vdash (\exists x_m)F_m(x_m) \supset Q$$

Since $F_m(y_m)$ is preceded by $(\exists x_m)F_m(x_m)$ in the original quasi-demonstration, we have also

$$\vdash (\exists x_m)F_m(x_m)$$

and finally by R 1

$$\vdash Q$$

Thus we have a demonstration that

$$(2) \qquad P_1, P_2, \ldots, P_n, S'_{k_1}, S'_{k_2}, \ldots, S'_{k_t}, F_1(y_1), F_2(y_2), \ldots, F_{m-1}(y_{m-1}) \vdash Q$$

in which R 2 is not used to quantify any variable free in any of the premisses $S'_{k_1}, S'_{k_2}, \ldots, S'_{k_t}, F_1(y_1), F_2(y_2), \ldots, F_{m-1}(y_{m-1})$. The foregoing argument can be used m times to obtain a demonstration that

$$(m) \qquad P_1, P_2, \ldots, P_n, S'_{k_1}, S'_{k_2}, \ldots, S'_{k_t} \vdash Q$$

in which R 2 is not used to quantify any variable free in any of the premisses $S'_{k_1}, S'_{k_2}, \ldots, S'_{k_t}$. Hence we can use the Deduction Theorem again to obtain

$$P_1, P_2, \ldots, P_n, S'_{k_1}, S'_{k_2}, \ldots, S'_{k_{t-1}} \vdash S'_{k_t} \supset Q$$

Since S_{j_t} preceded S_{k_t} in the original quasi-demonstration, we obtain also from those same premisses

$$\vdash S_{j_t}$$

Now by t uses of R 2 we obtain

$$\vdash S'_{k_t}$$

and finally by R 1

$$\vdash Q$$

Thus we have a demonstration that

$$(m+1) \qquad P_1, P_2, \ldots, P_n, S'_{k_1}, S'_{k_2}, \ldots, S'_{k_{t-1}} \vdash Q$$

in which R 2 is not used to quantify any variable free in any of the premisses $S'_{k_1}, S'_{k_2}, \ldots, S'_{k_{t-1}}$. The foregoing argument can be used t times to obtain a genuine demonstration that

$$(m+t) \qquad P_1, P_2, \ldots, P_n \vdash Q$$

in which there are no uses of R 2 to quantify variables free in $P_1, P_2, \ldots, P_n$ which were not quantified by R 2 in the original quasi-demonstration. This completes our proof of Metatheorem V.

The preceding discussion suffices to show that our first-order function calculus RS_1 *is* an axiomatization of the logic *used* in Chapter 4 and the first three sections of Chapter 5. That it contains no *more* than was used there can be shown by giving demonstrations (in the earlier sense of 'demonstration') of the postulates of RS_1.

The question naturally arises as to *how much* logic is contained in RS_1. We shall present a proof that RS_1 is deductively complete in the sense that all logically true propositions involving quantification of individual variables only are provable as theorems within it. This sense will be specified more precisely in Section 10.6. Before proving the completeness of our first-order function calculus, however, we shall develop some notions and Metatheorems concerning the subject of 'Normal Forms.'

10.5 Normal Forms

We begin by defining the notion of 'prenex normal form'. A *wff* is in *prenex normal form* if and only if it has the structure $(Qx_1)(Qx_2)\ldots(Qx_n)G$ where $x_1, x_2, \ldots, x_n$ are distinct individual variables, (Qx_i) is either $(\exists x_i)$ or (x_i), and G is a quantifier-free *wff* containing at least one occurrence of each of the x_i's. Some samples of formulas of RS_1 that are in prenex normal form are

$$(\boldsymbol{x})(\boldsymbol{A}^1(\boldsymbol{x}))$$
$$(\exists \boldsymbol{x})((\boldsymbol{B}^1(\boldsymbol{x})) \supset (\boldsymbol{C}^2(\boldsymbol{xy})))$$
$$(\boldsymbol{y})((\exists \boldsymbol{z})(\sim(\boldsymbol{A}^3(\boldsymbol{xyz}))))$$

We now define the term 'scope' as follows: If $(Qx)B$ is a well formed part (i.e., a *wff* that is a part) of a *wff* A, then the scope of that particular occurrence of (Qx) in A is the particular occurrence of B which immediately follows (Qx).[6] And we define the phrase 'initially placed' in this way: A quantifier is initially placed in a *wff* F if it either stands at the beginning of F or is preceded (except for parentheses) only by other quantifiers, and its scope extends to the end of F. It should be clear that an equivalent definition of 'prenex normal form' can be stated as: A *wff* is in *prenex normal form* if and only if all its quantifiers are initially placed, no two quantifiers are on the same variable, and every variable occurring in a quantifier occurs also at least once within the *scope* of that quantifier.

We can now state and prove our next Metatheorem.

METATHEOREM VI. *Given any wff F of RS_1, a formula P in prenex normal form can be found such that* $\vdash F \equiv P$.

Proof: If F is in prenex normal form, P is F and $\vdash F \equiv P$ by Ⓟ. If F is not in prenex normal form, it must either: (a) contain a quantifier that is not initially placed, or (b) contain two quantifiers on the same variable, or (c) contain a variable that occurs *only* in a quantifier.

[6]This definition provides a more precise formulation of the notion of scope that was discussed informally in Section 4.4. See pp. 86–87.

CASE 1. All quantifiers in F are initially placed.

Subcase 1.i. At least two quantifiers are on the same variable. Here F is of the form

$$(Qx_1)(Qx_2)\dots(Qx_i)\dots(Qx_i)\dots(Qx_n)G$$

If we denote all of F that immediately follows the first (Qx_i) by A, then x_i does not occur free in A, and by Th. 14 or Th. 15, $\vdash (Qx_i)A \equiv A$. Hence by the Rule of Replacement the first of any initially placed quantifiers on the same variable can be simply dropped out to yield an equivalent formula. So if all quantifiers of F are initially placed, then there is an equivalent formula all of whose quantifiers are initially placed and which has at most one quantifier on any variable.

Subcase 1.ii. There is a variable in F that occurs only in a quantifier. Here we have F of the form

$$(Qx_1)\dots(Qx_i)\dots(Qx_n)G$$

where x_i does not occur free in G. If we denote all of F that immediately follows (Qx_i) by A, then x_i does not occur free in A and by Th. 14 or Th. 15, $\vdash (Qx_i)A \equiv A$. Hence by the Rule of Replacement any initially placed quantifier on a variable which has no occurrence outside of that quantifier may simply be dropped out to yield an equivalent formula. So if all quantifiers of F are initially placed, then there is an equivalent formula, all of whose quantifiers are initially placed and that contains no variables whose only occurrences are in quantifiers.

CASE 2. Some quantifiers in F are not initially placed. In this case it is convenient to write F in unabbreviated form, so the only propositional calculus symbols in it are $\cdot$ and $\sim$. Next we consider a *wff* G such that no two quantifiers in G are on the same variable, no variable in G has both free and bound occurrences, and $\vdash F \equiv G$. If F is such a *wff*, let F be G; otherwise G is obtained from F by interchanging bound variables according to Theorems 25 and 26 and the Rule of Replacement.

Apart from parentheses, each symbol other than a quantifier that occurs to the left of a given quantifier (Qx_j) in a *wff* G can be counted as a "reason" why (Qx_j) is not initially placed in G, and each symbol that occurs to the right of (Qx_j) in G and does not lie within its scope can also be counted as a "reason" why (Qx_j) is not initially placed in G. It is obvious that any quantifier (Qx_j) occurring in a *wff* G is initially placed in G if and only if there are no "reasons" why it is not initially placed in G.

Given any such *wff* G, in which the quantifier on x_j is not initially placed and all quantifiers in G preceding the quantifier on x_j are initially placed in G, we can construct a *wff* G_1 such that: every initially placed quantifier in G is initially placed in G_1 also, $\vdash G \equiv G_1$, and there are fewer "reasons" why the quantifier on x_j is not initially placed in G_1 than there were "reasons" why the quantifier on x_j was not initially placed in G. The construction of G_1 proceeds as follows.

If (Qx_j) is the first quantifier in G that is not initially placed in G, then it can occur in only one of the following wf parts of G: $((Qx_j)(A))\cdot(B)$ or $(B)\cdot((Qx_j)(A))$ or $\sim((Qx_j)(A))$.

In the first case, since x_j has no free occurrence in B, we replace the wf part $((Qx_j(A))\cdot(B)$ by $(Qx_j)((A)\cdot(\mathrm{B}))$, using Theorem 16 or Theorem 23 and the Rule of Replacement, to obtain the *wff* G_1.

In the second case, we first use Ⓟ and the Rule of Replacement to replace the wf part $(B)\cdot((Qx_j)(A))$ by $((Qx_j)(A))\cdot(B)$, and then replace that as in the preceding case by $(Qx_j)((A)\cdot(B))$ to obtain the *wff* G_1.

In the third case we obtain the *wff* G_1 using Theorem 7 or Theorem 8 and the Rule of Replacement to replace the wf part $\sim((Qx_j)(A))$ by $(Q^*x_j)(\sim(A))$, where (Q^*x_j) is either (x_j) or $(\exists x_j)$ according as (Qx_j) is $(\exists x_j)$ or (x_j).

If the quantifier on x_j is not initially placed in G_1, we repeat the construction just described to obtain G_2, in which there are still fewer "reasons" why the quantifier on x_j is not initially placed. Since each *wff* contains only finitely many symbols, in any *wff*, there can be only finitely many "reasons" why any quantifier in it is not initially placed. Hence the construction described need be iterated only finitely many times to produce a *wff* G_k in which the quantifier on x_j is initially placed.

A new series of such constructions can be used for each quantifier that is not initially placed in the original *wff* F, until finally a *wff* G_K is obtained in which all quantifiers are initially placed and such that $\vdash F \equiv G_K$. And to G_K there is an equivalent *wff* in prenex normal form by Case 1.

This completes the proof of Metatheorem IV.

We define two new terms at this point: In any *wff* in prenex normal form

$$(Qx_1)(Qx_2)\ldots(Qx_n)G$$

the group of quantifiers $(Qx_1)(Qx_2)\ldots(Qx_n)$ is the *prefix* and the quantifier-free formula G is the *matrix*.

EXERCISES

Find a prenex normal form for

1. $(x)(\exists y)(\exists x)(y)[(z)G(z)\cdot F(x) \supset (\exists z)H(z)\cdot F(z)]$
*2. $[(x)G(x) \vee H(y)] \equiv [(y)(z)F(y, z)]$
3. $(x)[F(x) \supset G(x)] \supset [(x)F(x) \supset (x)G(x)]$
*4. $(y)(\exists x)F(x, y) \supset (\exists x)(y)F(x, y)$
5. $(x_1)(\exists x_2)(x_3)(\exists x_4)(x_5)[\mathrm{P}^6(x_1, \ldots, x_6) \supset (z_1)(\exists x)(\exists z)H(x_1, z)]$

A *wff* in prenex normal form may contain free variables. We wish to be able, without any loss in generality, to confine our attention to *wffs* containing no free variables. These will be referred to as *closed wffs* (abbreviated *cwffs*). To legitimize our concentrating on *cwffs*, we state and prove the following:

METATHEOREM VII. *For any wff F a cwff G may be found in prenex normal form such that* $\vdash F$ *if and only if* $\vdash G$.

Proof: By Metatheorem VI we can always find the prenex normal form P for any *wff* F such that $\vdash F \equiv P$.

CASE 1. If P is a *cwff*, then G is the same as P, and because $\vdash F \equiv G$, $\vdash F$ if and only if $\vdash G$.

CASE 2. If P contains n free variables $x_1, x_2, \ldots, x_n$, then G is the *cwff* $(x_1)(x_2) \ldots (x_n)P$. For if $\vdash F$, then $\vdash P$, and by R 2, $\vdash (x_n)P$, and by n uses of R 2, we have $\vdash G$. And if $\vdash G$, that is, $\vdash (x_1)(x_2) \ldots (x_n)P$, then since $\vdash (x_1)(x_2) \ldots (x_n)P \supset (x_2) \ldots (x_n)P$ by P 5, R 1 gives $\vdash (x_2) \ldots (x_n)P$, and by n uses of P 5 and R 1, we have $\vdash P$, and hence $\vdash F$.

We prove now a further result that gives for any *wff* a still more specialized normal form *wff* that is a theorem if and only if the original *wff* is a theorem.

METATHEOREM VIII. *For any wff F a cwff R may be found which is in prenex normal form and begins with an existential quantifier, such that* $\vdash F$ *if and only if* $\vdash R$.

Proof: By Metatheorem VII, for any *wff* F, there is a *cwff* G in prenex normal form such that $\vdash F$ if and only if $\vdash G$. Let $D(t)$ be a function of one variable t, such that neither D nor t occurs in G. Then by Ⓟ, $\vdash G \equiv \{G \cdot [D(t) \supset D(t)]\}$.

Hence

$\vdash G \supset \{G \cdot [D(t) \supset D(t)]\}$	by Ⓟ
$\vdash (\exists t)\{G \supset \{G \cdot [D(t) \supset D(t)]\}\}$	by DR 8, (**EG**)
$\vdash G \supset (\exists t)\{G \cdot [D(t) \supset D(t)]\}$	by Th. 19 and R.R.

We also have

$\vdash \{G \cdot [D(t) \supset D(t)]\} \supset G$	P 2
$\vdash (t)\{\{G \cdot [D(t) \supset D(t)]\} \supset G\}$	R 2
$\vdash (\exists t)\{G \cdot [D(t) \supset D(t)]\} \supset G$	Th. 21 and R.R.

Hence

$\vdash (\exists t)\{G \cdot [D(t) \supset D(t)]\} \equiv G$	by Ⓟ

Now the prenex normal form of $(\exists t)\{G \cdot [D(t) \supset D(t)]\}$ is the formula *of type R* that was desired. It is closed, it is in prenex normal form, it begins with an existential quantifier, and $\vdash F$ if and only if $\vdash R$. Where G is $(Qx_1)\ (Qx_2) \ldots (Qx_n)G'$, then

$$\vdash (\exists t)\{G \cdot [D(t) \supset D(t)]\} \text{ or}$$
$$\vdash (\exists t)\{[(Qx_1)(Qx_2) \ldots (Qx_n)G'] \cdot \sim[D(t) \cdot \sim D(t)]\}$$

has as its prenex normal form

$$\vdash (\exists t)(Qx_1)(Qx_2) \ldots (Qx_n)\{G' \cdot \sim[D(t) \cdot \sim D(t)]\}$$

by repeated uses of Th. 16 or Th. 23 and the Rule of Replacement.

By definition, any *cwff* in prenex normal form that begins with an existential quantifier will be said to be *of type R*.

EXERCISES

For each of the following *wffs*, construct a formula *of type R* that is a theorem if and only if the given *wff* is a theorem:

*1. $(x)(y)(z)\{[H(x) \vee H(y)] \vee H(z)\}$
2. $F(x_1, x_2, \ldots, x_n) \supset G(z)$
*3. $(\exists x)G(x, y) \vee F(z)$
4. $(\exists z)F(z, z) \supset (x)G(x, y)$
5. $(x)F(x, y) \supset G(z)$

Next, we present three new definitions, of which the third defines the 'Skolem normal form', which is a still more specialized type of well formed formula, about which we shall establish a further Metatheorem.

Df. The *rank of an occurrence of an existential quantifier* in a formula *of type R* is the number of universal quantifiers which precede that existential quantifier in the prefix.

Df. The *rank of a formula of type R* is the greatest of the ranks of the occurrences of the existential quantifiers of its prefix.

Df. A *wff F is in Skolem normal form* if and only if it is *of type R* and of rank 0.

METATHEOREM IX. *Given any wff F, we can find a formula F_0 in Skolem normal form such that $\vdash F$ if and only if $\vdash F_0$.*

Proof: For any *wff F* we can construct a formula *of type R*, call it F_1, such that $\vdash F$ if and only if $\vdash F_1$. (If F is already *of type R*, then $F = F_1$.) If F_1 is of rank 0, then it is in the desired Skolem normal form. If F_1 is of rank $k > 0$, then we show how to construct a formula *of type* R of rank less than k, call it F_2, such that $\vdash F_1$ if and only if $\vdash F_2$. We thus embark on a process that will yield formulas $F_3, F_4, F_5, \ldots$ all *of type* R and of decreasing rank. Ultimately, then, we arrive at a formula F_0 *of type* R and of rank 0 such that $\vdash F$ if and only if $\vdash F_0$.

If F_1 is not in Skolem normal form, then its rank is greater than 0, that is, F_1 has the structure

$$(\exists x_1)(\exists x_2)\ldots(\exists x_n)(y)(Qz_1)(Qz_2)\ldots(Qz_m)G$$

where $(Qz_1)(Qz_2), \ldots, (Qz_m)$ are quantifiers, of which at least one is an existential quantifier. We now introduce the following notations:

1. $B(x_1, x_2, \ldots, x_n, y) = \text{df } (Qz_1)(Qz_2)\ldots(Qz_m)G$. This we shall abbreviate as A.
2. $H(y) = \text{df } H(x_1, x_2, \ldots, x_n, y)$ where H is any $(n+1)$-place predicate variable (with arguments $x_1, x_2, \ldots, x_n, y$) such that H does not occur in G.
3. $H(t) = \text{df } H(x_1, x_2, \ldots, x_n, t)$, where t is a variable that does not occur in G.
4. $C = \text{df } (\exists x_1)(\exists x_2)\ \ldots\ (\exists x_n)\{(y)[(Qz_1)(Qz_2)\ \ldots\ (Qz_m)G \supset H(y)] \supset (t)H(t)\}$

First, we want to show that $\vdash F_1$ if and only if $\vdash C$. The two implications are

(i) If $\vdash F_1$, then $\vdash C$. Here we prove that $\vdash F_1 \supset C$.

$\vdash A \supset \{[A \supset H(y)] \supset H(y)\}$	Ⓟ
$\vdash (y)\{A \supset \{[A \supset H(y)] \supset H(y)\}\}$	R 2
$\vdash (y)A \supset (y)\{[A \supset H(y)] \supset H(y)\}$	DR 5
$\vdash (y)A \supset \{(y)[A \supset H(y)] \supset (y)H(y)\}$	Th. 2, Ⓟ
$\vdash (y)A \supset \{(y)[A \supset H(y)] \supset (t)H(t)\}$	Th. 25, R.R.
$\vdash (x_n)\{(y)A \supset \{(y)[A \supset H(y)] \supset (t)H(t)\}\}$	R 2
$\vdash (\exists x_n)(y)A \supset (\exists x_n)\{(y)[A \supset H(y)] \supset (t)H(t)\}$	Th. 4, R 1
....	
....	
$\vdash (\exists x_1)(\exists x_2)\ldots(\exists x_n)(y)A \supset (\exists x_1)(\exists x_2)\ldots(\exists x_n)\{(y)[A \supset H(y)] \supset (t)H(t)\}$	by n uses of R 2, Th. 4, R 1
$\vdash F_1 \supset C$	df.

Hence, if $\vdash F_1$, then $\vdash C$.

(ii) If $\vdash C$, then $\vdash F_1$. Here we must assume that $\vdash C$ and derive $\vdash F_1$.

$\vdash (\exists x_1) \ldots (\exists x_n)\{(y)[(Qz_1) \ldots (Qz_m)G \supset H(y)] \supset (t)H(t)\}$ C
$\vdash (\exists x_1) \ldots (\exists x_n)\{(y)[B(x_1, \ldots, x_n, y) \supset$
$H(x_1, \ldots, x_n, y)] \supset (t)H(x_1, \ldots, x_n, t)\}$ df.
$\vdash (\exists x_2) \ldots (\exists x_n)\{(y)[B(x_1', \ldots, x_n, y) \supset$
$H(x_1', \ldots, x_n, y)] \supset (t)H(x_1', \ldots, x_n, t)\}$ **EI**

. . . .

$\vdash (y)[B(x_1', \ldots, x_n', y) \supset H(x_1', \ldots, x_n', y)] \supset (t)H(x_1', \ldots, x_n', t)$ n uses of **EI**

Now we substitute $B(x_1', \ldots, x_n', y)$ for the predicate variable $H(x_1', \ldots, x_n', y)$[7]

$\vdash (y)[B(x_1', \ldots, x_n', y) \supset B(x_1', \ldots, x_n', y)] \supset (t)B(x_1', \ldots, x_n', t)$
$\vdash (y)[B(x_1', \ldots, x_n', y) \supset B(x_1', \ldots, x_n', y)] \supset (y)B(x_1', \ldots, x_n', y)$ Th. 25, R.R.
$\vdash (\exists y)\{[B(x_1', \ldots, x_n', y) \supset B(x_1', \ldots, x_n', y)] \supset$
$(y)B(x_1', \ldots, x_n', y)\}$ Th. 18, R.R.
$\vdash [B(x_1', \ldots, x_n', y) \supset B(x_1', \ldots, x_n', y)] \supset (y)B(x_1', \ldots, x_n', y)$ **EI**
$\vdash B(x_1', \ldots, x_n', y) \supset B(x_1', \ldots, x_n', y)$ Ⓟ
$\vdash (y)B(x_1', \ldots, x_n', y)$ R 1
$\vdash (\exists x_n)(y)B(x_1', \ldots, x_n, y)$ Dr 8, (**EG**)

. . . .

$\vdash (\exists x_1) \ldots (\exists x_n)(y)B(x_1, \ldots, x_n, y)$ n uses of **EG**
$\vdash (\exists x_1) \ldots (\exists x_n)(y)(Qz_1) \ldots (Qz_m)G$ df.
$\vdash F_1$ df.

Hence if $\vdash C$, then $\vdash F_1$.

C is not in prenex normal form. Its prenex normal form is constructed as follows, using the notation Q^* defined as follows:

If (Qz_i) is (z_i), then (Q^*z_i) is $(\exists z_i)$
If (Qz_i) is $(\exists z_i)$, then (Q^*z_i) is (z_i)

$$C \equiv (\exists x_1) \ldots (\exists x_n)\{(y)[(Qz_1)(Qz_2) \ldots (Qz_m)G \supset H(y)] \supset (t)H(t)\}$$

Since $H(y)$ does not contain any free z_i's,

$$\vdash (Qz_1) \ldots (Qz_m)G \supset H(y) \equiv (Q^*z_1) \ldots (Q^*z_m)[G \supset H(y)]$$

by m uses of Th. 18 or Th. 21. Hence by the Rule of Replacement,

$$C \equiv (\exists x_1) \ldots (\exists x_n)\{(y)(Q^*z_1) \ldots (Q^*z_m)[G \supset H(y)] \supset (t)H(t)\}$$

[7] The rule governing such substitution (rule of substitution for functional variables) is difficult and complicated to express. Since this is our only use of it, we shall not state it here. A discussion of alternative formulations can be found in A. Church, *Introduction to Mathematical Logic I*, Princeton University Press, Princeton, N.J., 1956, pp. 193 ff., 289, and 297.

Since $(t)H(t)$ does not contain any free z_i's,

$$\vdash \{(y)(Q^*z_1) \ldots (Q^*z_m)[G \supset H(y)] \supset (t)H(t)\} \equiv \{(\exists y)(Qz_1) \ldots (Qz_m)[[G \supset H(y)] \supset (t)H(t)]\}$$

by $m + 1$ uses of Th. 18 or Th. 21. Hence by the Rule of Replacement,

$$C \equiv (\exists x_1) \ldots (\exists x_n)(\exists y)(Qz_1) \ldots (Qz_m)[[G \supset H(y)] \supset (t)H(t)]$$

Since t is not free in $G \supset H(y)$, by Th. 20,

$$\vdash [[G \supset H(y)] \supset (t)H(t)] \equiv (t)[[G \supset H(y)] \supset H(t)],$$

hence by the Rule of Replacement,

$$C \equiv (\exists x_1) \ldots (\exists x_n)(\exists y)(Qz_1) \ldots (Qz_m)(t)[[G \supset H(y)] \supset H(t)]$$

The right-hand formula is the formula F_2 *of type R*, such that $\vdash F$ if and only if $\vdash F_2$. Now the rank of F_2 is lower than the rank of F_1, for their prefixes are the same except that F_2 has an additional universal quantifier in the final position. This does not affect the rank. Another exception is that one of the universal quantifiers ((y)) which preceded an existential quantifier has been replaced by an existential quantifier, which lowers the rank by one. If F_2 has rank 0 then it is the Skolem normal form of F. If it is not, then the argument can be repeated to obtain a formula F_3 *of type R* of still lower rank. Hence we can ultimately arrive at a formula *of type R* of rank 0, which is the desired Skolem normal form.

EXERCISES

Find a Skolem normal form for

*1. $(\exists x)(y)(\exists z)[F(x, y) \vee G(z)]$

2. $(x)F(x) \supset [(\exists y)G(y) \supset (z)H(z)]$

*3. $(x)(\exists y)F(x, y)$

4. $(x)(y)(\exists z)(\exists x)\{[[F(x) \cdot G(y)] \vee H(z)] \supset I(w)\}$

5. $(\exists x_1)(x_2)(x_3)(\exists x_4)F(x_1, x_2, x_3, x_4)$

10.6 Completeness of RS_1

We turn now to the problem of proving our first-order function calculus complete. There are several different senses of the term 'complete' which must be distinguished. A very strong kind of completeness is possessed by a system *all* of whose well formed formulas are provable as theorems. But this kind of completeness amounts to inconsistency, and is of course undesirable. Our first

Metatheorem, which established the consistency of our system, proved it to lack *this* kind of completeness. A somewhat weaker kind of completeness is possessed by a system each of whose *wffs* is either provable as a theorem or else has its denial provable as a theorem; that is, for every *wff* F either $\vdash F$ or $\vdash \sim F$. This kind of completeness is not desirable either, for on their normal interpretations, each of the following *wffs*

$$N_1\text{: } (\exists x)F(x) \supset (x)F(x)$$
$$N_2\text{: } (\exists x)(\exists y)\{F(x) \cdot F(y) \cdot [G(x) \equiv \sim G(y)]\} \supset (x)F(x)$$
$$N_3\text{: } (\exists x)(\exists y)(\exists z)\{F(x) \cdot F(y) \cdot F(z) \cdot [G(x) \equiv \sim G(y)] \cdot [H(y) \equiv \sim H(z)] \cdot [I(x) \equiv \sim I(z)]\} \supset (x)F(x)$$

. . . .

. . . .

asserts (respectively) that there is at most one individual, that there are at most two individuals, that there are at most three individuals, . . . But if we want our logical system to be applicable to *any* possible nonempty universe, *regardless* of the exact number of individuals it contains, there is no N_i such that we want either N_i or $\sim N_i$ to be provable as theorems. A different kind of completeness was proved for the propositional calculus presented in Chapter 8. That logistic system was proved to be complete in the sense that every *wff* that (on its normal interpretation) is a truth table tautology is provable as a theorem in the system. That kind of completeness is admirably suited for a propositional calculus, but it will not do for a functional calculus. A first-order function calculus that did not have the *wff*

$$(\exists x)(y)F(x, y) \supset (y)(\exists x)F(x, y)$$

provable as a theorem within it would be woefully unsatisfactory and 'incomplete', even though the formula in question is *not* a truth table tautology. The reason for regarding as unsatisfactory a logic in which the given expression is a *wff* but not provable is that (on its normal interpretation) the formula expresses a logical truth—where a 'logical truth' is a proposition that is true in (or of) *every possible nonempty universe.* The kind of completeness desired for our first-order function calculus can be expressed loosely by saying that every logical truth that can be expressed in the system is provable as a theorem within it. This notion, however, must be expressed much more precisely before it can be used in any proof of completeness.[8]

[8] The following discussion and proof of completeness is an adaptation to RS_1 of the completeness proof for an alternative first-order function calculus given by Leon Henkin in his article 'The Completeness of the First-Order Functional Calculus', *The Journal of Symbolic Logic*, vol. 14 (1949), pp. 159–166. Professor Henkin simplified the published proof and he communicated it to this writer in the spring of 1951. It is reproduced here with his kind permission.

The first proof of completeness for a first-order function calculus was published by Kurt Gödel, 'Die Vollstandigkeit der Axiome der logischen Funktionenkalkuls', *Monatshefte für Mathematik und Physik*, vol. 37 (1930), pp. 349–360.

Instead of continuing to speak of 'possible universes', we shall talk about *models*, where a model is any (nonempty) collection of elements each of which is thought of as an *individual*. And instead of speaking of our system of logic as being 'applied' to a 'possible universe', we shall speak of a model as constituting an 'interpretation' of our formal system. This last term will be given a precise meaning presently. With any given set of individuals that may be intended to serve as a *model*, we assume that we are given also the attributes that may belong to the individual elements, and the relations (binary, ternary, and so on) that may hold between (or among) them. Now we define '*interpretation of a wff S with respect to a given model*' as an assignment of meanings such that

1. To each propositional symbol in S we assign a truth value, either **T** or **F**.
2. To each individual constant in S, and to each variable with free occurrence in S, we assign an element of the model.
3. To each predicate symbol in S we assign an attribute or a binary or ternary or n-ary relation according as it is of degree (has right superscript) 1, 2, 3, or n.

This notion of interpretation of a *wff* with respect to a given model is incomplete, however, because no mention is made of what to do with the logical symbols $\sim$, $\cdot$, and the quantifier symbol (x). Our intended or *normal* interpretation of a *wff* S with respect to a given model is defined as an interpretation of S with respect to that model, subject to the following conditions:

(a) Any well formed part $\sim W$ is assigned the truth value **T** or **F** according as the well formed part W is assigned the truth value **F** or **T**.
(b) Any well formed part of the form $X \cdot Y$ is assigned the truth value **T** if and only if both X and Y are assigned the truth value **T**.
(c) Any well formed part $(x)R$ is assigned the truth value **T** if and only if R is assigned the value **T**, regardless of which element of the model is assigned to all free occurrences of x in R.
(d) Any well formed part $P^n(x_1, x_2, \ldots, x_n)$ is assigned the truth value **T** if and only if the elements of the model assigned to $x_1, x_2, \ldots, x_n$, in that order, stand to each other in the n-ary relation assigned to P^n.

So far we have defined 'interpretation' and 'normal interpretation' only for a *wff* with respect to a specified model. An *interpretation of a system with respect to a given model* is an assignment of meanings that provides interpretations for *all wffs* of the system, and a *normal interpretation of a system with respect to a given model* is an assignment of meanings which provides *normal* interpretations for all *wffs* of the system.

There are many interpretations, and even many normal interpretations of a formal system with respect to a given model. On some of these the value **T** may be assigned a *wff* that on another interpretation is assigned the truth value **F**.

It is obvious, however, that the postulates of our first-order function calculus will be assigned the truth value **T** on *any* normal interpretation of the system with respect to *any* model. A particular normal interpretation of a *wff* S with respect to a specified model will be said to *satisfy* S with respect to that model if by it S is assigned the truth value **T**. A *wff* S will be said to be *satisfiable with respect to a model* if there is a normal interpretation of S with respect to that model which assigns the truth value **T** to S. And a *wff* S will be said to be *satisfiable* if there is at least one model and a normal interpretation of S with respect to it which assigns the truth value **T** to S. 'Valid' is another term that it is important to introduce in this connection. A *wff* S is *valid with respect to a model* if every normal interpretation of S with respect to that model assigns the truth value **T** to S. It follows immediately from the definitions that the *wff* S is *valid* with respect to a given model if and only if every normal interpretation of S with respect to that model *satisfies* S. Finally, we define the term 'valid' by itself: a *wff* S is *valid* if it is valid with respect to every model. It follows immediately from the definitions that for any *wff* S, either S is valid or $\sim$S is satisfiable.

In the notion of a *model*, we have a precise formulation of our informal notion of a possible universe, and *normal interpretation* is what ought to be signified by the looser phrase 'intended application'. Consequently, the *validity* of a *wff* is a precise notion that serves the purpose suggested by the phrase 'logically true'. The completeness we wish to prove for our first-order function calculus RS_1 can now be defined precisely. A first-order function calculus is *complete* if and only if every valid *wff* is provable in it as a theorem.

As for a propositional calculus, so for a first-order function calculus: the converse of completeness is analyticity. An *analytic* first-order function calculus is one all of whose theorems are valid. It is obvious that all the postulates of RS_1 are valid, and that any *wffs* derived from them by R 1 and R 2 are valid also. Clearly RS_1 is analytic. The consistency of RS_1 is a consequence of its analyticity, just as the consistency of R.S. was shown to be a consequence of the analyticity of R.S. in Chapter 8.

In establishing the completeness of RS_1, it will be sufficient to confine our attention to *wffs* containing no free individual variables. It is easily demonstrated that to every *wff* S which contains free individual variables, there corresponds a *wff* S^c (called the closure of S) containing no free individual variables, such that $\vdash$ S if and only if $\vdash S^c$, and such that S is valid if and only if S^c is valid. Where S contains exactly n free variables, say $x_1, x_2, \ldots, x_{n-1}, x_n$ (in the order of their occurrence), then S^c will be $(x_1)(x_2) \ldots (x_{n-1})(x_n)$S. Now if $\vdash$ S then $\vdash (x_n)$S by R 2, and another application of R 2 gives $\vdash (x_{n-1})(x_n)$S. Finally, we have $\vdash S^c$ by n uses of R 2. Conversely, if $\vdash S^c$, that is, $\vdash (x_1)(x_2) \ldots (x_{n-1})(x_n)$S, then applying R 1 to it and the instance of P 5 which is $(x_1)(x_2) \ldots (x_{n-1})(x_n)\text{S} \supset (x_2) \ldots (x_{n-1})(x_n)$S will give $\vdash (x_2) \ldots (x_{n-1})(x_n)$S; so by n uses of R 1 and P 5, we obtain $\vdash$ S. If S is valid, every normal interpretation of S with respect to any model assigns to S the value **T** regardless of which elements of the model its free individual variables $x_1, x_2, \ldots, x_n$ are taken to denote. From this it follows that (x_n)S is valid, and $(x_{n-1})(x_n)$S is valid, . . . , and finally that S^c is valid. On the other hand, if S^c is valid, $(x_2)(x_3) \ldots (x_n)$S is assigned the value **T**, regardless of which element of the model x_1 is taken to denote, from which it follows that $(x_3) \ldots (x_n)$S is assigned the value **T** regardless of which elements x_1 and x_2 are taken to denote, from which it follows . . . finally that S is assigned the value **T** regardless of which elements of the model the free individual variables $x_1, x_2, \ldots, x_n$ are taken to denote, which means that S is valid. Hence with no real loss in generality we can confine our attention to closed *wffs* (*wffs* containing no free variables), which we shall refer to as *cwffs*.

To prove our system complete, it is useful to introduce the notion of a *set* of *cwffs*. We shall use the capital Greek letters *Gamma* and *Lambda*, with and without subscripts, to denote *sets* of *cwffs* of RS_1. We shall also make use of braces to represent sets of *cwffs:* for example, the set whose only member is the *cwff* S will be written $\{S\}$, the set whose only two members are the *cwffs* S_1 and S_2, will be written $\{S_1, S_2\}$, and so on. To assert that there is a demonstration of the *wff* Q from the closed well formed premisses $P_1, P_2, \ldots, P_n$, we may write *either*

$$P_1, P_2, \ldots, P_n \vdash Q$$

or

$$\{P_1, P_2, \ldots, P_n\} \vdash Q$$

If Γ is a set of *cwffs* and S is a *cwff* not belonging to Γ, then the new set containing all members of Γ *and* S will be written as $\Gamma \cup \{S\}$. To assert that there is a demonstration of Q from premisses belonging to $\Gamma \cup \{S\}$ we may write either

$$\Gamma \cup \{S\} \vdash Q \quad \text{or} \quad \Gamma, S \vdash Q$$

An alternative statement of the Deduction Theorem makes use of this new notation:

$$\text{If } \Gamma, P \vdash Q \quad \text{then} \quad \Gamma \vdash P \supset Q$$

If there is a demonstration of the *wff* Q from the premisses $P_1, P_2, \ldots, P_n$, then there is, of course, a demonstration of Q from any set of *cwffs* Γ which contains $\{P_1, P_2, \ldots, P_n\}$ as a subset (which will be written $\{P_1, P_2, \ldots, P_n\} \subset \Gamma$), no matter what other *cwffs* Γ may contain. The set Γ may even be infinite: There is nothing in our proof of the Deduction Theorem that requires any revision to accommodate this possibility. Of course, in any demonstration of Q from an infinite set of premisses only a finite number of them will actually be *used*, for a demonstration is always a *finite* sequence of *wffs*.

We defined a first-order function calculus to be *complete* if and only if every valid *wff* is provable in it as a theorem, and then showed that the first-order function calculus RS_1 is complete if and only if every valid *cwff* is provable in it as a theorem. To say that every valid *cwff* is provable in RS_1 as a theorem is, by transposition, to say that for any *cwff* S, if S is *not* a theorem, then S is *not* valid.

In defining the terms 'valid' and 'satisfiable', we remarked that for any *wff* S, either S is valid or $\sim$S is satisfiable. Hence to say that S is not valid is to say that $\sim$S is satisfiable. And so we can say that RS_1 is complete if and only if, for any *cwff* S, if S is not a theorem, then $\sim$S is satisfiable. We can establish this result by introducing a characteristic φ such that both

(1) If S is a *cwff* that is not a theorem, then S is a *cwff* having characteristic φ; and
(2) If S is a *cwff* having characteristic φ, then $\sim$S is satisfiable;

for from (1) and (2) the completeness of RS_1 will follow by a Hypothetical Syllogism.

We first define two attributes of *sets* of *cwffs*. A set Λ of *cwffs* is *inconsistent if* $\Lambda \vdash \sim P \cdot P$, otherwise Λ is *consistent*. It is obvious that *any wff* can be inferred from an inconsistent set, for where S is any *wff*, $\sim P \cdot P \vdash (\sim P \cdot P) \vee S$ by Ⓟ, which is logically equivalent to $\sim P \cdot P \vdash \sim(\sim P \cdot P) \supset S$. Since $\vdash \sim(\sim P \cdot P)$ by Ⓟ, we have by R 1 that $\sim P \cdot P \vdash S$, whence $\Lambda \vdash S$ where Λ is any inconsistent set of *cwffs* and S is any *wff* whatever. Now the characteristic φ referred to above can be defined as follows: A *cwff* S has characteristic φ if and only if $\{\sim S\}$ is a consistent set.

And we demonstrate (1) above as the following:

LEMMA: If S is a *cwff* that is not a theorem, then $\{\sim S\}$ is a consistent set.

Proof: We show that for any *cwff* S, if $\{\sim S\}$ is *not* a consistent set then S *is* a theorem. Let S be any *cwff* such that $\{\sim S\}$ is not consistent. If $\{\sim S\}$ is not consistent, then it is inconsistent, which means that

$$\sim S \vdash \sim P \cdot P$$

Because S is a *cwff*, we can use the Deduction Theorem to obtain

$$\vdash \sim S \supset \sim P \cdot P$$

from which

$$\vdash \sim(\sim P \cdot P) \supset \sim\sim S$$

follows by Ⓟ. We already have

$$\vdash \sim(\sim P \cdot P)$$

by Ⓟ, so by R 1 we obtain

$$\vdash \sim\sim S$$

which by Ⓟ gives

$$\vdash S$$

which is the desired result.

Instead of demonstrating (2) above, simply, as

If S is a *cwff* such that $\{\sim S\}$ is a consistent set, then $\sim S$ is satisfiable,

we prove a somewhat more general result, for the statement of which the term 'simultaneously satisfiable' must be defined. The definition is this: A set Λ of *cwffs* of a formal system is *simultaneously satisfiable* if and only if there is a model and a normal interpretation of all formulas of Λ with respect to that model which satisfies all the *wffs* of Λ. Now we state and prove

METATHEOREM X. *Any consistent set of cwffs of* RS_1 *is simultaneously satisfiable.*

Proof: Our proof of the simultaneous satisfiability of any arbitrary consistent set Λ of *cwffs* of RS_1 proceeds by constructing a larger set of *cwffs* which includes Λ as a subset, and then proving that this larger set is simultaneously satisfiable. This will, of course, establish the desired result for Λ.

We begin by considering a set of distinct symbols $u_1, u_2, u_3, \ldots$ which are different from any of those contained in RS_1. We denote by 'RS_1^*' the formal system obtained from RS_1 by adding to it the symbols $u_1, u_2, u_3, \ldots$ which serve as individual constants. The only difference between RS_1 and RS_1^* is that the latter has a larger number of individual symbols. In every other respect the two systems are identically the same. We assume—for definiteness—that all *cwffs* of RS_1^* have been ordered in a sequence, so we can refer to them as the first, the second, the third, and so on (there are many alternative ways of accomplishing this, of course).

Starting with any arbitrary consistent set Λ of *cwffs* of RS_1, we enlarge it by adding *cwffs* to it one at a time. The first *cwff* to be added is $(\exists x)Q_1 \supset Q_1^*$, where $(\exists x)Q_1$ is the first *cwff* of RS_1^*, which begins with an existential quantifier; and Q_1^* is the result of replacing all free occurrences of x in Q_1 by u_{i_1}, where u_{i_1} is the first u_i which has no occurrence in Q_1. We may call the resulting set 'Λ_1', where $\Lambda_1 =_{df} \Lambda \cup \{(\exists x)Q_1 \supset Q_1^*\}$, and can easily prove it to be consistent. For if Λ_1 were inconsistent, we should have Λ, $(\exists x)Q_1 \supset Q_1^* \vdash \sim P \cdot P$, and by the Deduction Theorem, $\Lambda \vdash [(\exists x)Q_1 \supset Q_1^*] \supset \sim P \cdot P$. Since Λ contains no occurrences of u_{i_1} nor any free occurrence of any individual variables, in the derivation of $[(\exists x)Q_1 \supset Q_1^*] \supset \sim P \cdot P$ from

Λ we could replace all occurrences of u_{i_1} by the free individual variable x to obtain $\Lambda \vdash [(\exists x)Q_1 \supset Q_1] \supset \sim P \cdot P$. From that we could successively obtain

$\Lambda \vdash (x)\{[(\exists x)Q_1 \supset Q_1] \supset \sim P \cdot P\}$	by R 2
$\Lambda \vdash (\exists x)[(\exists x)Q_1 \supset Q_1] \supset \sim P \cdot P$	by Th. 21 and R.R.
$\Lambda \vdash [(\exists x)Q_1 \supset (\exists x)Q_1] \supset \sim P \cdot P$	by Th. 19 and R.R.
$\Lambda \vdash [(\exists x)Q_1 \supset (\exists x)Q_1]$	by Ⓟ

and finally

$$\Lambda \vdash \sim P \cdot P$$

which is contrary to the assumption that Λ is consistent. Hence Λ_1 is a consistent set.

The next *cwff* to be added is $(\exists x)Q_2 \supset Q_2^*$, where $(\exists x)Q_2$ is the second *cwff* of RS_1^* which begins with an existential quantifier, and Q_2^* is the result of replacing all free occurrences of x in Q_2 by u_{i_2}, where u_{i_2} is the first u_i with no occurrence in either Q_2 or Λ_1. We now define Λ_2 as $\Lambda_1 \cup \{(\exists x)Q_2 \supset Q_2^*\}$, and it can be proved to be consistent in exactly the same way that Λ_1 was. We continue to add *cwffs* one at a time: to Λ_{j-1} we add $(\exists x)Q_j \supset Q_j^*$ where $(\exists x)Q_j$ is the *jth cwff* of RS_1^* which begins with an existential quantifier, and Q_j^* is the result of replacing all free occurrences of x in Q_j by u_{i_j}, where u_{i_j} is the first u_i with no occurrence either in Q_j or in Λ_{j-1}. Each Λ_j is consistent, each of them contains Λ as a subset, and each set is contained in all of the later sets of the sequence.

Next, we define the set Λ_ω as the sum or union of all sets in the preceding sequence. (The existence of the set Λ_ω is guaranteed by the Union Axiom. See page 184.) Any *cwff* of RS_1^* is a member of Λ_ω provided that it is a member of any of the sets Λ_i of the sequence. It is clear that $\Lambda \subset \Lambda_\omega$, and easily proved that Λ_ω is consistent. For if Λ_ω were inconsistent, there would be a demonstration of $\sim P \cdot P$ from a finite number of formulas of Λ_ω. Each of these would have to occur in Λ or some Λ_i for the first time and in every Λ_j thereafter, for Λ_ω contains no formulas except those which occur in Λ or Λ_1 or Λ_2 or Λ_3 or . . . , and we have remarked that $\Lambda \subset \Lambda_1 \subset \Lambda_2 \subset \Lambda_3 \subset \ldots$ Since there can be only a finite number of premisses used in any demonstration, there would have to be a first set Λ_i which contains all the premisses actually used in the alleged proof that $\Lambda_\omega \vdash \sim P \cdot P$. But that proof, then, would equally well be a proof that $\Lambda_i \vdash \sim P \cdot P$, contrary to the fact that each Λ_i is consistent. Hence Λ_ω is consistent.

Now we proceed to enlarge Λ_ω until we obtain a *maximal* consistent set Γ of *cwffs* of RS_1^*. Our method of 'enlarging' Λ_ω is by successively adding to it each *cwff* of RS_1^* that is consistent with the formulas of Λ_ω, together with those previously added to it. The resulting set Γ will be consistent, since any demonstration that $\Gamma \vdash \sim P \cdot P$ would make use of only a finite number of formulas of Γ, the last of which (in the order of their addition) could *not* have been added if the directions for adding formulas had been followed. From our definition of Γ, it follows for any *cwff* of RS_1^* which is not in Γ, that its addition to Γ would result in an inconsistent set; hence we call Γ a *maximal* consistent set.

It is obvious that Γ contains all the postulates of RS_1^* that are closed and the closures of those that are not, and also all theorems of RS_1^* that are closed, because if S is a *cwff* such that $\Gamma \vdash S$, then S is contained in Γ. For since Γ is maximal, the addition to it of any *cwff* S not already contained in it results in an inconsistent set, that is, for any *cwff* S not belonging to Γ, we have $\Gamma, S \vdash \sim P \cdot P$. From this, by the Deduction Theorem, we have $\Gamma \vdash S \supset \sim P \cdot P$. Now if we were also to have $\Gamma \vdash S$, then by R 1 we should have $\Gamma \vdash \sim P \cdot P$, contrary to the fact that Γ is consistent. It should also be observed that where S is any *cwff* of RS_1^*, either $\Gamma \vdash S$ or $\Gamma \vdash \sim S$. This follows from the fact that every *cwff* S of RS_1^* either is or is not contained in Γ. If S is in Γ, then obviously, $\Gamma \vdash S$. While if S is not in Γ, then $\Gamma, S \vdash \sim P \cdot P$ and by the Deduction Theorem, $\Gamma \vdash S \supset \sim P \cdot P$. From this, by Ⓟ, we have $\Gamma \vdash \sim(\sim P \cdot P) \supset \sim S$, and since we also have by Ⓟ that $\Gamma \vdash \sim(\sim P \cdot P)$, by R 1 we have $\Gamma \vdash \sim S$. From the preceding remarks it follows that for any *cwff* S, either S belongs to Γ or $\sim$S belongs to Γ.

To show that the formulas of Γ, and hence certainly those of Λ, are simultaneously satisfiable, we take as our model the set *I* of all individual constants of RS_1^*, regarded as *individuals*. These are all the individual constants of RS_1, a, b, c, a_1, b_1, c_1, a_2, b_2, c_2, . . . , together with all the symbols u_1, u_2, u_3, u_4, ... Our interpretation of the *cwffs* of Γ consists of the following assignment of meanings:

1. To each propositional symbol of RS_1^* we assign the truth value **T** if it belongs to Γ; otherwise it is assigned the truth value **F**.
2. To each individual constant of RS_1^* (considered as a symbol in an interpreted system) we assign itself (considered as an individual which belongs to the model *I*). There are no free individual variables in any *wff* of Γ, since these are all *cwffs;* consequently no attention need be paid to them here.
3. To each predicate symbol P^n we assign the n-ary relation (or attribute, if $n = 1$) among n individuals $a_1, a_2, \ldots, a_n$ of *I* of having their names, properly parenthesized, adjoined in that order to P^n, constitute a *cwff* of Γ, or $P^n(a_1, a_2, \ldots, a_n) \epsilon \Gamma$.

The interpretation is completed and made normal by adding these conditions:

(a) Any *cwff* $\sim$S is assigned the truth value **T** or **F** according as the *cwff* S is assigned the truth value **F** or **T**.
(b) Any *cwff* $X \cdot Y$ is assigned the truth value **T** if both X and Y are assigned the truth value **T**; otherwise it is assigned **F**.
(c) Any *cwff* $(x)R$ is assigned the truth value **T** if the truth value **T** is assigned to every result of replacing all free occurrences of x in R by an individual constant of RS_1^*; otherwise it is assigned **F**.

(d) Every *cwff* $P^n(a_1, a_2, \ldots, a_n)$ is assigned the truth value **T** if the elements of the model assigned to $a_1, a_2, \ldots, a_n$, in that order, stand to each other in the n-ary relation assigned to P^n, otherwise it is assigned **F**.

We now can prove that all *wffs* of Γ are satisfied on this interpretation. What we shall actually prove is the stronger result that every *cwff* S of RS_1^* is assigned the truth value **T** or **F** according as S belongs or does not belong to Γ. Our proof is by strong induction on the number of symbols in S, counting each propositional symbol, each *cwff* of the form $P^n(a_1, a_2, \ldots, a_n)$, each occurrence of a universal quantifier (x), and each occurrence of $\cdot$ and $\sim$ as a single symbol.

(α) In case $n = 1$, S is either a propositional symbol or a *cwff* of the form $P^n(a_1, a_2, \ldots, a_n)$. If S is a propositional symbol, by paragraph 1 of the original assignment of meanings, S is assigned the truth value **T** or **F** according as it is contained or not contained in Γ. If S is the *cwff* $P^n(a_1, a_2, \ldots, a_n)$ then by paragraph 3 and condition (d), S is assigned the truth value **T** or **F** according as it is contained or not contained in Γ.

(β) Here we assume that every *cwff* of RS_1^* which contains fewer than k symbols is assigned the truth value **T** or **F** according as it is contained or not contained in Γ. Now we consider any *cwff* S of RS_1^* that contains exactly $k > 1$ symbols. Here S must be either $\sim W$ or $X \cdot Y$ or $(x)R$.

CASE 1. S is $\sim W$.

If $\sim W$ is in Γ, then since Γ is a maximal consistent set, W is not in Γ. Here, by the β-case assumption, since W contains fewer than k symbols, W is assigned the truth value **F**. By condition (a), $\sim W$ must be assigned **T**. Hence if S is in Γ, it is assigned the truth value **T**.

If $\sim W$ is not in Γ, then W is in Γ. Here W is assigned **T**, so by condition (a), $\sim W$ must be assigned **F**. Hence if S is not in Γ, it is assigned the truth value **F**.

CASE 2. S is $X \cdot Y$.

If $X \cdot Y$ is in Γ, then by Ⓟ $\Gamma \vdash X$ and $\Gamma \vdash Y$, so X and Y are both in Γ. Since each contains fewer than k symbols, by the β-case assumption they are both assigned the truth value **T**, and by condition (b), S is also.

If $X \cdot Y$ is not in Γ, then not both X and Y can be in Γ (for if they were, then by Ⓟ $\Gamma \vdash X \cdot Y$ and $X \cdot Y$ would have to be contained in Γ), so not both X and Y are assigned the truth value **T**, and by condition (b), S is assigned the truth value **F**.

CASE 3. S is $(x)R$.

If $(x)R$ is in Γ then $\Gamma \vdash (x)R$, and where Q is any result of replacing all free occurrences of x in R by an individual constant, $\vdash (x)R \supset Q$ by P 5, so by R 1, $\Gamma \vdash Q$. Since Γ is a maximal consistent set, every such Q is contained in Γ, and since each of them contains fewer than k symbols, they are all assigned the truth value **T** by the β-case assumption. Now by condition (c), $(x)R$ is assigned **T** also. Hence if $(x)R$ is in Γ, then $(x)R$ is assigned the truth value **T**.

If $(x)R$ is not in Γ, then $\sim(x)R$ is in Γ, whence $\Gamma \vdash \sim(x)R$, and by Th. 7 and the Rule of Replacement, $\Gamma \vdash (\exists x)\sim R$. Since Γ is a maximal consistent set, it must contain $(\exists x)\sim R$. But $(\exists x)\sim R$ is a *cwff* of $RS_1{}^*$ which begins with an existential quantifier, and so the formula $(\exists x)\sim R \supset \sim R^*$ must have been added to some Λ_{j-1} to form Λ_j—where $\sim R^*$ is the result of replacing all free occurrences of x in $\sim R$ by u_{i_j}. Since the *cwff* $(\exists x)\sim R \supset \sim R^*$ belongs to some such Λ_j, it must belong to Γ. Hence, by R 1, $\Gamma \vdash \sim R^*$, so $\sim R^*$ belongs to Γ also. Since Γ is consistent, R^* does *not* belong to Γ, and since it contains fewer than k symbols, by the β-case assumption, it is assigned the truth value **F**. Hence by condition (c), $(x)R$ is assigned the truth value **F**.

This completes the induction, and proves that all formulas of Γ, and hence all those of Λ, are simultaneously satisfiable. Metatheorem X, together with the discussion which preceded it, completes our proof of

METATHEOREM XI. *RS_1 is complete.*

In the preceding section we showed that RS_1 is logically equivalent to the natural deduction techniques developed in Chapter 4. Hence our proof of Metatheorem XI validates the claim made at the end of Chapter 4 that the natural deduction techniques there developed 'permit the demonstration of all logically true propositions constructed out of truth-functional connectives and the quantifications of individual variables'.

10.7 RS_1 with Identity

We obtain a first-order calculus with identity by adding to RS_1 a new binary relation symbol, the predicate constant '$\boldsymbol{I}^2$', and additional axioms governing the new symbol to impose those conditions on it that serve to characterize identity. Our metalanguage expression for *wffs* containing the new identity symbol, for example, '$\boldsymbol{I}^2(\boldsymbol{xy})$', will be '$x = y$'. Although it is customary to use two or more special axioms or axiom schemata for identity, a single axiom schema is sufficient.[9] It is

P 6. $$(y)\{Fy \equiv (\exists x)[(x = y)\cdot Fx]\}$$

From this new postulate, in the presence of the other postulates, rules, theorems, derived rules, and metatheorems of RS_1 already at hand, we can derive the following results concerning identity.

[9]The idea of using this one axiom schema is credited to Hao Wang by W. V. O. Quine in *Set Theory and Its Logic*, Harvard University Press, Cambridge, Mass., 1963, p. 13.

THEOREM I–1: $\vdash [(x = y) \cdot Fx] \supset Fy$

Proof:	$(x = y) \cdot Fx \vdash (y)\{Fy \equiv (\exists x)[(x = y) \cdot Fx]\}$	P 6
	$(x = y) \cdot Fx \vdash Fy \equiv (\exists x)[(x = y) \cdot Fx]$	DR 7
	$(x = y) \cdot Fx \vdash (\exists x)[(x = y) \cdot Fx] \supset Fy$	Ⓟ
	$(x = y) \cdot Fx \vdash (x = y) \cdot Fx$	premiss
	$(x = y) \cdot Fx \vdash (\exists x)[(x = y) \cdot Fx]$	DR 8
	$(x = y) \cdot Fx \vdash \mathrm{Fy}$	R 1
	$\vdash [(x = y) \cdot Fx] \supset Fy$	D.T.

THEOREM I–2: $\vdash y = y$

Proof: We begin with the instance of P 6 in which 'Fz' is '$\sim(z = y)$':

$(y)[\sim(y = y) \equiv (\exists x)[(x = y) \cdot \sim(x = y)]]$	P 6
$\sim(y = y) \equiv (\exists x)[(x = y) \cdot \sim(x = y)]$	DR 7
$(y = y) \equiv \sim(\exists x)[(x = y) \cdot \sim(x = y)]$	Ⓟ
$\sim[(x = y) \cdot \sim(x = y)]$	Ⓟ
$(x)\sim[(x = y) \cdot \sim(x = y)]$	R 2
$\sim(\exists x)[(x = y) \cdot \sim(x = y)]$	Th. 8 & R.R.
$y = y$	Ⓟ

Theorems I–1 and I–2 are the usual axioms for identity.

THEOREM I–3: $\vdash (x)(y)[(x = y) \supset (y = x)]$

Proof: We begin with an instance of Th. I–1 in which 'Fz' is '$z = x$':

$[(x = y) \cdot (x = x)] \supset (y = x)$	Th. I–1
$(x = x) \supset [(x = y) \supset (y = x)]$	Ⓟ
$x = x$	Th. I–2
$(x = y) \supset (y = x)$	R 1
$(y)[(x = y) \supset (y = x)]$	R 2
$(x)(y)[(x = y) \supset (y = x)]$	R 2

THEOREM I–4: $\vdash (x)(y)(z)\{[(x = y) \cdot (y = z)] \supset (x = z)\}$

Proof: We begin with an instance of Th. I–1 in which 'Fu' is '$\sim(u = z)$'

$[(x = y) \cdot \sim(x = z)] \supset \sim(y = z)$	Th. I–1
$(x = y) \supset [\sim(x = z) \supset \sim(y = z)]$	Ⓟ
$(x = y) \supset [(y = z) \supset (x = z)]$	Ⓟ
$[(x = y) \cdot (y = z)] \supset (x = z)$	Ⓟ
$(z)\{[(x = y) \cdot (y = z)] \supset (x = z)\}$	R 2
$(y)(z)\{[(x = y) \cdot (y = z)] \supset (x = z)\}$	R 2
$(x)(y)(z)\{[(x = y) \cdot (y = z)] \supset (x = z)\}$	R 2

THEOREM I-5: $\vdash (x = y) \equiv (y = x)$

Proof:	$(x)(y)[(x = y) \supset (y = x)]$	Th. I-3
	$(y)[(x = y) \supset (y = x)]$	DR 7
	$(x = y) \supset (y = x)$	DR 7
	$(z)[(x = z) \supset (z = x)]$	DR 9
	$(w)(z)[(w = z) \supset (z = w)]$	DR 9
	$(z)[(y = z) \supset (z = y)]$	DR 7
	$(y = x) \supset (x = y)$	DR 7
	$[(x = y) \supset (y = x)] \cdot [(y = x) \supset (x = y)]$	Ⓟ
	$(x = y) \equiv (y = x)$	df.

The four 'Principles of Identity' stated in Section 5.4 can be established here as derived rules of inference.

DR I-1. $x = y \vdash y = x$

Proof:	$x = y$	premiss
	$y = x$	Th. I-5 & R.R.

DR I-2. $Fx, y = x \vdash Fy$

Proof:	$y = x$	premiss
	$x = y$	DR I-1
	Fx	premiss
	$(x = y) \cdot Fx$	Ⓟ
	$[(x = y) \cdot Fx] \supset Fy$	Th. I-1
	Fy	R 1

DR I-3. $Fx, \sim Fy \vdash \sim(y = x)$

Proof:	$[(x = y) \cdot Fx] \supset Fy$	Th. I-1
	$(x = y) \supset (Fx \supset Fy)$	Ⓟ
	Fx	premiss
	$\sim Fy$	premiss
	$\sim(Fx \supset Fy)$	Ⓟ
	$\sim(x = y)$	Ⓟ
	$\sim(y = x)$	Th. I-5 & R.R.

The fourth principle, $p \vdash \nu = \nu$, is already given by Theorem I-2.

A more complicated logistic system is required for the formalization of the logical principles involved in appraising arguments that concern attributes of attributes or of relations, or relations among attributes and relations, or the notions of *all* or *some* attributes or relations. Such logistic systems are generally called *extended function calculi*, and if they are consistent, they are demonstrably incomplete.[10] But these more advanced parts of symbolic logic lie beyond the scope of the present book.

10.8 First-Order Logic Including ZF Set Theory

The **ZF** Set Theory described in Chapter 7 can be formalized as a first-order theory. The only special mathematical symbol required is that for set membership. Or rather, since the first-order theory, as a logistic system, is uninterpreted, we can specify that the new and special symbol is the binary relation symbol $\in$, which we intend subsequently to interpret as set membership.

Indeed, if our purpose is simply to develop set theory, and thereby mathematics, in our first-order logic, our notations can be much more parsimonious. Let us begin again to develop a first-order logic with formalized Set Theory as our goal. Instead of the elaborate panoply of primitives enumerated in Section 10.1, let us begin with the one two-place predicate symbol $\in$, and a denumerably infinite list of individual variables

$$u,\ v,\ w,\ x,\ y,\ z,\ u_1,\ v_1,\ w_1,\ x_1,\ y_1,\ z_1,\ u_2,\ v_2,\ \ldots$$

Of course, we shall need the four logical symbols $\sim$, $\cdot$, (, and). Those are all the primitive or undefined terms we need.

We can define the additional logical symbols v, $\supset$, $\equiv$, and $\exists$, for our object logic, as in Section 10.1. Then we can define *wff* for our system recursively as follows:

(α) If x and y are individual symbols, then $x \in y$ is a *wff*.
(β) 1. If F is a *wff*, then $\sim(F)$ is a *wff*.
2. If F and G are *wffs*, then $(F)\cdot(G)$ is a *wff*.
3. If F is a *wff* and x is an individual variable, then $(x)(F)$ is a *wff*.

We assume the familiar five postulates of RS_1 together with the axiom schema for identity, P 6 of Section 10.7. We assume the two familiar rules of inference R 1 and R 2 of Section 10.1, and the definition of demonstration. It is easy enough to define $=$ in terms of $\in$ as in **ZF-1**, and then to define $\neq$ and $\notin$ as obvious abbreviations. All of the **ZF** axioms can be symbolized in this system. Then, we can either add the **ZF** axioms as additional axioms, or else take them all together as premisses to every argument whose conclusion is specifically set theoretic or mathematical.

[10] Proved by Kurt Gödel in 'Über formal unentscheidbare Sätze der Principia Mathematica und verwandter Systeme', *Monatshefte für Mathematik und Physik*, vol. 38 (1931), pp. 173–198, translated in *From Frege to Gödel, A Source Book in Mathematical Logic, 1879–1931*, ed. Jean Van Heijenoort, Harvard University Press, Cambridge, Mass., 1967, pp. 596–616. An informal exposition of Gödel's proof by Ernest Nagel and James R. Newman can be found in I. M. Copi and J. A. Gould, eds., *Contemporary Philosophical Logic*, St. Martin's Press, New York, 1978.

So far so good.

There is a new paradox lurking, however. We have seen, in our completeness proof for RS_1, how that system is satisfied by a denumerable model. On the assumption that the **ZF** Set Theory is consistent, it, too, has a denumerable model. This was proved by Thoralf Skolem. But the **ZF** system contains both an axiom of infinity and a power set axiom, which permits proof that there are nondenumerable sets. Here is the Skolem Paradox: A system which has a denumerable model contains a theorem that asserts the existence of nondenumerable sets.

How shall we understand this situation? It cannot be said that there is general agreement about what has gone wrong. Here, for what they may be worth, are two of the proposed resolutions.

One thing that can be said is that the set membership relation simply has not been (cannot be?) completely described or defined formally. So the binary relation in our model, which serves as interpretation for $\in$, is not set membership but something only formally parallel to it. In short, this proposed resolution says that the possibility of a denumerable model for a nondenumerable theory has shown only that our model is 'nonstandard', and is not what we were trying to talk about or specify.

Another way out of the Skolem Paradox is based on the realization that equinumerousness or equivalence of sets is established only by actually producing a function that is or effects a one-to-one correspondence between them. It may be that the function needed to map a set onto its power set simply cannot be defined *in the system.* If the set is denumerable, then its power set can be proved nondenumerable *in the system* only because the function that would enumerate its members does not exist *in the system.* It is as though the power set, seen from outside the system, *can* be proved equinumerous with the denumerable set, but to specify the one-to-one correspondence requires a different, more powerful system than **ZF**. This proposed resolution has more adherents than one might have thought. Intuitively, sets are the size they are, quite independently of what functions are available to map one onto another. This view seems to me to be an extreme kind of operationalism—which appears to suggest that if you cannot measure something, then it does not exist!

It is heartening to be able to end a textbook that has no prerequisites with a problem that has not yet been solved.

APPENDIX

A

Incompleteness of the Nineteen Rules

The nineteen Rules of Inference presented in Sections 3.1 and 3.2 of Chapter 3 are *incomplete,*[1] which is to say that there are truth functionally valid arguments whose validity cannot be proved using only those nineteen Rules. To discuss and establish this incompleteness, it is useful to introduce the notion of a characteristic that is 'hereditary with respect to a set of rules of inference'. We offer this definition: A characteristic Φ is *hereditary with respect to a set of rules of inference* if and only if whenever Φ belongs to one or more statements, it also belongs to every statement deduced from them by means of those Rules of Inference. For example, *truth* is a characteristic that is hereditary with respect to the nineteen Rules of Inference presented in the first two sections of Chapter 3. As we have already said, any conclusion must be true if it can be deduced from true premisses by means of our nineteen Rules of Inference. Indeed, we should not want to use any rules of inference with respect to which truth was *not* hereditary.

Now, to prove that a set of rules of inference is incomplete, we must find a characteristic Φ and a valid argument α such that

(1) Φ is hereditary with respect to the set of rules of inference; and
(2) Φ belongs to the premisses of α but not to the conclusion of α.

The characteristic *truth* is hereditary with respect to any set of rules of inference in which we may be seriously interested and, therefore, satisfies condition (1) above. But where α is a valid argument, it follows immediately from our definition of validity that *truth* can never satisfy condition (2) above. Hence to prove the incompleteness of our nineteen Rules we must find a characteristic other than *truth* that is hereditary with respect to our nineteen Rules, and can belong to the premisses but not to the conclusion of some valid argument α.

[1]The following proof of incompleteness was devised by my friend Professor Leo Simons of the University of Cincinnati. See Leo Simons, 'Logic Without Tautologies', *Notre Dame Journal of Formal Logic,* vol. 15 (1974), pp. 411–431, and 'More Logics Without Tautologies', ibid., vol. 19 (1978), pp. 543–557. For some discussion of the matter, see Robert L. Armstrong, 'A Question about Completeness', ibid., vol. 17 (1976), pp. 295–296, and Paul J. Campbell, 'An Answer to Armstrong's Question about Incompleteness in Copi', ibid., vol. 18 (1977), pp. 262–264.

To obtain such a characteristic, we introduce a three-element model in terms of which the symbols in our nineteen Rules can be interpreted. The three elements are the numbers **0**, **1**, and **2**, which play roles analogous to those of the truth values true (**T**) and false (**F**) introduced in Chapter 2. Every statement will have one of the three elements of the model assigned to it, and it will be said to assume, take on, or have one of the three *values* **0**, **1**, or **2**. Just as in Chapter 2 the statement variables p, q, r, . . . , were allowed to range over the two truth values **T** and **F**, so here we allow the statement variables p, q, r, . . . to range over the three values **0**, **1**, and **2**.

The five symbols '$\sim$', '$\cdot$', 'v', '$\supset$', and '$\equiv$' that occur in our nineteen Rules can be redefined for (or in terms of) our three-element model by the following three-valued tables:

p	$\sim p$
0	2
1	1
2	0

p	q	$p \cdot q$	$p \vee q$	$p \supset q$	$p \equiv q$
0	0	0	0	0	0
0	1	1	0	1	1
0	2	2	0	2	2
1	0	1	0	0	1
1	1	1	1	1	1
1	2	2	1	1	1
2	0	2	0	0	2
2	1	2	1	0	1
2	2	2	2	0	0

Alternative (but equivalent) analytical definitions can be given as follows, where 'min(x, y)' denotes the minimum of the numbers x and y, and 'max(x, y)' denotes the maximum of the numbers x and y.

$$\begin{aligned} \sim p &= 2 - p \\ p \cdot q &= \max(p, q) \\ p \vee q &= \min(p, q) \\ p \supset q &= \min(2 - p, q) \\ p \equiv q &= \max(\min(2 - p, q), \min(2 - q, p)) \end{aligned}$$

The desired characteristic Φ that is hereditary with respect to our nineteen Rules of Inference is the characteristic of having the value **0**. To prove that this characteristic is hereditary with respect to the nineteen Rules, it will suffice to show that it is hereditary with respect to each of the nineteen Rules. This can be shown for each rule by means of a three-valued table. For example, that having the value **0** is hereditary with respect to *Modus Ponens*, can be seen by examining the table above, which defines the value of '$p \supset q$' as a function of the values of 'p' and of 'q'. The two premisses 'p' and '$p \supset q$' both have the value **0** only in the first row, and there the conclusion 'q' has the value **0** also. Examining the same table shows that having the value **0** is hereditary also for Simplification, Conjunction, and Addition. Filling in additional columns for

'$\sim p$' and '$\sim q$' will show that having the value **0** is hereditary with respect to *Modus Tollens* and Disjunctive Syllogism also. That it is hereditary with respect to Hypothetical Syllogism can be shown by the following table:

p	q	r	$p \supset q$	$q \supset r$	$p \supset r$
0	0	0	0	0	0
0	0	1	0	1	1
0	0	2	0	2	2
0	1	0	1	0	0
0	1	1	1	1	1
0	1	2	1	1	2
0	2	0	2	0	0
0	2	1	2	0	1
0	2	2	2	0	2
1	0	0	0	0	0
1	0	1	0	1	1
1	0	2	0	2	1
1	1	0	1	0	0
1	1	1	1	1	1
1	1	2	1	1	1
1	2	0	1	0	0
1	2	1	1	0	1
1	2	2	1	0	1
2	0	0	0	0	0
2	0	1	0	1	0
2	0	2	0	2	0
2	1	0	0	0	0
2	1	1	0	1	0
2	1	2	0	1	0
2	2	0	0	0	0
2	2	1	0	0	0
2	2	2	0	0	0

Only in the first, tenth, nineteenth, twenty-second, twenty-fifth, twenty-sixth, and twenty-seventh rows do the two premisses '$p \supset q$' and '$q \supset r$' both have the value **0**, and in each of them the conclusion '$p \supset r$' has the value **0** also. Even larger tables would be needed to show that having the value **0** is hereditary with respect to Constructive Dilemma and Destructive Dilemma. Larger tables can be constructed, but they are not absolutely necessary, because the alternative analytical definitions on page 334 can be used to show that having the value **0** is hereditary with respect to the Dilemmas, as on page 336.

When we construct three-valued tables to verify that having the value **0** is hereditary with respect to replacement of statements by their logical equivalents, we notice that although the biconditionals themselves need not have the value **0**, the expressions flanking the equivalence sign necessarily have the same value. For example, in the table appropriate to the first of De Morgan's Theorems,

p	q	$\sim p$	$\sim q$	$p \cdot q$	$\sim(p \cdot q)$	$\sim p \vee \sim q$	$\sim(p \cdot q) \equiv (\sim p \vee \sim q)$
0	0	2	2	0	2	2	0
0	1	2	1	1	1	1	1
0	2	2	0	2	0	0	0
1	0	1	2	1	1	1	1
1	1	1	1	1	1	1	1
1	2	1	0	2	0	0	0
2	0	0	2	2	0	0	0
2	1	0	1	2	0	0	0
2	2	0	0	2	0	0	0

the equivalent expressions '$\sim(p \cdot q)$' and '$\sim p \vee \sim q$' have the same value in every row even though the statement of their equivalence fails to possess the value **0** in rows two, four, and five. It should be obvious, however, that having the value **0** is hereditary with respect to the replacement of all or part of any statement by any other statement that is equivalent to the part replaced.

Alternative proofs that having the value **0** is hereditary with respect to the nineteen Rules make use of our analytical definitions of the logical symbols. For example, that having the value **0** is hereditary with respect to Constructive Dilemma can be shown by the following argument. By assumption, '$(p \supset q) \cdot (r \supset s)$' and '$p \vee r$' both have the value **0**. Hence both '$p \supset q$' and '$r \supset s$' have the value **0**, so either $p = \mathbf{2}$ or $q = \mathbf{0}$ and either $r = \mathbf{2}$ or $s = \mathbf{0}$. Since '$p \vee r$' has the value **0**, either $p = \mathbf{0}$ or $r = \mathbf{0}$. If $p = \mathbf{0}$, then $p \neq \mathbf{2}$, whence $q = \mathbf{0}$, and if $r = \mathbf{0}$, then $r \neq \mathbf{2}$ whence $s = \mathbf{0}$. Therefore, either $q = \mathbf{0}$ or $s = \mathbf{0}$ whence '$q \vee s$' has the value **0**, which was to be shown.

Once it has been established that the characteristic of having the value **0** is hereditary with respect to the nineteen Rules, to prove the incompleteness of those Rules, one need only exhibit a valid argument whse premisses have the value **0**, but whose conclusion does not have the value **0**. Such an argument is

$$A \supset B$$
$$\therefore A \supset (A \cdot B)$$

whose validity is easily established by a truth table. Where 'A' has the value **1** and 'B' has the value **0**, the premiss '$A \supset B$' has the value $\mathbf{1} \supset \mathbf{0} = \mathbf{0}$ but the conclusion '$A \supset (A \cdot B)$' has the value $\mathbf{1} \supset (\mathbf{1} \cdot \mathbf{0}) = \mathbf{1} \supset \mathbf{1} = \mathbf{1}$. Therefore, the nineteen Rules of Inference are incomplete.

APPENDIX

B

Normal Forms and Boolean Expansions

In this Appendix we develop an alternative method for recognizing valid truth-functional arguments. Given a complicated statement form, one can often find a simpler one logically equivalent to it, to which the given form can be 'reduced' by algebraic operations. In treating the logic of compound statements and statement forms from the algebraic point of view, it is convenient to use a different notation for denial or negation. Here we symbolize the negation of an expression by drawing a bar above it. In this notation, the familiar truths of De Morgan are symbolized as $\overline{p \cdot q} \equiv (\bar{p} \vee \bar{q})$ and $\overline{p \vee q} \equiv (\bar{p} \cdot \bar{q})$. All of the statement forms treated here are truth-functional, so their status as tautologous, contradictory, or contingent remains unchanged when any part is replaced by an expression that is logically equivalent to the part replaced. Thus $p \vee \bar{p}$ is a tautology and remains one when p is replaced in it by $\bar{\bar{p}}$, since p and $\bar{\bar{p}}$ are logically equivalent by the principle of Double Negation.

Since $p \cdot (q \cdot r)$ and $(p \cdot q) \cdot r$ are logically equivalent, they may be indifferently written as $p \cdot q \cdot r$. Similarly, $p \vee (q \vee r)$ and $(p \vee q) \vee r$ may be written as $p \vee q \vee r$. The logical equivalences involved are principles of Association. The convention of dropping unnecessary parentheses permits us to enunciate generalized statements of De Morgan's Theorems as $\overline{p_1 \cdot p_2 \cdot p_3 \cdot \ldots \cdot p_n} \equiv (\bar{p}_1 \vee \bar{p}_2 \vee \bar{p}_3 \vee \ldots \vee \bar{p}_n)$ and $\overline{p_1 \vee p_2 \vee p_3 \vee \ldots \vee p_n} \equiv (\bar{p}_1 \cdot \bar{p}_2 \cdot \bar{p}_3 \cdot \ldots \cdot \bar{p}_n)$. The two tautologies $[p \cdot (q \vee r)] \equiv [(p \cdot q) \vee (p \cdot r)]$ and $[p \vee (q \cdot r)] \equiv [(p \vee q) \cdot (p \vee r)]$ are principles of Distribution. Generalized principles of Distribution may be expressed as

$$[p \cdot (q_1 \vee q_2 \vee \ldots \vee q_n)] \equiv [(p \cdot q_1) \vee (p \cdot q_2) \vee \ldots \vee (p \cdot q_n)]$$

and

$$[p \vee (q_1 \cdot q_2 \cdot \ldots \cdot q_n)] \equiv [(p \vee q_1) \cdot (p \vee q_2) \cdot \ldots \cdot (p \vee q_n)]$$

The capital Greek letters pi and sigma are often used to express generalized logical products (conjunctions) and generalized logical sums (disjunctions). We define

$$\prod_{i=1}^{n} p_i = \text{df}\, (p_1 \cdot p_2 \cdot \ldots \cdot p_n)$$

and

$$\sum_{i=1}^{n} p_i = \text{df}\,(p_1 \vee p_2 \vee \ldots \vee p_n)$$

Given these notations we can express our generalized De Morgan Theorems as

$$\overline{\prod_{i=1}^{n} p_i} \equiv \sum_{i=1}^{n} \overline{p_i} \quad \text{and} \quad \overline{\sum_{i=1}^{n} p_i} \equiv \prod_{i=1}^{n} \overline{p_i}$$

and the generalized Distribution principles as

$$p \cdot \sum_{i=1}^{n} q_i \equiv \sum_{i=1}^{n} (p \cdot q_i) \quad \text{and} \quad p \vee \prod_{i=1}^{n} q_i \equiv \prod_{i=1}^{n} (p \vee q_i)$$

It is convenient to have four other principles of logical equivalence available for purposes of algebraic manipulation and transformation.

First, the principles of Commutation expressed as

$$(p \cdot q) \equiv (q \cdot p) \quad \text{and} \quad (p \vee q) \equiv (q \vee p)$$

Second, the principles of Tautology, which, stated as

$$p \equiv (p \vee p) \quad \text{and} \quad p \equiv (p \cdot p)$$

assure us that any statement, wherever it may occur, is replaceable by the disjunction (conjunction) both of whose disjuncts (conjuncts) are the same as the given statement, and vice versa.

Third is the principle that any statement p is logically equivalent to the conjunction of itself with any tautology of the form $q \vee \overline{q}$, that is

$$p \equiv [p \cdot (q \vee \overline{q})]$$

Our fourth and final principle is the logical equivalence

$$p \equiv [p \vee (q \cdot \overline{q})]$$

which permits us to interchange p and $p \vee (q \cdot \overline{q})$ wherever either of them occurs.

It is clear that by invoking the defining equivalences

$$(p \supset q) \equiv (\overline{p} \vee q) \quad \text{and} \quad (p \equiv q) \equiv [(p \cdot q) \vee (\overline{p} \cdot \overline{q})]$$

material conditionals and biconditionals can always be eliminated in favor of, or expressed in terms of, conjunctions, disjunctions, and negations. Moreover, by repeated uses of De Morgan's Theorems and Double Negation, any formula can be replaced by a logically equivalent one in which no negation symbol applies to a compound part. Thus

$$\overline{\bar{p} \cdot [(\overline{\bar{q} \vee r}) \cdot (s \cdot \bar{t})]}$$

is transformed by repeated applications of De Morgan's principles and Double Negation as follows:

$$\bar{\bar{p}} \vee [\overline{(\overline{\bar{q} \vee r}) \cdot (s \cdot \bar{t})}]$$

$$p \vee [(\overline{\overline{\bar{q} \vee r}}) \vee (\overline{s \cdot \bar{t}})]$$

$$p \vee [(\bar{q} \vee r) \vee (\bar{s} \vee \bar{\bar{t}})]$$

$$p \vee [(\bar{q} \vee r) \vee (\bar{s} \vee t)]$$

which becomes, by the principle of Association,

$$p \vee \bar{q} \vee r \vee \bar{s} \vee t$$

In this expression, the negation symbol is applied only to the single variables q and s.

The principle of Distribution permits us to change any given formula into an equivalent one in which conjunction symbols occur—if at all—only between single variables or their negations, or else into a different equivalent form in which disjunction symbols occur—if at all—only between single variables or their negations. Thus the expression

(1) $$(p \vee q) \cdot (r \vee s)$$

in which the conjunction symbol occurs between disjunctions, reduces by repeated applications of the principles of Distribution, Commutation, and Association into

(2) $$(p \cdot r) \vee (q \cdot r) \vee (p \cdot s) \vee (q \cdot s)$$

in which conjunction symbols connect single variables only. And the expression

(3) $$(p \cdot q) \vee (r \cdot s)$$

in which the disjunction symbol occurs between conjunctions, reduces by repeated applications of the other parts of the principles of Distribution,

Commutation, and Association into

(4) $$(p \vee r) \cdot (q \vee r) \cdot (p \vee s) \cdot (q \vee s)$$

in which disjunction symbols connect single variables only.

Two 'normal forms' can be defined. A formula is in *conjunctive normal form* when, in addition to statement variables, it contains no symbols other than those for conjunction, disjunction, and negation; negation symbols apply only to single variables; and no disjunct is a conjunction, that is, disjunction symbols occur only between single variables or their negations. Thus (1) and (4) above are in *conjunctive normal form.* A formula is in *disjunctive normal form* when, in addition to statement variables, it contains no symbols other than those for conjunction, disjunction, and negation: negation symbols apply only to single variables; and no conjunct is a disjunction, that is, conjunction symbols occur only between single variables or their negations. Thus (2) and (3) above are in *disjunctive normal form.*

Let us examine the moderately complicated formula $(p \supset \bar{q}) \supset (p \equiv \bar{q})$ to see what gain in perspicuity is achieved by reducing it to disjunctive normal form. By replacing the material implication symbols by their definitions, the initial expression is reduced first to

$$(\bar{p} \vee \bar{q}) \supset (p \equiv \bar{q})$$

and then to

$$(\overline{\bar{p} \vee \bar{q}}) \vee (p \equiv \bar{q})$$

Replacing the material equivalence symbol by its definition, we obtain

$$(\overline{\bar{p} \vee \bar{q}}) \vee [(p \cdot \bar{q}) \vee (\bar{p} \cdot \bar{\bar{q}})]$$

Applying De Morgan's principle, we obtain

$$(\bar{\bar{p}} \cdot \bar{\bar{q}}) \vee [(p \cdot \bar{q}) \vee (\bar{p} \cdot \bar{\bar{q}})]$$

which, by the principles of Double Negation and Association, becomes

$$(p \cdot q) \vee (p \cdot \bar{q}) \vee (\bar{p} \cdot q)$$

Although it is already in disjunctive normal form, it can be further simplified by the following transformations. First, we apply the principle of Tautology to replace the first disjunct by its 'double', obtaining

$$(p \cdot q) \vee (p \cdot q) \vee (p \cdot \bar{q}) \vee (\bar{p} \cdot q)$$

Then we rearrange the terms by simply commuting or interchanging the second and third disjuncts, to get

$$(p \cdot q) \vee (p \cdot \overline{q}) \vee (p \cdot q) \vee (\overline{p} \cdot q)$$

Using the principle of Distribution on the first pair of disjuncts, we obtain

$$[p \cdot (q \vee \overline{q})] \vee (p \cdot q) \vee (\overline{p} \cdot q)$$

and using it and Commutation on the two right-hand disjuncts, we get

$$[p \cdot (q \vee \overline{q})] \vee [(p \vee \overline{p}) \cdot q]$$

Now, by the principle that the conjunction of any statement p with a tautology of the form $q \vee \overline{q}$ is logically equivalent to the statement itself, we obtain first

$$p \vee [(p \vee \overline{p}) \cdot q]$$

and finally

$$p \vee q$$

which is logically equivalent to, but much simpler than the form with which we began.[1]

The term 'normal form' is sometimes reserved for more specific types of expressions. These more specific types are also called 'Boolean Expansions' or 'Boolean normal forms', after the British logician George Boole (1815–1864). A formula containing the variables $p, q, r, \ldots$ is said to be in *disjunctive Boolean normal form*, or a *disjunctive Boolean Expansion*, when it is in disjunctive normal form, when every disjunct contains exactly one occurrence of every variable (either the variable or its negation), when the variables occur in alphabetical order in each disjunct, and when no two disjuncts are the same. The disjunctive Boolean Expansion of $p \vee q$ is formed by first replacing p by $p \cdot (q \vee \overline{q})$ and q by $(p \vee \overline{p}) \cdot q$ to obtain the equivalent expression $p \cdot (q \vee \overline{q}) \vee (p \vee \overline{p}) \cdot q$, then using the Disjunction and Association rules to obtain $(p \cdot q) \vee (p \cdot \overline{q}) \vee (p \cdot q) \vee (\overline{p} \cdot q)$, and finally cancelling out any repetitions of the same disjunct by the rules of Commutation and Tautology, which results in $(p \cdot q) \vee (p \cdot \overline{q}) \vee (\overline{p} \cdot q)$. The disjunctive Boolean Expansion of any noncontradictory formula can be obtained by the same general method.

A formula containing the variables $p, q, r, \ldots$ is said to be in *conjunctive*

[1] An application of normal forms to electrical circuits (parallel connections are represented by 'v' and series connections are represented by '·') has been made by Claude E. Shannon, 'A Symbolic Analysis of Relay and Switching Circuits', *Transactions of the American Institute of Electrical Engineers*, vol. 57 (1938), pp. 713–723.

An informal account of the above and several other interesting applications can be found in John E. Pfeiffer's article 'Symbolic Logic', in *Scientific American*, vol. 183, no. 6 (December 1950). See also Martin Gardner, 'Logic Machines', ibid., vol. 186, no. 3 (March 1952).

Boolean normal form, or a *conjunctive Boolean Expansion,* provided that it is in conjunctive normal form, every conjunct contains exactly one occurrence of every variable (either the variable or its negation), the variables occur in alphabetical order in each conjunct, and no two conjuncts are the same. The conjunctive Boolean Expansion of $p \cdot q$ is formed by first replacing p by $p \vee (q \cdot \bar{q})$ and q by $(p \cdot \bar{p}) \vee q$ to obtain $[p \vee (q \cdot \bar{q})] \cdot [(p \cdot \bar{p}) \vee q]$. Then, using the Distribution and Association rules to obtain $(p \vee q) \cdot (p \vee \bar{q}) \cdot (p \vee q) \cdot (\bar{p} \vee q)$, and, finally, cancelling out any repetition of the same conjunct by the rules of Commutation and Tautology results in $(p \vee q) \cdot (p \vee \bar{q}) \cdot (\bar{p} \vee q)$. The conjunctive Boolean Expansion of any nontautologous formula is easily obtained by the same general method and can be used for deciding whether the original formula is contradictory.

Any conjunctive Boolean Expansion containing n variables and 2^n conjuncts is reducible to an explicit contradiction. Thus the conjunctive Boolean Expansion $(p \vee q) \cdot (p \vee \bar{q}) \cdot (\bar{p} \vee q) \cdot (\bar{p} \vee \bar{q})$ is equivalent by the rule of Distribution to $[p \vee (q \cdot \bar{q})] \cdot [\bar{p} \vee (q \cdot \bar{q})]$, which is equivalent to the explicit contradiction $p \cdot \bar{p}$. It is equally obvious that the conjunctive Boolean Expansion of any contradiction will contain 2^n conjuncts if it involves n variables. Hence the general rule is that a statement form containing n variables is contradictory if and only if it has a conjunctive Boolean Expansion that contains 2^n conjuncts.

The negation of a conjunctive Boolean Expansion is reducible by repeated applications of De Morgan's Theorem and Double Negation to a logically equivalent disjunctive Boolean Expansion that contains the same variables and has the same number of disjuncts as the original conjunctive Boolean Expansion has conjuncts. (Thus the conjunctive Boolean Expansion $(p \vee q \vee r) \cdot (p \vee q \vee \bar{r}) \cdot (\bar{p} \vee \bar{q} \vee r)$ has its negation $\overline{(p \vee q \vee r) \cdot (p \vee q \vee \bar{r}) \cdot (\bar{p} \vee \bar{q} \vee r)}$ logically equivalent to the disjunctive Boolean Expansion $(\bar{p} \cdot \bar{q} \cdot \bar{r}) \vee (\bar{p} \cdot \bar{q} \cdot r) \vee (p \cdot q \cdot \bar{r})$.) Since the negation of a contradiction is a tautology, and a conjunctive Boolean Expansion containing n variables is a contradiction if and only if it contains 2^n conjuncts, it follows that a disjunctive Boolean Expansion containing n variables is a tautology if and only if it contains 2^n disjuncts.

That such a disjunctive Boolean Expansion must be tautologous can perhaps be seen more clearly by the following considerations. In the first place, since we are concerned with truth-functional compounds only, to speak of substituting statements for the statement variables of a statement form is equivalent to speaking of an assignment of truth values to the variables of that statement form. Now just as each row of a truth table with n initial or guide columns represents a different assignment of truth values to the n statement variables involved, so each disjunct of the Boolean Expansion represents a different assignment of truth values to the variables they contain. And just as the 2^n rows of a truth table having n initial or guide columns represent all possible assignments of truth values to its variables, so the 2^n disjuncts of a disjunctive Boolean Expansion represent all possible assignments of truth values to its variables. Since the 2^n disjuncts represent all possible assignments of truth values to its variables, at least one of them must be true. And since a disjunctive

Boolean Expansion asserts only that at least one of its disjuncts is true, any disjunctive Boolean Expansion containing n variables and 2^n disjuncts is tautologous. This point is made in somewhat different terms in Chapter 8, Section 8.6, and again in Chapter 9, Section 9.2.

It was pointed out in Chapter 2 that a truth-functional argument is valid if and only if its corresponding conditional statement (whose antecedent is the conjunction of the argument's premisses and whose consequent is the argument's conclusion) is a tautology. Since counting the number of disjuncts of its disjunctive Boolean Expansion permits us to decide whether or not a given form is a tautology, this provides us with an alternative method of deciding the validity of arguments. Thus the argument form $p \vee q, \sim p \therefore q$ is proved valid by constructing the disjunctive Boolean Expansion of its corresponding conditional $[(p \vee q) \cdot \bar{p}] \supset q$, and observing that the number of its disjuncts is 2^2.

Since the negation of a tautology is a contradiction, an argument is valid if and only if the negation of its corresponding conditional is a contradiction. Hence another method of deciding the validity of an argument is to form the conjunctive Boolean Expansion of the negation of its corresponding conditional and count the number of its conjuncts. If it contains n distinct variables and has 2^n conjuncts, then the argument is valid; otherwise it is invalid.

EXERCISES

For each of the following, find as simple an equivalent as you can:

*1. $p \equiv \bar{p}$
2. $p \vee (q \cdot \bar{p})$
3. $p \vee (p \cdot q)$
4. $q \cdot (p \vee \bar{q})$
*5. $p \equiv (q \cdot \bar{p})$
6. $p \cdot (p \vee q)$
7. $p \cdot q \cdot \bar{p}$
8. $(p \vee q) \cdot q$
9. $p \vee (q \cdot r \cdot \bar{q})$
*10. $p \supset [(p \supset q) \supset q]$
11. $(p \vee q) \cdot (q \vee r) \cdot (p \vee r) \cdot r$
12. $(p \cdot q) \vee (q \cdot r) \vee (p \cdot r) \vee r$
13. $p \cdot \{p \supset [q \cdot (q \supset r)]\}$
14. $p \vee \{\bar{p} \supset [q \vee (\bar{q} \supset r)]\}$
*15. $(p \vee q) \supset [(p \supset q) \supset \overline{(q \vee \bar{q})}]$
16. $(p \cdot \bar{q}) \supset [(p \supset q) \supset \overline{(p \cdot \bar{q})}]$
17. $\{p \cdot [(q \cdot r) \vee \bar{p}]\} \vee \{(p \cdot q) \cdot [r \supset (\bar{p} \cdot \bar{r})]\} \vee \{p \cdot [\bar{p} \supset (\bar{p} \cdot q \cdot \bar{r})]\}$
18. $(p \cdot q) \vee (\bar{p} \cdot \bar{q} \cdot r) \vee (\bar{p} \cdot q) \vee (p \cdot \bar{q} \cdot r) \vee (p \cdot \bar{q}) \vee (p \cdot q \cdot r) \vee (\bar{p} \cdot q \cdot r)$

APPENDIX

C

The Ramified Theory of Types[1]

The *simple* theory of types, which was expounded briefly on pages 153–154, suffices to eliminate such logical paradoxes as that of the alleged attribute *impredicable*. But there is another kind of paradox or contradiction that is not prevented by the simple theory of types. One example of this other kind is the paradox of the liar, which was discussed on page 214. These two kinds of paradoxes were first explicitly distinguished by F. P. Ramsey in 1926.[2] Since then those of the first kind have been known as 'logical paradoxes', the second kind either as 'epistemological paradoxes' or, more usually, as 'semantical paradoxes'.

A singularly clear example of a semantical paradox is due to Kurt Grelling.[3] The Grelling paradox can be stated informally as follows. Some words designate attributes that are exemplified by the words themselves, for example, 'English' is an English word, and 'short' is a short word. Other words designate attributes not exemplified by the words themselves, for example, 'French' is not a French word, and 'long' is not a long word. We shall use the word 'heterological' to designate the attribute a word has just in case it designates an attribute not exemplified by the word itself. Thus the words 'French' and 'long' are both heterological. The contradiction now arises when we ask if the word 'heterological' is heterological. If it is heterological, then it designates an attribute that it does not exemplify, and since it designates heterological, it is not, itself, heterological. But if it is not heterological, then the attribute that it designates *is* exemplified by itself, so that it *is* heterological. Here the contradiction is made explicit: if it is, then it isn't, and if it isn't, then it is.

A somewhat more formal derivation of the Grelling paradox, patterned after

[1]The reader who desires a more comprehensive discussion of this topic should consult Chapter 3 of I. M. Copi, *The Theory of Logical Types*, Routledge & Kegan Paul Ltd., London, 1971, pp. 76–114.

[2]F. P. Ramsey, 'The Foundations of Mathematics', *Proceedings of the London Mathematical Society*, 2nd series, vol. 25 (1926), pp. 338–384. Reprinted in *The Foundations of Mathematics*, Harcourt Brace Jovanovich, New York, and Kegan Paul, Trench, Trubner & Co. Ltd., London, 1931, pp. 1–61.

[3]K. Grelling and L. Nelson, 'Bemerkungen zu den Paradoxien von Russell und Burali-Forti', *Abhandlung der Fries'schen Schule*, n.s. vol. 2 (1907–1908), pp. 300–324.

Ramsey's version of it,[4] is the following. Where 'Des' designates the name relation, so that 's designates ϕ' is symbolized as 'sDesϕ', we begin with the definition

$$\text{Het(s)} = \text{df } (\exists\phi)\text{:sDes}\phi\cdot\text{sDes}\psi \equiv_{\psi} \psi = \phi\cdot\sim\phi(\text{s})^{5}$$

CASE 1

1. Het('Het') $\supset$:$(\exists\phi)$:'Het'Des$\phi\cdot$'Het'Des$\psi \equiv_{\psi} \psi = \phi\cdot\sim\phi$('Het')
2. $\supset$:'Het'Des$\phi\cdot$'Het'Des$\psi \equiv_{\psi} \psi = \phi\cdot\sim\phi$('Het')
3. $\supset$:'Het'Des$\phi\cdot$'Het'DesHet $\equiv$ Het $= \phi\cdot\sim\phi$('Het')
4. $\supset$:'Het'DesHet $\equiv$ Het $= \phi\cdot\sim\phi$('Het')
5. $\supset$:Het $= \phi\cdot\sim\phi$('Het')
6. $\supset$:$\sim$Het('Het')

CASE 2

1. $\sim$Het('Het') $\supset$:$(\phi)\sim$['Het'Des$\phi\cdot$'Het'Des$\psi \equiv_{\psi} \psi = \phi\cdot\sim\phi$('Het')]
2. $\supset$:'Het'DesHet $\supset$ $\sim$['Het'Des$\psi \equiv_{\psi}$ ψ = Het$\cdot\sim$Het('Het')]
3. $\supset$:'Het'Des$\psi \equiv_{\psi} \psi$ = Het$\cdot\supset\cdot$Het('Het')
4. $\supset$:'Het'Des$\psi \equiv_{\psi} \psi$ = Het (assuming 'Het' univocal)
5. $\supset$:Het('Het')

Hence

$$\text{Het('Het')} \equiv \sim\text{Het('Het')}$$

which is a contradiction.

The derivation of the preceding contradiction did not violate any restrictions imposed by the simple theory of logical types, so that theory does not prevent such contradictions. Contradictions of this second kind can be eliminated, however, by the adoption of a more complicated version of the theory of logical types, known as the 'ramified' (or 'branching') theory of logical types. Now, according to the simple theory of logical types, all entities are divided into different logical types, of which the lowest contains all individuals, the next lowest consists of all attributes of individuals (designated by functions of individuals), the next, of all attributes of attributes of individuals (designated by functions of functions of individuals), and so on. There are also hierarchies of relations among individuals and attributes, and of attributes of relations, but they need not be considered here. Where we use lower-case letters to denote individuals and capital letters to designate attributes, then we can represent

[4] Op cit., p. 358 and pp. 369–372; reprint p. 27 and pp. 42–46.

[5] Where writing the *psi* as a subscript to the equivalence symbol abbreviates the universal quantification of the equivalence with respect to *psi*, so that 'sDes$\psi \equiv_{\psi} \psi = \phi$' is an abbreviation for '(ψ)[sDes$\psi \equiv \psi = \phi$]'.

this hierarchy as follows, where the subscript attached to a function indicates its proper level in the hierarchy:

. . . .
. . . .
. . . .

type 3: $F_3, G_3, H_3, \ldots$
type 2: $F_2, G_2, H_2, \ldots$
type 1: $F_1, G_1, H_1, \ldots$
type 0: $a, b, c, \ldots, x, y, z$

Only a function of type 1 can significantly be predicated of an individual, and in general, a function of type i can be significantly predicated of a function of type j if and only if $i = j + 1$.

The preceding hierarchy presents a rough 'picture' of the simple theory of logical types. Now the ramified theory divides each type above level zero into a further hierarchy. Thus all functions of type 1, which may be predicated of individuals, are divided into different *orders* in the following fashion.

All propositional functions of type 1 that contain either no quantifiers or else quantifiers on individual variables only are said to be *first-order* functions. For example, $F_1(x)$ and $(y)[F_1(y) \supset G_1(x)]$ are first-order functions of type 1. First-order functions will have a left superscript 1 attached to indicate their position in the hierarchy of orders. Thus all first-order functions of type 1 may be listed as ${}^1F_1, {}^1G_1, {}^1H_1, \ldots$ Next, all propositional functions of type 1 that contain quantifiers on first-order functions, but no quantifiers on any other functions, are *second-order* functions. Examples of second-order functions of type 1 are $({}^1F_1)[{}^1F_1(x) \equiv {}^1F_1(a)]$ and $(\exists {}^1G_1)(\exists y)[{}^1G_1(y) \supset {}^1H_1(x)]$. Second-order functions will have a left superscript 2 attached, and all second-order functions of type 1 may be listed as ${}^2F_1, {}^2G_1, {}^2H_1, \ldots$ In general, an nth order function of type 1 will contain quantifiers on functions of order $n - 1$, but will have no quantifiers on functions of order m where $m \geq n$.

The ramified theory of logical types can be described compendiously by means of the following two-dimensional array:

	Order 1	*Order* 2	*Order* 3	. . .
	.	.	.	
	.	.	.	
	.	.	.	
type 3:	${}^1F_3, {}^1G_3, {}^1H_3, \ldots;$	${}^2F_3, {}^2G_3, {}^2H_3, \ldots;$	${}^3F_3, {}^3G_3, {}^3H_3, \ldots;$	. . .
type 2:	${}^1F_2, {}^1G_2, {}^1H_2, \ldots;$	${}^2F_2, {}^2G_2, {}^2H_2, \ldots;$	${}^3F_2, {}^3G_2, {}^3H_2, \ldots;$	. . .
type 1:	${}^1F_1, {}^1G_1, {}^1H_1, \ldots;$	${}^2F_1, {}^2G_1, {}^2H_1, \ldots;$	${}^3F_1, {}^3G_1, {}^3H_1, \ldots;$	. . .

Just as the simple hierarchy of types prevents us from speaking about all functions or attributes and permits us to speak only about all functions of individuals or all functions of functions of individuals, and so forth, so the hierarchy of orders prevents us from speaking about all functions or attributes of a given type. Such ranking permits us to speak only about all first-order functions of a given type, or all second-order functions of a given type, and so forth. Thus we cannot, according to the ramified theory of types, say that Bob has all of Al's good qualities, which would ordinarily be symbolized as

$$(F_1)\{[G_2(F_1) \cdot F_1(a)] \supset F_1(b)\}$$

Instead, we can say that Bob has all of Al's good *first-order* qualities, symbolized as

$$({}^1F_1)\{[{}^1G_2({}^1F_1) \cdot {}^1F_1(a)] \supset {}^1F_1(b)\}$$

or that Bob has all of Al's good *second-order* qualities, symbolized as

$$({}^2F_1)\{[{}^1G_2({}^2F_1) \cdot {}^2F_1(a)] \supset {}^2F_1(b)\}$$

or that Bob has all of Al's good *n*th order qualities, for some specified n. It should be noticed that the attribute of having all of Al's first-order attributes, symbolized

$$({}^1F_1)[{}^1F_1(a) \supset {}^1F_1(x)]$$

is a *second-order* attribute.

The preceding formulation of the ramified theory of logical types is crude, but it suffices to eliminate the Grelling paradox. In Case 1, the step from 2 to 3 is rejected on the grounds that the function 'Het' is of higher order than the function variable 'ψ' and may not be instantiated in its place. In Case 2, the step from 1 to 2 is likewise rejected on the grounds that the function 'Het' is of higher order than the function variable 'ϕ' and may not be instantiated in its place.

Another feature of the ramified theory of logical types is that it divides *propositions* into a hierarchy of propositions of different orders. Just as any function (of any type) is a first-order function if it makes no reference to any totality of functions of that type (i.e., if it contains no quantifier on any function variable of that type), so any proposition is said to be a *first-order proposition* if it makes no reference to any totality of propositions (i.e., if it contains no quantifier on any propositional variable). In general, a proposition is of order $n + 1$ if it contains a quantifier on a propositional variable of order n but contains no quantifier on any propositional variable of order m where $m \geq n$.[6] The restriction here is that we can never refer to *all* propositions, but only to all propositions of this or that specified order. Thus we cannot say that 'None

[6]In *Principia Mathematica*, propositions are also divided into different orders on the basis of differences in the orders of the functions they contain. We shall ignore that subtlety in the present exposition, however.

of the propositions uttered by Smith tends to incriminate him', which we might partially symbolize as

$$(p)[(\text{Smith utters } p) \supset \sim(p \text{ tends to incriminate Smith})]$$

but can say instead either that 'None of the first-order propositions uttered by Smith tends to incriminate him', or that 'None of the second-order propositions uttered by Smith tends to incriminate him', and so forth. We would partially symbolize the second of these alternative propositions as

$$({}^2p)[(\text{Smith utters } {}^2p) \supset \sim({}^2p \text{ tends to incriminate Smith})]$$

That proposition contains a quantifier on a propositional variable of order 2, and is, therefore, a proposition of order 3.

By dividing propositions into different orders and permitting reference only to propositions of some specified order or orders, the ramified theory of logical types effectively prevents the paradox of the liar. Any version of that paradox involves the assertion that *all propositions satisfying a certain condition are false,* where the assertion, itself, is a proposition that satisfies that condition. (The condition in question might be *being asserted by the speaker,* or *being written in a specified location,* or *being uttered by a Cretan,* etc.) The paradox is fully explicit when the assertion in question is the *only* proposition that satisfies the specified condition, for in that case, if it is true, then it is false, and if it is false, then it is true. Such a contradiction is prevented by the ramified theory of logical types in the following way. The assertion can refer only to all propositions of a certain order, so it can assert only for some specified n that *all* nth *order propositions satisfying a certain condition are false.* But no paradox can arise here, because the italicized sentence expresses a proposition of order $n + 1$, and even if it satisfies the specified condition, it is not an nth order proposition and, therefore, does not assert its own falsehood.

The ramified theory of logical types, including both the hierarchy of types and the hierarchy of orders, was recommended by Russell and Whitehead not only for 'its ability to solve certain contradictions', but also for having 'a certain consonance with common sense which makes it inherently credible'.[7] They claimed to have deduced the theory from what they called the 'vicious-circle principle', of which one of their formulations was: 'Whatever involves *all* of a collection must not be one of the collection'. But that principle is not obvious, and their putative deduction is not convincing.[8] The chief merit of the theory

[7] A. N. Whitehead and B. Russell, 'Introduction to the first edition', *Principia Mathematica,* chap. II.

[8] In his 'Foundations of Mathematics', F. P. Ramsey remarked on the 'rather sloppy way' in which the type theory was deduced from the 'vicious-circle principle' (p. 356, p. 24 of reprint), and challenged the validity of that principle itself, holding that '. . . we may refer to a man as the tallest in a group, thus identifying him by means of a totality of which he is himself a member without there being any vicious circle' (p. 368, p. 41 of reprint).

of logical types, at least of its ramified version, would seem to be its prevention of the paradoxes.

There are difficulties as well as advantages with the ramified theory of logical types. One of these concerns the notion of identity. The usual definition of identity (of individuals) is

$$(x = y) = \text{df}\, (F_1)[F_1(x) \equiv F_1(y)]$$

from which definition all of the usual attributes of the identity relation can be deduced. But that definition violates the ramified theory of logical types, since reference is made in it to *all* functions of type 1. Were we to replace it by the definition

$$(x = y) = \text{df}\, ({}^1F_1)[{}^1F_1(x) \equiv {}^1F_1(y)]$$

which provides that x and y are identical if they have all of their *first-order* attributes in common, the possibility arises that x and y might be identical (in this sense) and still have different *second-order* attributes. For any n, defining identity as the sharing of all attributes of order n would permit "identical" individuals to differ with respect to attributes of order m where $m > n$. If we accept the restrictions of the ramified theory of logical types, then we cannot define the identity relation, and even if we accept the relation of identity as primitive or undefined, we could not state all the rules for its usage.

Other disadvantages of the ramified theory of logical types are more technical, and will merely be mentioned. Mathematicians wish to establish their theorems for *all* functions (of numbers), but they cannot do so if they are bound by the hierarchy of orders. Moreover, certain existence theorems in analysis, such as that of the Least Upper Bound, cannot be proved within the restrictions of the ramified theory of types. The Cantorean theory of the infinite, which is basic to nearly all of modern mathematics, cannot be established within the rigid framework of the ramified type theory. And even the principle of mathematical induction must be abandoned in its full generality, for its complete statement is prevented by the ramified theory of logical types.

To relax the excessive restrictions of the hierarchy of orders, the 'Axiom of Reducibility' was introduced. That axiom asserts that to any function of any order and any type there corresponds a formally equivalent *first-order* function of the same type (two functions are *formally equivalent* when, for any admissible argument, they are either both true or both false). In the presence of this axiom, the identity relation *can* be satisfactorily defined in terms of first-order functions and all of the disadvantages mentioned in the preceding paragraph disappear. The question naturally arises: Does the Axiom of Reducibility relax the restrictions of the ramified theory of logical types sufficiently to permit the reintroduction of the paradoxes? If *all* of the semantical terms such as 'true', 'false', and 'designates' are added to a logical system like that of *Principia*

Mathematica, which contains both the ramified type theory and the Axiom of Reducibility, and names for all functions are added also, then at least some of the paradoxes reappear.[9] On the other hand, if the names of *some* functions are left out, then even in the presence of the semantical terms, the paradoxes do not seem to be derivable even with the aid of the Axiom of Reducibility.[10]

It may not be out of place here to indicate briefly how the 'levels of language' method of avoiding the semantical paradoxes[11] is remarkably similar to the ramified type theory's hierarchy of orders.[12] Confining our remarks to the Grelling paradox, we note that this paradox does not arise in an object language (like the extended function calculus, for example) when we assume that there are no symbols in it that designate symbols. Now does it arise in the metalanguage of that object language. Since the metalanguage contains synonyms for all symbols of the object language and names for all symbols of the object language as well as its own variables and the name relation (which we write as 'Des'), the symbol 'Het' can be defined in it. By definition:

$$\text{Het}(v)\cdot \equiv :\cdot(\exists\phi):v\text{Des}\phi\cdot v\text{Des}\psi \equiv_{\psi} \psi = \phi\cdot\sim\phi(v)$$

Yet the Grelling cannot be derived in the metalanguage, because although there is a symbol for the function Het, there is no symbol for the name of that function. In other words, in the metalanguage, although we could substitute 'Het' for 'ϕ', we cannot substitute "Het" for 'v' because "Het" is not a symbol of the metalanguage.

So far, there are no complications. The paradox does not arise in the object language because that structure contains no names for the symbols in it. Nor does the paradox arise in the metalanguage, because there is no name for the function symbol 'Het' in that language. The *threat* of the Grelling arises only in the meta-metalanguage. *If* certain safeguards or refinements of the levels of language theory are ignored, the paradox seems to be derivable. Ignoring those safeguards, we have, by definition, in the meta-metalanguage,

$$\text{Het}(v)\cdot \equiv :\cdot(\exists\phi):v\text{Des}\phi\cdot v\text{Des}\psi \equiv_{\psi} \psi = \phi\cdot\sim\phi(v)$$

Since the meta-metalanguage *does* contain a name for the function symbol 'Het', we substitute that name, "Het", for the variable 'v' and obtain the contradiction, as we did in our first derivation.

[9] See the author's article on 'The Inconsistency or Redundancy of *Principia Mathematica*' in *Philosophy and Phenomenological Research,* vol. 11, no. 2 (December 1950), pp. 190–199.

[10] See the review of the preceding article by Alonzo Church, *The Journal of Symbolic Logic,* vol. 16, no. 2 (June 1951), pp. 154–155.

[11] First suggested by Bertrand Russell in his Introduction to L. Wittgenstein's *Tractatus logico-philosophicus,* Harcourt Brace Jovanovich, New York, and Kegan Paul, Trench, Trubner & Co. Ltd., London, 1922, p. 23.

[12] The remainder of this appendix is reprinted by permission from 'The Inconsistency or Redundancy of *Principia Mathematica*', *Philosophy and Phenomenological Research,* vol. 11, University of Buffalo, Buffalo, N. Y., 1950.

The way in which the safeguards of the levels of language theory serve to prevent this contradiction makes it very similar to the theory of orders. The definition of Het in the meta-metalanguage that was written above requires the addition of subscripts to resolve its ambiguity. Once these ambiguities are pointed out and resolved, the contradiction vanishes.

In the first place, the meta-metalanguage contains two symbols for the name relation, 'Des_1' and 'Des_2'.[13] The first of these is the meta-metalanguage's synonym for the name relation in the metalanguage. The full sentence

$$\upsilon \mathrm{Des}_1 \phi$$

asserts that the function symbol denoted by 'υ' is a symbol of the object language and designates the function ϕ. The second has no synonym in the metalanguage. The full sentence

$$\upsilon \mathrm{Des}_2 \phi$$

asserts that the function symbol denoted by 'υ' is a symbol of the metalanguage and designates the function ϕ. These are quite different.

And in the second place, the meta-metalanguage contains two symbols 'Het_1' and 'Het_2', between whose meanings there is significant difference. The first of these is the meta-metalanguage's synonym for the function symbol 'Het' of the metalanguage. The full sentence

$$\mathrm{Het}_1(\upsilon)$$

asserts that the function symbol denoted by 'υ' is a symbol of the object language and designates in the object language an attribute that it does not possess. The second has no synonym in the metalanguage. The full sentence

$$\mathrm{Het}_2(\upsilon)$$

asserts that the function symbol denoted by 'υ' is a symbol of the metalanguage and designates in the metalanguage an attribute that it does not possess. Their definitions are different:

$$\mathrm{Het}_1(\upsilon) = \mathrm{df}\ (\exists\phi){:}\upsilon\mathrm{Des}_1\phi \cdot \upsilon\mathrm{Des}_1\psi \equiv_{\psi} \psi = \phi \cdot {\sim}\phi(\upsilon)$$

and

$$\mathrm{Het}_2(\upsilon) = \mathrm{df}\ (\exists\phi){:}\upsilon\mathrm{Des}_2\phi \cdot \upsilon\mathrm{Des}_2\psi \equiv_{\psi} \psi = \phi \cdot {\sim}\phi(\upsilon)$$

It is clear that we cannot define Het_1 in terms of Des_2 because the values of the arguments of the two functions are terms of different languages, of the object

[13] As was suggested by Ramsey in a somewhat different context, op. cit., p. 370, in reprint, p. 43.

language for Het_1 and of the metalanguage for Des_2. For the same reason, Het_2 cannot be defined in terms of Des_1.

No version of the Grelling can be derived from the definition of Het_1, because the only values of its arguments are terms of the object language, and there is no term of the object language analogous to either 'Het_1' or 'Het_2'. The only possibility lies in the direction of deriving a contradiction from the definition of Het_2, and this is thwarted by something remarkably like *Principia's* theory of orders.

In the definition of Het_2 we cannot substitute for 'v' the name of the symbol for that function, because although the function symbol 'Het_2' occurs in the meta-metalanguage, no *name* of that function symbol occurs there. The best we can do is to substitute the meta-metalanguage's name of the function symbol of the metalanguage that is synonymous with 'Het_1', for which we *do* have a name in the meta-metalanguage (call it "Het"). Making the substitution, we have

$$\text{Het}_2(\text{`Het'}) \cdot \equiv : \cdot (\exists\phi) : \text{`Het'}\text{Des}_2\phi \cdot \text{`Het'}\text{Des}_2\psi \equiv_\psi \psi = \phi \cdot \sim\phi(\text{`Het'})$$

If we endeavor to deduce a contradiction from this equivalence by an argument parallel to earlier versions, we are unable to do so. There is a choice of function symbol to substitute for the generalized function variable 'ψ', for there are two function symbols in our meta-metalanguage that look promising: 'Het_1' and 'Het_2'.

If we substitute 'Het_1', we obtain

$$\text{Het}_2(\text{`Het'}) \cdot \supset : \text{`Het'}\text{Des}_2\phi \cdot \text{`Het'}\text{Des}_2\text{Het}_1 \equiv \text{Het}_1 = \phi \cdot \sim\phi(\text{`Het'})$$

Since "Het" is the meta-metalanguage's name for the function symbol of the metalanguage that is synonymous with 'Het_1', we have

$$\text{`Het'}\text{Des}_2\text{Het}_1 \cdot \text{`Het'}\text{Des}_2\text{Het}_1 \equiv \text{Het}_1 = \phi$$

and consequently

$$\text{Het}_2(\text{`Het'}) \supset \sim\text{Het}_1(\text{`Het'})$$

But this is no part of any contradiction. It is antecedently known on independent grounds that if any term satisfies Het_2, it is in the metalanguage and not in the object language, whereas only terms of the object language satisfy Het_1.

On the other hand, if we substitute 'Het_2', we obtain

$$\text{Het}_2(\text{`Het'}) \supset : \text{`Het'}\text{Des}_2\phi \cdot \text{`Het'}\text{Des}_2\text{Het}_2 \equiv \text{Het}_2 = \phi \cdot \sim\phi(\text{`Het'})$$

From this, *if* "Het'Des_2Het_2' were true, we should indeed be able to obtain a contradiction. But "Het'Des_2Het_2' is *not* true, because the argument "Het" denotes a symbol of the metalanguage, whereas the attribute denoted by the argument 'Het_2' is not denoted by any symbol of the metalanguage. In other words, "Het'Des_2Het_2' is false because 'Het_2' is a symbol of the meta-metalanguage which has no synonym in the metalanguage.

This is very like the theory of orders, because the contradiction is evaded by arranging that certain symbols of the meta-metalanguage are defined *over certain ranges*. Thus 'Des_1' is defined over a narrower range than 'Des_2', and 'Het_1' is defined over a narrower range than 'Het_2'; Des_1 and Het_1 are satisfied only by symbols of the object language, Des_2 and Het_2 are satisfied only by symbols of the metalanguage, which is a wider and more inclusive language. Not only is the levels of language theory remarkably analogous to the theory of orders, but where each metalanguage is conceived as actually containing the object language with which it deals,[14] it can be identified with the Russellian theory of orders *as applied to symbols* rather than to the functions they denote.

In spite of the indicated similarities, there are fundamental differences between the two theories. Most significant is that unlike the ramified type theory, the levels of language device for avoiding paradoxes does not jeopardize the derivation of any parts of classical mathematics. Thus no need arises for any analogue to the reducibility axiom.

[14] This conception is recommended by A. Tarski in his 'Semantic Conception of Truth', *Philosophy and Phenomenological Research*, vol. 4 (1944), pp. 341–375. Reprinted in H. Feigl and W. Sellars, *Readings in Philosophical Analysis*, Appleton-Century-Crofts, New York, 1949, pp. 52–84. See especially p. 350 in vol. 4 and pp. 60–61 in reprint.

Solutions to Selected Exercises

Exercises on pages 14–15:

I. **1.** **F** **5.** **T** **10.** **F** **15.** **F**
II. **1.** **F** **5.** **T** **10.** **T** **15.** **F**
III. **1.** $(A \cdot B) \vee C$ **5.** $(C \vee D) \cdot \sim(C \cdot D)$
IV. **1.** $S \cdot W$ **5.** $G \cdot S$

Exercises on pages 18–19:

I. **1.** **T** **5.** **F** **10.** **T** **15.** **T**
II. **1.** **T** **5.** **F** **10.** **F** **15.** **F**
III. **1.** $(A \cdot C) \supset \sim D$ **5.** $\sim A \supset \sim(C \vee D)$ **10.** $(A \supset \sim C) \cdot (\sim C \supset D)$

Exercises on pages 25–26:

I. **a.** 4 is the specific form of **a.** **e.** 13 has **e** as a substitution instance. **j.** 10 has **j** as a substitution instance, and 20 is the specific form of **j**.
II. **1.** Valid **5.** Invalid—shown by row 2. **10.** Invalid—shown by row 1. **15.** Valid.
III. **1.** Invalid—shown by row 1. **5.** Valid. **10.** Invalid—shown by row 3. **15.** Invalid—shown by row 12.

Exercises on pages 30–31:

I. **1.** Contingent—**F** in row 1, **T** in row 2. **5.** Tautologous.
II. **1.** Not equivalent—shown by rows 2 and 3. **5.** Equivalent.

Exercises on pages 34–38:

I. **1.** Simplification (Simp.). **5.** Hypothetical Syllogism (H.S.) **10.** Constructive Dilemma (C.D.)

II.

1.		**5.**		**10.**	
3.	2, Simp.	4.	3, Simp.	7.	1, 6, M.T.
4.	1, 3, M.P.	5.	2, 4, M.P.	8.	2, 6, M.T.
5.	3, Simp.	6.	4, Simp.	9.	7, 8, Conj.
6.	4, 5, M.P.	7.	6, Add.	10.	3, 9, M.P.
7.	6, Simp.	8.	1, 7, M.P.	11.	4, 10, C.D.
8.	7, Add.	9.	8, 5, H.S.	12.	5, 11, D.D.

III. 1.

1.	$A \supset B$	
2.	$C \supset D$	
3.	$(\sim B \vee \sim D) \cdot (\sim A \vee \sim B)$	$/ \therefore \sim A \vee \sim C$
4.	$(A \supset B) \cdot (C \supset D)$	1, 2, Conj.
5.	$\sim B \vee \sim D$	3, Simp.
6.	$\sim A \vee \sim C$	4, 5, D.D.

5.

1.	$(R \supset \sim S) \cdot (T \supset \sim U)$	
2.	$(V \supset \sim W) \cdot (X \supset \sim Y)$	
3.	$(T \supset W) \cdot (U \supset S)$	
4.	$V \vee R$	$/ \therefore \sim T \vee \sim U$
5.	$V \supset \sim W$	2, Simp.
6.	$R \supset \sim S$	1, Simp.
7.	$(V \supset \sim W) \cdot (R \supset \sim S)$	5, 6, Conj.
8.	$\sim W \vee \sim S$	7, 5, C.D.
9.	$\sim T \vee \sim U$	3, 8, D.D.

IV. 1.

1.	$(A \vee G) \supset S$	
2.	$A \cdot T \quad / \therefore S$	
3.	A	2, Simp.
4.	$A \vee G$	3, Add.
5.	S	1, 4, M.P.

5.

1.	$(\sim K \cdot P) \supset (B \vee R)$	
2.	$\sim K \supset (B \supset D)$	
3.	$K \vee (R \supset E)$	
4.	$\sim K \cdot P$	$/ \therefore D \vee E$
5.	$\sim K$	4, Simp.
6.	$B \supset D$	2, 5, M.P.
7.	$R \supset E$	3, 5, D.S.
8.	$(B \supset D) \cdot (R \supset E)$	6, 7, Conj.
9.	$B \vee R$	1, 4, M.P.
10.	$D \vee E$	8, 9, C.D.

10.

1.	$T \vee (E \supset D)$	
2.	$T \supset C$	
3.	$(E \supset G) \supset (D \supset I)$	
4.	$(\sim T \cdot \sim C) \supset (D \supset G)$	
5.	$\sim C$	
6.	$\sim I \vee \sim G$	$/ \therefore \sim D \vee \sim E$
7.	$\sim T$	2, 5, M.T.
8.	$E \supset D$	1, 7, D.S.
9.	$\sim T \cdot \sim C$	7, 5, Conj.
10.	$D \supset G$	4, 9, M.P.
11.	$E \supset G$	8, 10, H.S.
12.	$D \supset I$	3, 11, M.P.
13.	$(D \supset I) \cdot (E \supset G)$	12, 11, Conj.
14.	$\sim D \vee \sim E$	13, 6, D.D.

Exercises on pages 43–47:

I. **1.** Commutation (Com.) **5.** Tautology (Taut.) **10.** Commutation (Com.) **15.** Distribution (Dist.)

II. **1.**
3. 2, Add.
4. 3, De M.
5. 1, 4, M.T.
6. 5, De M.
7. 6, Com.
8. 7, Simp.

5.
4. 1, Exp.
5. 2, Impl.
6. 5, Taut.
7. 4, 6, M.T.
8. 7, De M.
9. 3, 8, D.D.
10. 9, Impl.

III. **1.**

1.	$\sim A$	$/\therefore A \supset B$
2.	$\sim A \vee B$	1, Add.
3.	$A \supset B$	2, Impl.

5.

1.	$K \supset L$	$/\therefore K \supset (L \vee M)$
2.	$\sim K \vee L$	1, Impl.
3.	$(\sim K \vee L) \vee M$	2, Add.
4.	$\sim K \vee (L \vee M)$	3, Assoc.
5.	$K \supset (L \vee M)$	4, Impl.

10.

1.	$A \supset \sim(B \supset C)$	
2.	$(D \cdot B) \supset C$	
3.	D	$/\therefore \sim A$
4.	$D \supset (B \supset C)$	2, Exp.
5.	$B \supset C$	4, 3, M.P.
6.	$\sim\sim(B \supset C)$	5, D.N.
7.	$\sim A$	1, 6, M.T.

15.

1.	$(\sim U \vee V) \cdot (U \vee W)$	
2.	$\sim X \supset \sim W$	$/\therefore V \vee X$
3.	$\sim U \vee V$	1, Simp.
4.	$U \supset V$	3, Impl.
5.	$W \supset X$	2, Trans.
6.	$(U \supset V) \cdot (W \supset X)$	4, 5, Conj.
7.	$(U \vee W) \cdot (\sim U \vee V)$	1, Com.
8.	$U \vee W$	7, Simp.
9.	$V \vee X$	6, 8, C.D.

20.

1.	$O \supset (P \supset Q)$	
2.	$P \supset (Q \supset R)$	$/\therefore O \supset (P \supset R)$
3.	$(O \cdot P) \supset Q$	1, Exp.
4.	$(P \cdot Q) \supset R$	2, Exp.
5.	$(Q \cdot P) \supset R$	4, Com.
6.	$Q \supset (P \supset R)$	5, Exp.
7.	$(O \cdot P) \supset (P \supset R)$	3, 6, H.S.
8.	$O \supset [P \supset (P \supset R)]$	7, Exp.
9.	$O \supset [(P \cdot P) \supset R]$	8, Exp.
10.	$O \supset (P \supset R)$	9, Taut.

25.

1.	$J \cdot (K \cdot L)$	
2.	$(K \cdot J) \supset [M \vee (N \vee O)]$	
3.	$\sim M \cdot (\sim P \cdot \sim O) /\therefore N$	
4.	$(J \cdot K) \cdot L$	1, Assoc.
5.	$J \cdot K$	4, Simp.
6.	$K \cdot J$	5, Com.
7.	$M \vee (N \vee O)$	2, 6, M.P.
8.	$\sim M$	3, Simp.
9.	$N \vee O$	7, 8, D.S.
10.	$O \vee N$	9, Com.
11.	$(\sim M \cdot \sim P) \cdot \sim O$	3, Assoc.

12.	$\sim O \cdot (\sim M \cdot \sim P)$	11, Com.
13.	$\sim O$	12, Simp.
14.	N	10, 13, D.S.

IV. **1.**

1.	$S \supset G$	
2.	$\sim S \supset E$	$/\therefore G \vee E$
3.	$\sim G \supset \sim S$	1, Trans.
4.	$\sim G \supset E$	3, 2, H.S.
5.	$\sim\sim G \vee E$	4, Impl.
6.	$G \vee E$	5, D.N.

5.

1.	$(G \cdot L) \vee (W \cdot T)$	
2.	$\sim G$	$/\therefore T$
3.	$\sim G \vee \sim L$	2, Add.
4.	$\sim (G \cdot L)$	3, De M.
5.	$W \cdot T$	1, 4, D.S.
6.	$T \cdot W$	5, Com.
7.	T	6, Simp.

10.

1.	$B \vee T$	
2.	$(B \vee C) \supset (L \cdot M)$	
3.	$\sim L$	$/\therefore T$
4.	$\sim L \vee \sim M$	3, Add.
5.	$\sim (L \cdot M)$	4, De M.
6.	$\sim (B \vee C)$	2, 5, M.T.
7.	$\sim B \cdot \sim C$	6, De M.
8.	$\sim B$	7, Simp.
9.	T	1, 8, D.S.

15.

1.	$(W \supset M) \cdot (I \supset E)$	
2.	$W \vee I$	
3.	$(W \supset \sim E) \cdot (I \supset \sim M)$	$/\therefore E \equiv \sim M$
4.	$M \vee E$	1, 2, C.D.
5.	$\sim E \vee \sim M$	3, 2, C.D.
6.	$E \supset \sim M$	5, Impl.
7.	$\sim\sim M \vee E$	4, D.N.
8.	$\sim M \supset E$	7, Impl.
9.	$(E \supset \sim M) \cdot (\sim M \supset E)$	6, 8, Conj.
10.	$E \equiv \sim M$	9, Equiv.

21.

1.	$(T \supset E) \cdot (A \supset L)$	$/\therefore (T \vee A) \supset (E \vee L)$
2.	$T \supset E$	1, Simp.
3.	$\sim T \vee E$	2, Impl.
4.	$(\sim T \vee E) \vee L$	3, Add.
5.	$\sim T \vee (E \vee L)$	4, Assoc.
6.	$(E \vee L) \vee \sim T$	5, Com.
7.	$(A \supset L) \cdot (T \supset E)$	1, Com.
8.	$A \supset L$	7, Simp.
9.	$\sim A \vee L$	8, Impl.
10.	$(\sim A \vee L) \vee E$	9, Add.
11.	$\sim A \vee (L \vee E)$	10, Assoc.
12.	$\sim A \vee (E \vee L)$	11, Com.
13.	$(E \vee L) \vee \sim A$	12, Com.
14.	$[(E \vee L) \vee \sim T] \cdot [(E \vee L) \vee \sim A]$	6, 13, Conj.
15.	$(E \vee L) \vee (\sim T \cdot \sim A)$	14, Dist.
16.	$(\sim T \cdot \sim A) \vee (E \vee L)$	15, Com.
17.	$\sim (T \vee A) \vee (E \vee L)$	16, De M.
18.	$(T \vee A) \supset (E \vee L)$	17, Impl.

Exercises on page 49:

1.

A	B	C	D
F	**T**	**F**	**F**

5.

	T	U	V	W	X
	T	**T**	**T**	**T**	**F**
or	**F**	**F**	**T**	**F**	**T**

Exercises on page 52:

21.

1.	$(T \supset E) \cdot (A \supset L)$	$/\therefore (T \vee A) \supset (E \vee L)$
2.	$T \vee A$	$/\therefore E \vee L$
3.	$E \vee L$	1, 2, C.D.

Exercises on page 54:

3.

1.	$(H \supset I) \cdot (J \supset K)$				
2.	$(I \vee K) \supset L$				
3.	$\sim L$	$/\therefore \sim(H \vee J)$			
4.	$\sim(I \vee K)$	2, 3, M.T.	4′.	$\sim\sim(H \vee J)$	I.P.
5.	$\sim I \cdot \sim K$	4, De M.	5′.	$H \vee J$	4′, D.N.
6.	$\sim I$	5, Simp.	6′.	$I \vee K$	1, 5′, C.D.
7.	$H \supset I$	1, Simp.	7′.	L	2, 6′, M.P.
8.	$\sim H$	7, 6, M.T.	8′.	$L \cdot \sim L$	7′, 3, Conj.
9.	$(J \supset K) \cdot (H \supset I)$	1, Com.			
10.	$J \supset K$	9, Simp.			
11.	$\sim K \cdot \sim I$	5, Com.			
12.	$\sim K$	11, Simp.			
13.	$\sim J$	10, 12, M.T.			
14.	$\sim H \cdot \sim J$	8, 13, Conj.			
15.	$\sim(H \vee J)$	14, De M.			

5.

1.	$(V \supset \sim W) \cdot (X \supset Y)$				
2.	$(\sim W \supset Z) \cdot (Y \supset \sim A)$				
3.	$(Z \supset \sim B) \cdot (\sim A \supset C)$				
4.	$V \cdot X$	$/\therefore \sim B \cdot C$			
5.	$V \supset \sim W$	1, Simp.	5′.	$\sim(\sim B \cdot C)$	I.P.
6.	V	4, Simp.	6′.	$\sim\sim B \vee \sim C$	5′, De M.
7.	$\sim W$	5, 6, M.P.	7′.	$\sim Z \vee \sim\sim A$	3, 6′, D.D.
8.	$\sim W \supset Z$	2, Simp.	8′.	$\sim\sim W \vee \sim Y$	2, 7′, D.D.
9.	Z	8, 7, M.P.	9′.	$\sim V \vee \sim X$	1, 8′, D.D.
10.	$Z \supset \sim B$	3, Simp.	10′.	$\sim(V \cdot X)$	9′, De M.
11.	$\sim B$	10, 9, M.P.	11′.	$(V \cdot X) \cdot \sim(V \cdot X)$	4, 10′, Conj.
12.	$(X \supset Y) \cdot (V \supset W)$	1, Com.			
13.	$X \supset Y$	12, Simp.			
14.	$X \cdot V$	4, Com.			
15.	X	14, Simp.			
16.	Y	13, 15, M.P.			
17.	$(Y \supset \sim A) \cdot (\sim W \supset Z)$	2, Com.			
18.	$Y \supset \sim A$	17, Simp.			
19.	$\sim A$	18, 16, M.P.			
20.	$(\sim A \supset C) \cdot (Z \supset \sim B)$	3, Com.			
21.	$\sim A \supset C$	20, Simp.			
22.	C	21, 19, M.P.			
23.	$\sim B \cdot C$	11, 22, Conj.			

Exercises on page 56:

I. 1.

1.	P	$/\therefore Q \supset P$	(C.P.)
2.	$P \vee \sim Q$	1, Add.	
3.	$\sim Q \vee P$	2, Com.	
4.	$Q \supset P$	3, Impl.	

5.

1.	$\sim\sim P$	$/\therefore P$	(C.P.)
2.	P	1, D.N.	

10. 1. $A \supset B$ $\quad /\therefore A \supset (A \cdot B)$ (C.P.)
2. A $\quad /\therefore A \cdot B$ (C.P.)
3. B 1, 2, M.P.
4. $A \cdot B$ 2, 3, Conj.

15. 1. $A \supset (B \cdot C)$ $\quad /\therefore [B \supset (D \cdot E)] \supset (A \supset D)$ (C.P.)
2. $B \supset (D \cdot E)$ $\quad /\therefore A \supset D$ (C.P.)
3. A $\quad /\therefore D$ (C.P.)
4. $B \cdot C$ 1, 3, M.P.
5. B 4, Simp.
6. $D \cdot E$ 2, 5, M.P.
7. D 6, Simp.

II. **1.** 1. $\sim[(A \supset B) \vee (A \supset \sim B)]$ $\quad /\therefore (A \supset B) \vee (A \supset \sim B)$ (I.P.)
2. $\sim(A \supset B) \cdot \sim(A \supset \sim B)$ 1, De M.
3. $\sim(\sim A \vee B) \cdot \sim(\sim A \vee \sim B)$ 2, Impl.
4. $(\sim\sim A \cdot \sim B) \cdot (\sim\sim A \cdot \sim\sim B)$ 3, De M.
5. $(\sim\sim A \cdot \sim B) \cdot (\sim\sim B \cdot \sim\sim A)$ 4, Com.
6. $[(\sim\sim A \cdot \sim B) \cdot \sim\sim B] \cdot \sim\sim A$ 5, Assoc.
7. $[\sim\sim A \cdot (\sim B \cdot \sim\sim B)] \cdot \sim\sim A$ 6, Assoc.
8. $[(\sim B \cdot \sim\sim B) \cdot \sim\sim A] \cdot \sim\sim A$ 7, Com.
9. $(\sim B \cdot \sim\sim B) \cdot (\sim\sim A \cdot \sim\sim A)$ 8, Assoc.
10. $\sim B \cdot \sim\sim B$ 9, Simp.

5. 1. $\sim[(A \supset B) \vee (\sim A \supset C)]$ $\quad /\therefore (A \supset B) \vee (\sim A \supset C)$ (I.P.)
2. $\sim(A \supset B) \cdot \sim(\sim A \supset C)$ 1, De M.
3. $\sim(\sim A \vee B) \cdot \sim(\sim\sim A \vee C)$ 2, Impl.
4. $(\sim\sim A \cdot \sim B) \cdot (\sim\sim\sim A \cdot \sim C)$ 3, De M.
5. $\sim\sim A \cdot [\sim B \cdot (\sim\sim\sim A \cdot \sim C)]$ 4, Assoc.
6. $\sim\sim A \cdot [(\sim\sim\sim A \cdot \sim C) \cdot \sim B]$ 5, Com.
7. $\sim\sim A \cdot [\sim\sim\sim A \cdot (\sim C \cdot \sim B)]$ 6, Assoc.
8. $(\sim\sim A \cdot \sim\sim\sim A) \cdot (\sim C \cdot \sim B)$ 7, Assoc.
9. $\sim\sim A \cdot \sim\sim\sim A$ 8, Simp.

Exercises on page 61:

1. 1. $A \supset B$
2. $B \supset [(C \supset \sim\sim C) \supset D]$ $\quad /\therefore A \supset D$
→3. A
4. B 1, 3, M.P.
5. $(C \supset \sim\sim C) \supset D$ 2, 4, M.P.
→6. C
7. $\sim\sim C$ 6, D.N.
8. $C \supset \sim\sim C$ 6–7, C.P.
9. D 5, 8, M.P.
10. $A \supset D$ 3–9, C.P.

Exercises on pages 69–71:

I. **1.** $(x)(Sx \supset Rx)$ **5.** $(x)(Sx \supset Ex)$
10. $(x)(Dx \supset Bx)$ **15.** $(x)[Cx \supset (\sim Wx \vee Sx)]$
20. $(x)[(Ix \cdot Ux) \supset (Px \vee Fx)]$

25. $(x)[(Ax \cdot Ox) \supset (Dx \supset \sim Rx)]$

30. $(x)[(Hx \cdot Gx) \supset Tx]$ **35.** $(x)[(Gx \cdot Hx) \supset Tx]$ **40.** $(x)[(Tx \cdot Hx) \supset Gx]$

II. **1.** $(x)(Cx \supset Bx)$ **5.** $(x)[Mx \supset (Vx \cdot Fx)]$

Exercises on pages 76–77:

I. **1.**

1.	$(x)(Ax \supset Bx)$	
2.	$\sim Bt$	$/\therefore \sim At$
3.	$At \supset Bt$	1, **UI**
4.	$\sim At$	3, 2, M.T.

5.

1.	$(x)(Kx \supset Lx)$	
2.	$(x)[(Kx \cdot Lx) \supset Mx]$	$/\therefore (x)(Kx \supset Mx)$
3.	Ky	
4.	$Ky \supset Ly$	1, **UI**
5.	Ly	4, 3, M.P.
6.	$Ky \cdot Ly$	3, 6, Conj.
7.	$(Ky \cdot Ly) \supset My$	2, **UI**
8.	My	7, 6, M.P.
9.	$Ky \supset My$	3–8, C.P.
10.	$(x)(Kx \supset Mx)$	9, **UG**

II. **1.**

1.	$(x)(Ax \supset Bx)$	
2.	$\sim Bc$	$/\therefore \sim Ac$
3.	$Ac \supset Bc$	1, **UI**
4.	$\sim Ac$	3, 2, M.T.

5.

1.	$(x)(Lx \supset Mx)$	
2.	$(\exists x)(Lx \cdot Nx)$	$/\therefore (\exists x)(Nx \cdot Mx)$
3.	$La \cdot Na$	2, **EI**
4.	$La \supset Ma$	1, **UI**
5.	La	3, Simp.
6.	Ma	4, 5, M.P.
7.	$Na \cdot La$	3, Com.
8.	Na	7, Simp.
9.	$Na \cdot Ma$	8, 6, Conj.
10.	$(\exists x)(Nx \cdot Mx)$	9, **EG**

10.

1.	$(x)(Dx \supset Ex)$	
2.	$(x)(Ex \supset Fx)$	
3.	$(x)(Fx \supset Gx)$	$/\therefore (x)(Dx \supset Gx)$
4.	$Dy \supset Ey$	1, **UI**
5.	$Ey \supset Fy$	2, **UI**
6.	$Fy \supset Gy$	3, **UI**
7.	$Dy \supset Fy$	4, 5, H.S.
8.	$Dy \supset Gy$	7, 6, H.S.
9.	$(x)(Dx \supset Gx)$	8, **UG**

15.

1.	$(x)[(Bx \vee Vx) \supset (Ox \cdot Dx)]$	$/\therefore (x)(Bx \supset Dx)$
2.	By	
3.	$(By \vee Vy) \supset (Oy \cdot Dy)$	1, **UI**
4.	$By \vee Vy$	2, Add.
5.	$Oy \cdot Dy$	3, 4, M.P.
6.	$Dy \cdot Oy$	5, Com.
7.	Dy	6, Simp.
8.	$By \supset Dy$	2–7, C.P.
9.	$(x)(Bx \supset Dx)$	8, **UG**

20.

1.	$(x)\{(Bx \vee Wx) \supset [(Ax \vee Fx) \supset Sx]\}$ $/ \therefore (x)[Bx \supset (Ax \supset Sx)]$	
→ 2.	By	
→ 3.	Ay	
4.	$(By \vee Wy) \supset [(Ay \vee Fy) \supset Sy]$	1, **UI**
5.	$By \vee Wy$	2, Add.
6.	$(Ay \vee Fy) \supset Sy$	4, 5, M.P.
7.	$Ay \vee Fy$	3, Add.
8.	Sy	6, 7, M.P.
9.	$Ay \supset Sy$	3–8, C.P.
10.	$By \supset (Ay \supset Sy)$	2–9, C.P.
11.	$(x)[Bx \supset (Ax \supset Sx)]$	10, **UG**

Exercises on pages 82–83:

I. **1.** $(\exists x)(Ax \cdot Bx)$ $\boxed{a, c}$ $(Aa \cdot Ba) \vee (Ac \cdot Bc)$
Ac — Ac
$\therefore Bc$ — $\therefore Bc$

Aa	*Ac*	*Ba*	*Bc*
T	T	T	F

5. $(\exists x)(Kx \cdot Lx)$ $\boxed{a, b}$ $(Ka \cdot La) \vee (Kb \cdot Lb)$
$(\exists x)(\sim Kx \cdot \sim Lx)$ — $(\sim Ka \cdot \sim La) \vee (\sim Kb \cdot \sim Lb)$
$\therefore (\exists x)(Lx \cdot \sim Kx)$ — $\therefore (La \cdot \sim Ka) \vee (Lb \cdot \sim Kb)$

	Ka	*Kb*	*La*	*Lb*
	T	F	T	F
or	F	T	F	T

II. **1.** $(x)(Ax \supset Bx)$ $\boxed{j}$ $Aj \supset Bj$
Bj — Bj
$\therefore Aj$ — $\therefore Aj$

Aj	*Bj*
F	T

5. $(x)(Kx \supset \sim Lx)$ $\boxed{a, b}$ $(Ka \supset \sim La) \cdot (Kb \supset \sim Lb)$
$(\exists x)(Mx \cdot Lx)$ — $(Ma \cdot La) \vee (Mb \cdot Lb)$
$\therefore (x)(Kx \supset \sim Mx)$ — $\therefore (Ka \supset \sim Ma) \cdot (Kb \supset \sim Mb)$

	Ka	*Kb*	*La*	*Lb*	*Ma*	*Mb*
	T	F	F	T	T	T
or	F	T	T	F	T	T

III. **1.** $(x)(Ax \supset Bx)$ $\boxed{j}$ $Aj \supset Bj$
Bj — Bj
$\therefore Aj$ — $\therefore Aj$

Aj	*Bj*
F	T

5.

1.	$(x)(Vx \supset Cx)$	
2.	$(\exists x)(Rx \cdot \sim Cx)$	$/ \therefore (\exists x)(Rx \cdot \sim Vx)$
3.	$Ra \cdot \sim Ca$	2, **EI**
4.	$Va \supset Ca$	1, **UI**

5. $\sim Ca \cdot Ra$	3, Com.
6. $\sim Ca$	5, Simp.
7. $\sim Va$	4, 6, M.T.
8. Ra	3, Simp.
9. $Ra \cdot \sim Va$	8, 7, Conj.
10. $(\exists x)(Rx \cdot \sim Vx)$	9, **EG**

10. $(x)[Tx \supset (Ox \vee Ex)]$
$(x)[(Ox \cdot Tx) \supset \sim Ex]$
$(\exists x)(Tx \cdot Ex)$
$\therefore (\exists x)(Tx \cdot Ox)$

[a]

$Ta \supset (Oa \vee Ea)$
$(Oa \cdot Ta) \supset \sim Ea$
$Ta \cdot Ea$
$\therefore Ta \cdot Oa$

Ta	*Oa*	*Ea*
T	**F**	**T**

Exercises on page 88:

1. $(\exists x)Dx \supset (\exists x)(Py \cdot By)$
5. $(x)[Bx \supset (Yx \supset Rx)]$
10. $(x)\{(Ox \cdot Px) \supset [(y)(My \supset \sim Py) \vee Mx]\}$
15. $(x)\{[Sx \cdot (y)(Sy \supset Wy)] \supset Wx\}$

Exercises on page 92:

I. 1. *d* and *f*
5. *none*
II. 1. *a*, *b*, and *c*
2. *g*

Exercises on pages 100–101:

3. The use of **UG** at line 4 is wrong, because ν ('*x*') occurs free in the assumption '$Fx \equiv Gy$' within whose scope the premiss $\Phi\nu$ ('$Fx \equiv Gy$') lies.

6. There are three mistakes here:
a. The use of **UG** at line 5 is wrong, because ν ('*z*') does not occur free in $\Phi\nu$ ('$(\exists x)[(Fx \cdot Gz) \supset Hy]$') at all places that μ ('*y*') occurs free in $\Phi\mu$ ('$(\exists x)[(Fx \cdot Gy) \supset Hy]$').
b. The use of **UG** at line 5 is wrong, because the variable ν ('*z*') occurs free in line 2 within whose scope $\Phi\nu$ ('$(\exists x)[(Fx \cdot Gz) \supset Hy]$') lies.
c. The use of **UI** at line 7 is wrong, because ν ('*x*') does not occur free in $\Phi\nu$ ('$(\exists x)[(Fx \cdot Gx) \supset Hx]$') at all places that μ ('*y*') occurs free in $\Phi\mu$ ('$(\exists x)[(Fx \cdot Gy) \supset Hy]$').

9. There are three (or four) mistakes here:
Line 3, '$Fx \cdot Gy$', is claimed at line 5 to be a $\Phi\nu$ corresponding to the $\Phi\mu$ ('$Fx \cdot Gx$') in line 1. This claim is false regardless of whether we take ν to be '*x*' or to be '*y*'. So in either case we note that
a. The use of **EI** at line 5 is wrong, because ν ('*x*' **or** '*y*') does not occur free in $\Phi\nu$ ('$Fx \cdot Gy$') at all places that μ ('*x*') occurs free in $\Phi\mu$ ('$Fx \cdot Gx$').
If we take ν in $\Phi\nu$ ('$Fx \cdot Gy$') to be '*x*', then we note that
b. The use of **EI** at line 5 is wrong, because ν ('*x*') occurs free in *p* ('*Fx*'), the last line lying within the scope of the assumption $\Phi\nu$ ('$Fx \cdot Gy$').
In any case, there are two additional mistakes present:
c. The use of **EI** at line 8 is wrong, because ν ('*x*') occurs free in line 5 which precedes $\Phi\nu$ ('$\sim Fx \cdot Gx$').
d. The use of **EI** at line 8 is wrong, because ν ('*x*') occurs free in *p* ('$\sim Fx$'), the last line lying within the scope of the assumption $\Phi\nu$ ('$\sim Fx \cdot Gx$').

Exercises on pages 103–105:

I. 1.

1.	$(x)(Ax \supset Bx)$	$/\therefore (x)(Bx \supset Cx) \supset (Ak \supset Ck)$
→ 2.	$(x)(Bx \supset Cx)$	
→ 3.	Ak	
4.	$Ak \supset Bk$	1, **UI**
5.	Bk	4, 3, M.P.
6.	$Bk \supset Ck$	2, **UI**
7.	Ck	6, 5, M.P.
8.	$Ak \supset Ck$	3–7, C.P.
9.	$(x)(Bx \supset Cx) \supset (Ak \supset Ck)$	2–8, C.P.

5.

1.	$(\exists x)Lx \supset (y)My$	$/\therefore (x)[Lx \supset (y)My]$
→ 2.	Lx	
3.	$(\exists x)Lx$	2, **EG**
4.	$(y)My$	1, 3, M.P.
5.	$Lx \supset (y)My$	2–4, C.P.
6.	$(x)[Lx \supset (y)My]$	5, **UG**

10.

1.	$(\exists x)Ax \supset (y)(By \supset Cy)$	
2.	$(\exists x)Dx \supset (\exists y)By$	$/\therefore (\exists x)(Ax \cdot Dx) \supset (\exists y)Cy$
→ 3.	$(\exists x)(Ax \cdot Dx)$	
→ 4.	$Ax \cdot Dx$	
5.	Ax	4, Simp.
6.	$(\exists x)Ax$	5, **EG**
7.	$(y)(By \supset Cy)$	1, 6, M.P.
8.	Dx	4, Simp.
9.	$(\exists x)Dx$	8, **EG**
10.	$(\exists y)By$	2, 9, M.P.
→ 11.	By	
12.	$By \supset Cy$	7, **UI**
13.	Cy	12, 11, M.P.
14.	$(\exists y)Cy$	13, **EG**
15.	$(\exists y)Cy$	10, 11–14, **EI**
16.	$(\exists y)Cy$	3, 4–5, **EI**
17.	$(\exists x)(Ax \cdot Dx) \supset (\exists y)Cy$	3–16, C.P.

15.

1.	$(x)\{Ox \supset [(y)(Py \supset Qy) \supset Rx]\}$	
2.	$(x)\{Rx \supset [(y)(Py \supset Sy) \supset Tx]\}$	$/\therefore (y)[Py \supset (Qy \cdot Sy)] \supset (x)(Ox \supset Tx)$
→ 3.	$(y)[Py \supset (Qy \cdot Sy)]$	
→ 4.	Ox	
5.	$Ox \supset [(y)(Py \supset Qy) \supset Rx]$	1, **UI**
6.	$(y)(Py \supset Qy) \supset Rx$	5, 4, M.P.
7.	$Py \supset (Qy \cdot Sy)$	3, **UI**
8.	$(Py \supset Qy) \cdot (Py \supset Sy)$	7, Dist.
9.	$Py \supset Qy$	8, Simp.
10.	$(y)(Py \supset Qy)$	9, **UG**
11.	Rx	6, 10, M.P.
12.	$Rx \supset [(y)(Py \supset Sy) \supset Tx]$	2, **UI**
13.	$(y)(Py \supset Sy) \supset Tx$	12, 11, M.P.
14.	$Py \supset Sy$	8, Simp.
15.	$(y)(Py \supset Sy)$	14, **UG**
16.	Tx	13, 15, M.P.
17.	$Ox \supset Tx$	4–16, C.P.
18.	$(x)(Ox \supset Tx)$	17, **UG**
19.	$(y)[Py \supset (Qy \cdot Sy)] \supset (x)(Ox \supset Tx)$	3–18, C.P.

II. 4.

1.	$(\exists x)Gx \supset (y)[Cy \supset Gy]$	
2.	$(\exists x)(Px \cdot Tx) \supset (y)(Gy \supset Ty)$	$/\therefore (\exists x)[Px \cdot (Tx \cdot Gx)] \supset (y)(Cy \supset Ty)$
→ 3.	$(\exists x)[Px \cdot (Tx \cdot Gx)]$	
→ 4.	$Pu \cdot (Tu \cdot Gu)$	
5.	Gu	4, Simp.
6.	$(\exists x)Gx$	5, **EG**
7.	$Pu \cdot Tu$	4, Simp.
8.	$(\exists x)(Px \cdot Tx)$	7, **EG**
9.	$(y)(Cy \supset Gy)$	1, 6, M.P.
10.	$(y)(Gy \supset Ty)$	2, 8, M.P.
11.	$Cw \supset Gw$	9, **UI**
12.	$Gw \supset Tw$	10, **UI**
13.	$Cw \supset Tw$	11, 12, H.S.
14.	$(y)(Cy \supset Ty)$	13, **UG**
15.	$(y)(Cy \supset Ty)$	3, 4–14, **EI**
16.	$(\exists x)[Px \cdot (Tx \cdot Gx)] \supset (y)(Cy \supset Ty)$	3–15, C.P.

8.

1.	$(x)[Rx \supset (Sx \vee Mx)]$	
2.	$(x)[(Ux \cdot Rx) \supset \sim Sx]$	$/\therefore (x)(Ux \supset Rx) \supset (y)(Uy \supset My)$
→ 3.	$(x)(Ux \supset Rx)$	
→ 4.	Uz	
5.	$Uz \supset Rz$	3, **UI**
6.	Rz	5, 4, M.P.
7.	$Uz \cdot Rz$	4, 6, Conj.
8.	$(Uz \cdot Rz) \supset \sim Sz$	2, **UI**
9.	$\sim Sz$	8, 7, M.P.
10.	$Rz \supset (Sz \vee Mz)$	1, **UI**
11.	$Sz \vee Mz$	10, 6, M.P.
12.	Mz	11, 9, D.S.
13.	$Uz \supset Mz$	4–12, C.P.
14.	$(y)(Uy \supset My)$	13, **UG**
15.	$(x)(Ux \supset Rx) \supset (y)(Uy \supset My)$	3–14, C.P.

12.

1.	$(x)\{Lx \supset [(y)(Py \supset Vy) \supset Mx]\}$	
2.	$(\exists z)(Pz \cdot Vz) \supset (y)(Py \supset Vy)$	$/\therefore (\exists x)Lx \supset [(\exists z)(Pz \cdot Vz) \supset (\exists y)My]$
→ 3.	$(\exists x)Lx$	
→ 4.	$(\exists z)(Pz \cdot Vz)$	
5.	$(y)(Py \supset Vy)$	2, 4, M.P.
→ 6.	Lu	
7.	$Lu \supset [(y)(Py \supset Vy) \supset Mu]$	1, **UI**
8.	$(y)(Py \supset Vy) \supset Mu$	7, 6, M.P.
9.	Mu	8, 5, M.P.
10.	$(\exists y)My$	9, **EG**
11.	$(\exists y)My$	3, 6–10, **EI**
12.	$(\exists z)(Pz \cdot Vz) \supset (\exists y)My$	4–11, C.P.
13.	$(\exists x)Lx \supset [(\exists z)(Pz \cdot Vz) \supset (\exists y)My]$	3–12, C.P.

Exercises on page 108:

1. $(x)(y)[Ax \supset (By \vee Cy)]$

$(z)\{[(y)By \vee (y)Cy] \supset Dz\}$ $\boxed{a, b}$

$\therefore (\exists x)(\exists z)(Ax \supset Dz)$

$\{[Aa \supset (Ba \vee Ca)] \cdot [Aa \supset (Bb \vee Cb)]\} \cdot \{[Ab \supset (Ba \vee Ca)] \cdot [Ab \supset (Bb \vee Cb)]$

$\{[(Ba \cdot Bb) \vee (Ca \cdot Cb)] \supset Da\} \cdot \{[(Ba \cdot Bb) \vee (Ca \cdot Cb)] \supset Db\}$

$\therefore [(Aa \supset Da) \vee (Aa \supset Db)] \vee [(Ab \supset Da) \vee (Ab \supset Db)]$

	Aa	*Ab*	*Ba*	*Bb*	*Ca*	*Cb*	*Da*	*Db*
	T	**T**	**T**	**F**	**F**	**T**	**F**	**F**
or	**T**	**T**	**F**	**T**	**T**	**F**	**F**	**F**

5. $(x)(\exists y)(Fx \equiv Gy)$

$\therefore (\exists y)(x)(Fx \equiv Gy)$ $\boxed{a, b}$

$[(Fa \equiv Ga) \vee (Fa \equiv Gb)] \cdot [(Fb \equiv Ga) \vee (Fb \equiv Gb)]$

$\therefore [(Fa \equiv Ga) \cdot (Fb \equiv Ga)] \vee [(Fa \equiv Gb) \cdot (Fb \equiv Gb)]$

	Fa	Fb	Ga	Gb
	T	**F**	**T**	**F**
or	**T**	**F**	**F**	**T**
or	**F**	**T**	**T**	**F**
or	**F**	**T**	**F**	**T**

Exercises on page 115:

1.

1.	Fy	
2.	$(\exists x)Fx$	1, **EG**
3.	$Fy \supset (\exists x)Fx$	1–2, C.P.
4.	$(y)[Fy \supset (\exists x)Fx]$	3, **UG**

5.

1.	$(\exists x)(Fx \cdot Gx)$	
2.	$Fy \cdot Gy$	
3.	Fy	2, Simp.
4.	$(\exists x)Fx$	3, **EG**
5.	Gy	2, Simp.
6.	$(\exists x)Gx$	5, **EG**
7.	$(\exists x)Fx \cdot (\exists x)Gx$	4, 6, Conj.
8.	$(\exists x)Fx \cdot (\exists x)Gx$	1, 2–7, **EI**
9.	$(\exists x)(Fx \cdot Gx) \supset [(\exists x)Fx \cdot (\exists x)Gx]$	1–8, **C.P.**

10.

1.	$\sim(\exists y)[(\exists x)Fx \supset Fy]$	
2.	$(y)\sim[(\exists x)Fx \supset Fy]$	1, **QN**
3.	$\sim[(\exists x)Fx \supset Fy]$	2, **UI**
4.	$(\exists x)Fx \cdot \sim Fy$	3, Impl.
5.	$\sim Fy$	4, Simp.
6.	$(x)\sim Fx$	5, **UG**
7.	$\sim(\exists x)Fx$	6, **QN**
8.	$(\exists x)Fx$	4, Simp.
9.	$(\exists x)Fx \vee (\exists y)[(\exists x)Fx \supset Fy]$	8, Add.
10.	$(\exists y)[(\exists x)Fx \supset Fy]$	9, 7, D.S.
11.	$\sim(\exists y)[(\exists x)Fx \supset Fy] \supset (\exists y)[(\exists x)Fx \supset Fy]$	1–10, C.P.
12.	$(\exists y)[(\exists x)Fx \supset Fy] \vee (\exists y)[(\exists x)Fx \supset Fy]$	11, Impl., D.N.
13.	$(\exists y)[(\exists x)Fx \supset Fy]$	12, Taut.

15.

1.	$(\exists x)Fx \supset (y)Gy$	
2.	Fx	
3.	$(\exists x)Fx$	2, **EG**
4.	$(y)Gy$	1, 4, M.P.
5.	Gy	4, **UI**
6.	$Fx \supset Gy$	2–5, C.P.
7.	$(y)(Fx \supset Gy)$	6, **UG**
8.	$(x)(y)(Fx \supset Gy)$	7, **UG**
9.	$[(\exists x)Fx \supset (y)Gy] \supset (x)(y)(Fx \supset Gy)$	1–8, C.P.

→ 1.	$(x)(y)(Fx \supset Gy)$	
← 2.	$(\exists x)Fx$	
→ 3.	Fx	
4.	$(y)(Fx \supset Gy)$	1, **UI**
5.	$Fx \supset Gy$	4, **UI**
6.	Gy	5, 3, M.P.
7.	$(y)Gy$	6, **UG**
8.	$(y)Gy$	2, 3–7, **EI**
9.	$(\exists x)Fx \supset (y)Gy$	2–8, C.P.
10.	$(x)(y)(Fx \supset Gy) \supset [(\exists x)Fx \supset (y)Gy]$	1–9, C.P.

Exercises on pages 127–130:

I. **1.** Every dog has his day.
5. All things come to him who waits.
10. A person is judged by the company he keeps.
15. It's an ill wind that blows nobody any good.
20. Nobody borrows everything from everybody.

II. **1.** $(x)[(Dx \cdot Mx) \supset (y)(Ty \supset \sim Txy)]$
5. $(x)\{[Bx \cdot (y)(Txy \supset Ly)] \supset (z)[Txz \supset (u)\sim Buz]\}$
10. $(x)\{Px \supset [(\exists y)Axy \supset (z)(Pz \supset Ezx)]\}$
15. $(x)\{Px \supset \{(\exists y)\{Dy \cdot Hxy \cdot (z)[(Pz \cdot Vzx) \supset Byz]\} \supset (\exists u)(Nu \cdot Mxu)]\}\}$
20. $(\exists x)\{Sx \cdot (y)[Py \supset (\exists z)Byzx]\}$
25. $(x)\{Sx \supset (y)\{Py \supset (\exists z)[(\exists w)Bwzx \cdot \sim Byzx]\}\}$
30. $(x)\{Px \supset (y)[Cy \supset (\exists z)(Bzx \cdot \sim Dxzy)]\}$
35. $(x)\{Cx \supset (y)[(\exists u)(\exists v)Dyuv \supset (\exists z)(\exists w)(Dyzw \cdot \sim Dyzx)]\}$
40. $(\exists x)\{Px \cdot (\exists y)\{Cy \cdot (z)[(\exists u)Dxzu \supset Dxzy]\}\}$

III. **1.** $(x)\{\{Px \cdot (\exists y)[My \cdot (\exists z)(Bz \cdot Bzy \cdot Sxz)]\} \supset$
$(\exists u)[Mu \cdot (\exists v)(Bv \cdot Bvx \cdot Suv)]\}$
5. $(x)\{\{Mx \cdot (y)[(Ry \cdot Byx) \supset Sxy]\} \supset$
$(z)[(Mz \cdot Pxz) \supset Hxz]\}$

Exercises on pages 132–134:

I. 1.

1.	$(\exists x)(y)[(\exists z)Ayz \supset Ayx]$	
2.	$(y)(\exists z)Ayz$	$/\therefore (\exists x)(y)Ayx$
→ 3.	$(y)[(\exists z)Ayz \supset Ayx]$	
4.	$(\exists z)Ayz \supset Ayx$	3, **UI**
5.	$(\exists z)Ayz$	2, **UI**
6.	Ayx	4, 5, M.P.
7.	$(y)Ayx$	6, **UG**
8.	$(\exists x)(y)Ayx$	7, **EG**
9.	$(\exists x)(y)Ayx$	1, 3–8, **EI**

5.

1.	$(\exists x)[Hx \cdot (y)(Iy \supset Jxy)]$	$/\therefore (x)(Hx \supset Ix) \supset (\exists y)(Iy \cdot Jyy)$
2.	$(x)(Hx \supset Ix)$	
3.	$Hx \cdot (y)(Iy \supset Jxy)$	
4.	Hx	3, Simp.
5.	$Hx \supset Ix$	2, **UI**
6.	Ix	5, 4, M.P.
7.	$(y)(Iy \supset Jxy)$	3, Simp.
8.	$Ix \supset Jxx$	7, **UI**
9.	Jxx	8, 6, M.P.
10.	$Ix \cdot Jxx$	6, 9, Conj.
11.	$(\exists y)(Iy \cdot Jyy)$	10, **EG**
12.	$(\exists y)(Iy \cdot Jyy)$	1, 3–11, **EI**
13.	$(x)(Hx \supset Ix) \supset (\exists y)(Iy \cdot Jyy)$	2–12, C.P.

II. 4.

1.	$(x)\{(\exists y)(Byb \cdot Lxyb) \supset Fx\}$	
2.	$(\exists x)(Cxb \cdot Lxab)$	$/\therefore (x)(Cxb \supset \sim Fx) \supset \sim Bab$
3.	$(x)(Cxb \supset \sim Fx)$	
4.	$Cxb \cdot Lxab$	
5.	$Cxb \supset \sim Fx$	3, **UI**
6.	Cxb	4, Simp.
7.	$\sim Fx$	5, 6, M.P.
8.	$(\exists y)(Byb \cdot Lxyb) \supset Fx$	1, **UI**
9.	$\sim(\exists y)(Byb \cdot Lxyb)$	8, 7, M.T.
10.	$(y)(\sim Byb \vee \sim Lxyb)$	9, **QN**, De M.
11.	$\sim Bab \vee \sim Lxab$	10, **UI**
12.	$Lxab$	4, Simp.
13.	$\sim Bab$	11, 12, Com., D.N., D.S.
14.	$\sim Bab$	2, 4–13, **EI**
15.	$3 \supset 14$	3–14, C.P.

6.

1.	$(x)\{(\exists y)[(\exists z)(Gz \cdot \sim Rz \cdot Sxzy)] \supset Cx\}$	
2.	$(z)[(Wz \cdot Orz) \supset (Slzr \vee Smzr)]$	$/\therefore (\exists z)(Wz \cdot Orz \cdot Gz \cdot \sim Rz) \supset [(u)(\sim Smur) \supset Cl]$
3.	$(\exists z)(Wz \cdot Orz \cdot Gz \cdot \sim Rz)$	
4.	$(u)(\sim Smur)$	
5.	$Wz \cdot Orz \cdot Gz \cdot \sim Rz$	
6.	$Wz \cdot Orz$	5, Simp.
7.	$(Wz \cdot Orz) \supset (Slzr \vee Smzr)$	2, **UI**
8.	$Slzr \vee Smzr$	7, 6, M.P.
9.	$\sim Smzr$	4, **UI**
10.	$Slzr$	8, 9, Com., D.S.
11.	$Gz \cdot \sim Rz$	5, Simp.
12.	$Gz \cdot \sim Rz \cdot Slzr$	11, 10, Conj.
13.	$(\exists z)(Gz \cdot \sim Rz \cdot Slzr)$	12, **EG**
14.	$(\exists y)[(\exists z)(Gz \cdot \sim Rz \cdot Slzy)]$	13, **EG**
15.	$(\exists y)[(\exists z)(Gz \cdot \sim Rz \cdot Slzy)] \supset Cl$	1, **UI**
16.	Cl	15, 14, M.P.
17.	Cl	3, 5–16, **EI**
18.	$4 \supset 17$	4–17, C.P.
19.	$3 \supset [18]$	3–18, C.P.

9.

1.	$(x)[(Px \cdot \sim Rxx) \supset (y)(Py \supset \sim Ryx)]$	
2.	$(y)\{Py \supset (x)[(Px \cdot \sim Ryx) \supset \sim Hyx]\}$ $\quad /\therefore (x)\{[Px \cdot (z)(Pz \supset \sim Rxz)] \supset (y)(Py \supset \sim Hyx)\}$	
→ 3.	$Px \cdot (z)(Pz \supset \sim Rxz)$	
→ 4.	Py	
5.	$(z)(Pz \supset \sim Rxz)$	3, Simp.
6.	$Px \supset \sim Rxx$	5, **UI**
7.	Px	3, Simp.
8.	$\sim Rxx$	6, 7, M.P.
9.	$Px \cdot \sim Rxx$	7, 8, Conj.
10.	$(Px \cdot \sim Rxx) \supset (y)(Py \supset \sim Ryx)$	1, **UI**
11.	$(y)(Py \supset \sim Ryx)$	10, 9, M.P.
12.	$Py \supset \sim Ryx$	11, **UI**
13.	$\sim Ryx$	12, 4, M.P.
14.	$Px \cdot \sim Ryx$	7, 13, Conj.
15.	$Py \supset (x)[(Px \cdot \sim Ryx) \supset \sim Hyx]$	2, **UI**
16.	$(x)[(Px \cdot \sim Ryx) \supset \sim Hyx]$	15, 4, M.P.
17.	$(Px \cdot \sim Ryx) \supset \sim Hyx$	16, **UI**
18.	$\sim Hyx$	17, 14, M.P.
19.	$4 \supset 18$	4–18, C.P.
20.	$(y)(19)$	19, **UG**
21.	$3 \supset 20$	3–20, C.P.
22.	$(x)\{21\}$	21, **UG**

Exercises on pages 139-140:

3.

1.	$(x)[Fx \supset (y)(Sy \supset Oxy)]$ $\quad /\therefore (y)[Sy \supset (x)(Fx \supset \sim Oyx)]$	
2.	$(x)(y)(Oxy \supset \sim Oyx)$ (auxiliary premiss)	
→ 3.	Sy	
→ 4.	Fx	
5.	$Fx \supset (y)(Sy \supset Oxy)$	1, **UI**
6.	$(y)(Sy \supset Oxy)$	5, 4, M.P.
7.	$Sy \supset Oxy$	6, **UI**
8.	Oxy	7, 3, M.P.
9.	$(y)(Oxy \supset \sim Oyx)$	2, **UI**
10.	$Oxy \supset \sim Oyx$	9, **UI**
11.	$\sim Oyx$	10, 8, M.P.
12.	$Fx \supset \sim Oyx$	4–11, C.P.
13.	$(x)(Fx \supset \sim Oyx)$	12, **UG**
14.	$Sy \supset (x)(Fx \supset \sim Oyx)$	3–13, C.P.
15.	$(y)[Sy \supset (x)(Fx \supset \sim Oyx)]$	14, **UG**

5.

1.	$(x)\{\{Px \cdot (\exists y)[Py \cdot (\exists z)(Cz \cdot Cyz) \cdot Nxy]\} \supset Ux\}$	$/\therefore (x)\{[Px \cdot (\exists z)(Bz \cdot Cxz)] \supset Ux\}$
2.	$(z)(Bz \supset Cz)$	(auxiliary premisses)
3.	$(x)(Nxx)$	(auxiliary premisses)
→ 4.	$Px \cdot (\exists z)(Bz \cdot Cxz)$	
5.	$(\exists z)(Bz \cdot Cxz)$	4, Simp.
→ 6.	$Bz \cdot Cxz$	
7.	Bz	6, Simp.
8.	$Bz \supset Cz$	2, **UI**
9.	Cz	8, 7, M.P.
10.	Cxz	6, Simp.
11.	$Cz \cdot Cxz$	9, 10, Conj.
12.	$(\exists z)(Cz \cdot Cxz)$	11, **EG**
13.	Px	4, Simp.
14.	Nxx	3, **UI**
15.	$Px \cdot (\exists z)(Cz \cdot Cxz) \cdot Nxx$	13, 12, 14, Conj.
16.	$(\exists y)[Py \cdot (\exists z)(Cz \cdot Cyz) \cdot Nxy]$	15, **EG**
17.	$Px \cdot (\exists y)[Py \cdot (\exists z)(Cz \cdot Cyz) \cdot Nxy]$	13, 16, Conj.
18.	$\{Px \cdot (\exists y)[Py \cdot (\exists z)(Cz \cdot Cyz) \cdot Nxy]\} \supset Ux$	1, **UI**
19.	Ux	18, 17, M.P.
20.	Ux	5, 6–19, **EI**
21.	$[Px \cdot (\exists z)(Bz \cdot Cxz)] \supset Ux$	4–20, C.P.
22.	$(x)\{[Px \cdot (\exists z)(Bz \cdot Cxz)] \supset Ux\}$	21, **UG**

7.

1.	$(x)[Vx \supset (y)(Oyx \supset \sim Ixy)]$	
2.	$(y)(x)[(Rx \cdot Oyx) \supset Tyx]$	$/\therefore (y)[(\exists x)(Vx \cdot Oyx) \supset (\exists x)(Tyx \cdot \sim Ixy)]$
3.	$(x)(Vx \supset Rx)$ (auxiliary premiss)	
→ 4.	$(\exists x)(Vx \cdot Oyx)$	
→ 5.	$Vx \cdot Oyx$	
6.	Vx	5, Simp.
7.	$Vx \supset Rx$	3, **UI**
8.	Rx	7, 6, M.P.
9.	Oyx	5, Simp.
10.	$Rx \cdot Oyx$	8, 9, Conj.
11.	$(x)[(Rx \cdot Oyx) \supset Tyx]$	2, **UI**
12.	$(Rx \cdot Oyx) \supset Tyx$	11, **UI**
13.	Tyx	12, 10, M.P.
14.	$Vx \supset (y)(Oyx \supset \sim Ixy)$	1, **UI**
15.	$(y)(Oyx \supset \sim Ixy)$	14, 6, M.P.
16.	$Oyx \supset \sim Ixy$	15, **UI**
17.	$\sim Ixy$	16, 9, M.P.
18.	$Tyx \cdot \sim Ixy$	17, 13, Conj.
19.	$(\exists x)(Tyx \cdot \sim Ixy)$	18, **EG**
20.	$(\exists x)(Tyx \cdot \sim Ixy)$	4, 5–19, **EI**
21.	$(\exists x)(Vx \cdot Oyx) \supset (\exists x)(Tyx \cdot \sim Ixy)$	4–20, C.P.
22.	$(y)[(\exists x)(Vx \cdot Oyx) \supset (\exists x)(Tyx \cdot \sim Ixy)]$	21, **UG**

Exercises on pages 149–150:

2.

1.	$(\exists x)\{Px \cdot Sx \cdot (y)[(Py \cdot Sy) \supset x = y] \cdot Lx\}$	$/\therefore (x)[(Px \cdot Sx) \supset Lx]$
→ 2.	$Pz \cdot Sz$	
→ 3.	$Px \cdot Sx \cdot (y)[(Py \cdot Sy) \supset x = y] \cdot Lx$	
4.	$(y)[(Py \cdot Sy) \supset x = y]$	3, Simp.
5.	$(Pz \cdot Sz) \supset x = z$	4, **UI**
6.	$x = z$	5, 2, M.P.
7.	Lx	3, Simp.
8.	Lz	6, 7, Id.
9.	Lz	1, 3–8, **EI**
10.	$(Pz \cdot Sz) \supset Lz$	2–9, C.P.
11.	$(x)[(Px \cdot Sx) \supset Lx]$	10, **UG**

4.

1.	$(\exists x)\{Px \cdot (y)[(Py \cdot x \neq y) \supset Fxy] \cdot Sx\}$	$/\therefore (y)[(Py \cdot \sim Sy) \supset (\exists x)(Px \cdot Fxy)]$
→ 2.	$Py \cdot \sim Sy$	
→ 3.	$Px \cdot (y)[(Py \cdot x \neq y) \supset Fxy] \cdot Sx$	
4.	Sx	3, Simp.
5.	$\sim Sy$	2, Simp.
6.	$x \neq y$	4, 5, Id.
7.	Py	2, Simp.
8.	$Py \cdot x \neq y$	7, 6, Conj.
9.	$(y)[(Py \cdot x \neq y) \supset Fxy]$	3, Simp.
10.	$(Py \cdot x \neq y) \supset Fxy$	9, **UI**
11.	Fxy	10, 8, M.P.
12.	Px	3, Simp.
13.	$Px \cdot Fxy$	12, 11, Conj.
14.	$(\exists x)(Px \cdot Fxy)$	13, **EG**
15.	$(\exists x)(Px \cdot Fxy)$	1, 3–14, **EI**
16.	$(Py \cdot \sim Sy) \supset (\exists x)(Px \cdot Fxy)$	2–15, C.P.
17.	$(y)[(Py \cdot \sim Sy) \supset (\exists x)(Px \cdot Fxy)]$	16, **UG**

6.

1.	$(x)\{Fx \supset (y)[(Fy \cdot Lxy) \supset Sxy]\}$ $/\therefore (\exists x)\{Fx \cdot (y)[(Fy \cdot x \neq y) \supset Lxy]\} \supset (\exists x)\{Fx \cdot (y)[(Fy \cdot x \neq y) \supset Sxy]\}$	
→ 2.	$(\exists x)\{Fx \cdot (y)[(Fy \cdot x \neq y) \supset Lxy]\}$	
→ 3.	$Fx \cdot (y)[(Fy \cdot x \neq y) \supset Lxy]$	
4.	Fx	3, Simp.
5.	$(y)[(Fy \cdot x \neq y) \supset Lxy]$	3, Simp.
6.	$Fx \supset (y)[(Fy \cdot Lxy) \supset Sxy]$	1, **UI**
7.	$(y)[(Fy \cdot Lxy) \supset Sxy]$	6, 4, M.P.
8.	$(Fy \cdot Lxy) \supset Sxy$	7, **UI**
9.	$(Fy \cdot x \neq y) \supset Lxy$	5, **UI**
10.	$Lxy \supset (Fy \supset Sxy)$	8, Com., Exp.
11.	$(Fy \cdot x \neq y) \supset (Fy \supset Sxy)$	9, 10, H.S.
12.	$(Fy \cdot x \neq y \cdot Fy) \supset Sxy$	11, Exp.
13.	$(Fy \cdot x \neq y) \supset Sxy$	12, Com., Taut.
14.	$(y)[(Fy \cdot x \neq y) \supset Sxy]$	13, **UG**
15.	$Fx \cdot (y)[(Fy \cdot x \neq y) \supset Sxy]$	4, 14, Conj.
16.	$(\exists x)\{Fx \cdot (y)[(Fy \cdot x \neq y) \supset Sxy]\}$	15, **EG**
17.	$(\exists x)\{Fx \cdot (y)[(Fy \cdot x \neq y) \supset Sxy]\}$	2, 3–16, **EI**
18.	$2 \supset 17$	2–17, C.P.

Exercises on pages 155–156:

I. 3. $(x)(y)[x \neq y \supset (\exists F)(Fx \cdot \sim Fy)]$

6. $(\exists x)\{Fxd \cdot (y)(Fyd \supset x = y) \cdot (G)[(Gx \cdot VG) \supset Gd] \cdot (H)[(Hx \cdot FH) \supset \sim Hd]\}$

9. $(x)\{[Mx \cdot (F)(VF \supset Fx)] \supset Vx\} \cdot (\exists x)[Mx \cdot Vx \cdot (\exists F)(VF \cdot \sim Fx)]$

II. 2.

→ 1. $(\exists x)(\exists F)Fx$	
→ 2. $(\exists F)Fx$	
→ 3. Fx	
4. $(\exists x)Fx$	3, **EG**
5. $(\exists F)(\exists x)Fx$	4, **EG**
6. $(\exists F)(\exists x)Fx$	2, 3–5, **EI**
7. $(\exists F)(\exists x)Fx$	1, 2–6, **EI**
8. $1 \supset 7$	1–7, C.P.

→ 1. $(\exists F)(\exists x)Fx$	
→ 2. $(\exists x)Fx$	
→ 3. Fx	
4. $(\exists F)Fx$	3, **EG**
5. $(\exists x)(\exists F)Fx$	4, **EG**
6. $(\exists x)(\exists F)Fx$	2, 3–5, **EI**
7. $(\exists x)(\exists F)Fx$	1, 2–6, **EI**
8. $1 \supset 7$	1–7, C.P.

6.

→ 1. $(x)(y)(z)[(Rxy \cdot Ryz) \supset Rxz] \cdot (x) \sim Rxx$	
2. $(x)(y)(z)[(Rxy \cdot Ryz) \supset Rxz]$	1, Simp.
3. $(y)(z)[(Rxy \cdot Ryz) \supset Rxz]$	2, **UI**
4. $(z)[(Rxy \cdot Ryz) \supset Rxz]$	3, **UI**
5. $(Rxy \cdot Ryx) \supset Rxx$	4, **UI**
6. $(x) \sim Rxx$	1, Simp.
7. $\sim Rxx$	6, **UI**
8. $\sim (Rxy \cdot Ryx)$	5, 7, M.T.
9. $\sim Rxy \vee \sim Ryx$	8, De M.
10. $Rxy \supset \sim Ryx$	9, Impl.
11. $(y)(Rxy \supset \sim Ryx)$	10, **UG**
12. $(x)(y)(Rxy \supset \sim Ryx)$	11, **UG**
13. $\{1\} \supset 12$	1–12, C.P.
14. $(R)\{13\}$	13, **UG**

10.

1. $(x)(y)[(x = y) \equiv (F)(Fx \equiv Fy)]$	$/ \therefore (x)(x = x)$
→ 2. Fx	
3. $\sim\sim Fx$	2, D.N.
4. Fx	3, D.N.
5. $Fx \supset Fx$	2–4, C.P.
6. $(Fx \supset Fx) \cdot (Fx \supset Fx)$	5, Taut.
7. $Fx \equiv Fx$	6, Equiv.
8. $(F)(Fx \equiv Fx)$	7, **UG**
9. $(y)[(x = y) \equiv (F)(Fx \equiv Fy)]$	1, **UI**
10. $(x = x) \equiv (F)(Fx \equiv Fx)$	9, **UI**
11. $[(x = x) \supset (F)(Fx \equiv Fx)] \cdot [(F)(Fx \equiv Fx) \supset (x = x)]$	10, Equiv.
12. $(F)(Fx \equiv Fx) \supset (x = x)$	11, Simp.
13. $x = x$	12, 8, M.P.
14. $(x)(x = x)$	13, **UG**

Exercises on pages 175–176:

TH. 2. There is at most one entity 1 in **C** such that $\alpha \cap 1 = \alpha$

Proof: Suppose there are entities 1_1 and 1_2 such that $\alpha \cap 1_1 = \alpha$, $\alpha \cap 1_2 = \alpha$. Then $1_1 \cap 1_2 = 1_1$ and $1_2 \cap 1_1 = 1_2$. $1_1 \cap 1_2 = 1_2 \cap 1_1$ by Ax. 6. Hence $1_1 = 1_2$ by two substitutions of equals for equals.

TH. 7. $0 \neq 1$

Proof: Suppose $0 = 1$. Then by Ax. 10 there is an α in **C** such that $\alpha \neq 0$ and $\alpha \neq 1$. However,

$\alpha \cap 1 = \alpha \cap 0$	by assumption that $0 = 1$.
$\alpha = \alpha \cap 0$	Ax. 4
$\alpha = 0$	Th. 6

contrary to $\alpha \neq 0$. Therefore, the supposition that $0 = 1$ must be false, so $0 \neq 1$.

TH. 12. $\alpha \cup (\beta \cup \gamma) = (\alpha \cup \beta) \cup \gamma$

Proof: LEMMA 1. $\alpha \cup (\alpha \cap \beta) = \alpha$

Proof:

$\alpha \cup (\alpha \cap \beta) = (\alpha \cap 1) \cup (\alpha \cap \beta)$	Ax. 4
$= \alpha \cap (1 \cup \beta)$	Ax. 8
$= \alpha \cap (\beta \cup 1)$	Ax. 5
$= \alpha \cap 1$	Th. 5
$= \alpha$	Ax. 4

LEMMA 2. $\alpha \cap (\alpha \cup \beta) = \alpha$

Proof:

$\alpha \cap (\alpha \cup \beta) = (\alpha \cup 0) \cap (\alpha \cup \beta)$	Ax. 3
$= \alpha \cup (0 \cap \beta)$	Ax. 7
$= \alpha \cup (\beta \cap 0)$	Ax. 6
$= \alpha \cup 0$	Th. 6
$= \alpha$	Ax. 3

LEMMA 3. If $\alpha \cap \gamma = \alpha$ and $\beta \cap \gamma = \beta$ then $(\alpha \cup \beta) \cap \gamma = \alpha \cup \beta$

Proof:

$(\alpha \cup \beta) \cap \gamma = \gamma \cap (\alpha \cup \beta)$	Ax. 6
$= (\gamma \cap \alpha) \cup (\gamma \cap \beta)$	Ax. 8
$= (\alpha \cap \gamma) \cup (\beta \cap \gamma)$	Ax. 6
$= \alpha \cup \beta$	hypothesis

A. $\alpha \cap [\alpha \cup (\beta \cup \gamma)] = \alpha$	L. 2
$\beta \cap [\alpha \cup (\beta \cup \gamma)] = (\beta \cap \alpha) \cup [\beta \cap (\beta \cup \gamma)]$	Ax. 8
$= (\beta \cap \alpha) \cup \beta$	L. 2
$= \beta \cup (\beta \cap \alpha)$	Ax. 5
B. $\beta \cap [\alpha \cup (\beta \cup \gamma)] = \beta$	L. 1
C. $(\alpha \cup \beta) \cap [\alpha \cup (\beta \cup \gamma)] = \alpha \cup \beta$	by L. 3 from *A*, *B*.
$\gamma \cap [\alpha \cup (\beta \cup \gamma)] = \gamma \cap [(\beta \cup \gamma) \cup \alpha]$	Ax. 5
$= [\gamma \cap (\beta \cup \gamma)] \cup (\gamma \cap \alpha)$	Ax. 8
$= [\gamma \cap (\gamma \cup \beta)] \cup (\gamma \cap \alpha)$	Ax. 5
$= \gamma \cup (\gamma \cap \alpha)$	L. 2
D. $\gamma \cap [\alpha \cup (\beta \cup \gamma)] = \gamma$	L. 1
E. $[(\alpha \cup \beta) \cup \gamma] \cap [\alpha \cup (\beta \cup \gamma)] = (\alpha \cup \beta) \cup \gamma$	by L. 3 from *C*, *D*.
$\alpha \cap [(\alpha \cup \beta) \cup \gamma] = [\alpha \cap (\alpha \cup \beta)] \cup (\alpha \cap \gamma)$	Ax. 8
$= \alpha \cup (\alpha \cap \gamma)$	L. 2

F. $\alpha \cap [(\alpha \cup \beta) \cup \gamma] = \alpha$	L. 1
$\beta \cap [(\alpha \cup \beta) \cup \gamma] = [\beta \cap (\alpha \cup \beta)] \cup (\beta \cap \gamma)$	Ax. 8
$= [\beta \cap (\beta \cup \alpha)] \cup (\beta \cap \gamma)$	Ax. 5
$= \beta \cup (\beta \cap \gamma)$	L. 2
G. $\beta \cap [(\alpha \cup \beta) \cup \gamma] = \beta$	L. 1
$\gamma \cap [(\alpha \cup \beta) \cup \gamma] = \gamma \cap [\gamma \cup (\alpha \cup \beta)]$	Ax. 5
H. $\gamma \cap [(\alpha \cup \beta) \cup \gamma] = \gamma$	L. 2
I. $(\beta \cup \gamma) \cap [(\alpha \cup \beta) \cup \gamma] = \beta \cup \gamma$	by L. 3 from *G, H.*
J. $[\alpha \cup (\beta \cup \gamma)] \cap [(\alpha \cup \beta) \cup \gamma] = \alpha \cup (\beta \cup \gamma)$	by L. 3 from *F, I.*
$\alpha \cup (\beta \cup \gamma) = [\alpha \cup (\beta \cup \gamma)] \cap [(\alpha \cup \beta) \cup \gamma]$	*J*
$= [(\alpha \cup \beta) \cup \gamma] \cap [\alpha \cup (\beta \cup \gamma)]$	Ax. 6
$\alpha \cup (\beta \cup \gamma) = (\alpha \cup \beta) \cup \gamma$	*E*

TH. 20. If $\alpha \cap \beta \neq 0$ and $\beta \cap -\gamma = 0$, then $\alpha \cap \gamma \neq 0$

Proof: $(\alpha \cap \gamma) \cap \beta = \alpha \cap (\gamma \cap \beta)$	Th. 13
$= \alpha \cap (\beta \cap \gamma)$	Ax. 6
$= (\alpha \cap \beta) \cap \gamma$	Th. 13
$= [(\alpha \cap \beta) \cap \gamma] \cup 0$	Ax. 3
$= [(\alpha \cap \beta) \cap \gamma] \cup (\alpha \cap 0)$	Th. 6
$= [(\alpha \cap \beta) \cap \gamma] \cup [\alpha \cap (\beta \cap -\gamma)]$	hypothesis
$= [(\alpha \cap \beta) \cap \gamma] \cup [(\alpha \cap \beta) \cap -\gamma]$	Th. 13
$= \alpha \cap \beta$	Th. 11
$(\alpha \cap \gamma) \cap \beta \neq 0$	hypothesis
$\alpha \cap \gamma \neq 0$	Th. 10

Exercises on pages 217–218:

5. Not a *wff*. **10.** *wff*.

Exercises on page 220:

5. $\sim((P)\cdot(\sim(\sim((Q)\cdot(\sim(P))))))$

10. $(\sim((\sim((P)\cdot(\sim(\sim(P)))))\cdot(\sim(\sim(P)))))\cdot(\sim((\sim(P))\cdot(\sim(\sim((P)\cdot(\sim(\sim(P))))))))$

Exercises on page 222:

1. $f_2(P, Q)$ is expressed as $\sim(P\cdot\sim Q)$
$f_5(P,Q)$ is expressed as $\sim(P\cdot Q)\cdot\sim(P\cdot\sim Q)$
$f_{11}(P,Q)$ is expressed as $\sim(P\cdot Q)\cdot\sim(P\cdot\sim Q)\cdot\sim(\sim P\cdot Q)$

Exercises on pages 226–227:

1. S(v,∼) is functionally complete, for it contains the same '∼' as does R.S., and the '·' of R.S. is definable in S(v,∼) as $P\cdot Q = \text{df} \sim(\sim P \vee \sim Q)$. Hence any truth function expressible in R.S. is expressible in S(v,∼), whose functional completeness now follows from that of R.S.

2. S(⊃,·) is functionally incomplete, for it contains no *wff* that can express a truth function whose value is *false* when all its arguments have the value *true.* Proof by strong induction on the number of symbols in the *wff* g(*P, Q, R,* ...) of S(⊃,·).
α) If g(*P, Q, R,* ...) contains just one symbol, it is either *P* or *Q* or *R* or .... If these are all *true,* then g(*P, Q, R,* ...) cannot have the value *false.*
β) Assume that any *wff* containing $<m$ symbols cannot be *false* when all its

arguments are *true*. Now consider any *wff* $g(P, Q, R, \ldots)$ containing exactly m symbols ($m > 1$). It must be either $g_1(P, Q, R, \ldots) \supset g_2(P, Q, R, \ldots)$ or $g_1(P, Q, R, \ldots) \cdot g_2(P, Q, R, \ldots)$. But $g_1(P, Q, R, \ldots)$ and $g_2(P, Q, R, \ldots)$ each contains $< m$ symbols, hence they must have the value *true* if all their arguments are *true*. And because $\mathbf{T} \supset \mathbf{T}$ and $\mathbf{T} \cdot \mathbf{T}$ are both $\mathbf{T}$, $g(P, Q, R, \ldots)$ cannot be *false* when all of its arguments have the value *true*. Hence no *wff* of $S(\supset, \cdot)$ can be *false* when its arguments are all *true*, so $S(\supset, \cdot)$ is functionally incomplete.

6. $S(\supset, +)$ is functionally complete, for the '$\sim$' of R.S. is definable in $S(\supset, +)$ as $\sim P = \text{df } P + (P \supset P)$, and the '$\cdot$' of R.S. is then definable in $S(\supset, +)$ as $P \cdot Q = \text{df } \sim[P \supset (P + Q)]$. Hence any truth function expressible in R.S. is expressible in $S(\supset, +)$, whose functional completeness now follows from that of R.S.

Exercises on page 237:

1. The same models that establish the independence of the H.A. postulates on pages 263–265 serve to establish the independence of the P_N axioms.

Exercises on pages 243–244:

TH. 6. $\vdash (R \sim\sim P) \supset (PR)$

Proof:	$\vdash \sim\sim P \supset P$	Th. 2
	$\vdash (R \sim\sim P) \supset (PR)$	DR 3

DR 6. $P \supset Q, Q \supset R \vdash P \supset R$

Proof:	$P \supset P$	Th. 7
	$P \supset Q$	pr.
	$Q \supset R$	pr.
	$P \supset R$	DR 5

TH. 10. $\vdash (PQ)R \supset P(QR)$

Proof:	$\vdash (PQ)R \supset PQ$	Ax. 2
	$\vdash PQ \supset P$	Ax. 2
	$\vdash (PQ)R \supset P$	DR 6
	$\vdash PQ \supset QP$	Th. 8
	$\vdash QP \supset Q$	Ax. 2
	$\vdash PQ \supset Q$	DR 6
	$\vdash (PQ)R \supset QR$	DR 7, Cor. 1
	$\vdash (PQ)R \supset P(QR)$	DR 8

DR 10. $P \supset R, Q \supset R \vdash (P \vee Q) \supset R$

Proof:	$P \supset R$	pr.
	$Q \supset R$	pr.
	$(P \vee Q) \supset (R \vee R)$	DR 9
	$\sim R \supset \sim R \sim R$	Ax. 1
	$[\sim R \supset \sim R \sim R] \supset [\sim(\sim R \sim R) \supset \sim\sim R]$	Th. 5
	$\sim(\sim R \sim R) \supset \sim\sim R$	R 1
	$(R \vee R) \supset \sim\sim R$	df.
	$\sim\sim R \supset R$	Th. 2
	$(P \vee Q) \supset R$	DR 5

TH. 14. $\vdash [PQ \supset R] \supset [P \supset (Q \supset R)]$

Proof:	$\vdash \sim\sim(Q{\sim}R) \supset (Q{\sim}R)$	Th. 2
	$\vdash P{\sim}{\sim}(Q{\sim}R) \supset P(Q{\sim}R)$	DR 7, Cor. 2
	$\vdash P(Q{\sim}R) \supset (PQ){\sim}R$	Th. 10, Cor.
	$\vdash P{\sim}{\sim}(Q{\sim}R) \supset (PQ){\sim}R$	DR 6
	$\vdash \{P{\sim}{\sim}(Q{\sim}R) \supset (PQ){\sim}R\} \supset \{\sim[(PQ){\sim}R] \supset \sim[P{\sim}{\sim}(Q{\sim}R)]\}$	Th. 5
	$\vdash \sim[(PQ){\sim}R] \supset \sim[P{\sim}{\sim}(Q{\sim}R)]$	R 1
	$\vdash [PQ \supset R] \supset [P \supset (Q \supset R)]$	df.

Exercises on page 247:

TH. 17, COR. $\vdash P \supset (P \vee Q)$

Proof:	$\vdash P \supset (Q \vee P)$	Th. 17
	$\vdash (Q \vee P) \supset (P \vee Q)$	Th. 11
	$\vdash P \supset (P \vee Q)$	DR 6

DR 14. $P, Q \vdash PQ$

Proof:	$P \supset (Q \supset PQ)$	Th. 15
	P	pr.
	$Q \supset (PQ)$	R 1
	Q	pr.
	PQ	R 1

Exercises on pages 249-250:

DR 20. $(P \supset Q)\cdot(R \supset S), P \vee R \vdash Q \vee S$

Proof:	$(P \supset Q)\cdot(R \supset S)$	pr.
	$P \supset Q$	DR 19
	$R \supset S$	DR 19, Cor.
	$(P \vee R) \supset (Q \vee S)$	DR 9
	$P \vee R$	pr.
	$Q \vee S$	R 1

TH. 30. $\vdash P(Q \vee R) \equiv PQ \vee PR$

Proof:	$\vdash PQ \supset P$	Ax. 2
	$\vdash PR \supset P$	Ax. 2
	$\vdash (PQ \vee PR) \supset P$	DR 10
	$\vdash PQ \supset QP$	Th. 8
	$\vdash QP \supset Q$	Ax. 2
	$\vdash PQ \supset Q$	DR 6
	$\vdash PR \supset RP$	Th. 8
	$\vdash RP \supset R$	Ax. 2
	$\vdash PR \supset R$	DR 6
	$\vdash (PQ \vee PR) \supset (Q \vee R)$	DR 9
	$\vdash (PQ \vee PR) \supset P(Q \vee R)$	DR 8
	$\vdash P(Q \vee R) \supset (Q \vee R)P$	Th. 8

$\vdash (Q \vee R)P \supset QP \vee RP$	Th. 18
$\vdash P(Q \vee R) \supset QP \vee RP$	DR 6
$\vdash QP \equiv PQ$	Th. 21
$\vdash P(Q \vee R) \supset PQ \vee RP$	MT IV, Cor.
$\vdash RP \equiv PR$	Th. 21
$\vdash P(Q \vee R) \supset PQ \vee PR$	MT IV, Cor.
$\vdash P(Q \vee R) \equiv PQ \vee PR$	DR 14, df.

Exercises on page 276:

The H.A. system has already been proved to be a model system of logic. Because P_N has the same primitive symbols as H.A., it, too, is functionally complete. The analyticity of P_N is shown by using truth tables to show that all four of its axioms are tautologies, and then showing that any *wff* that follows from tautologies by repeated applications of the P_N Rule must be a tautology also. Because its Axioms 1, 2, and 4 are the same as the Postulates 1, 2, and 4 of H.A., and because its Rule is the same as the rule R′ 1 of H.A., and because the primitives of the two systems are the same, we can prove the deductive completeness of P_N by deriving P 3 of H.A. from the four Axioms of P_N. First, we establish a derived rule for P_N:

DR 1. $P \supset Q, Q \supset R \vdash_{P_N} P \supset R$

Demonstration:		
	1. $(Q \supset R) \supset [(\sim P \vee Q) \supset (\sim P \vee R)]$	Ax. 4
	2. $Q \supset R$	premiss
	3. $(\sim P \vee Q) \supset (\sim P \vee R)$	Rule
	4. $(P \supset Q) \supset (P \supset R)$	df.
	5. $P \supset Q$	premiss
	6. $P \supset R$	Rule

TH. 1. $\vdash_{P_N} (P \vee Q) \supset (Q \vee P)$

Proof:		
	1. $[Q \supset (Q \vee P)] \supset \{(P \vee Q) \supset [P \vee (Q \vee P)]\}$	Ax. 4
	2. $Q \supset (Q \vee P)$	Ax. 2
	3. $(P \vee Q) \supset [P \vee (Q \vee P)]$	Rule
	4. $[P \vee (Q \vee P)] \supset [Q \vee (P \vee P)]$	Ax. 3
	5. $(P \vee Q) \supset [Q \vee (P \vee P)]$	DR 1
	6. $[(P \vee P) \supset P] \supset \{[Q \vee (P \vee P)] \supset (Q \vee P)\}$	Ax. 4
	7. $(P \vee P) \supset P$	Ax. 1
	8. $[Q \vee (P \vee P)] \supset (Q \vee P)$	Rule
	9. $(P \vee Q) \supset (Q \vee P)$	DR 1

Exercises on page 280:

1. 1. *CAppp*
2. *CpApq*
3. *CApqAqp*
4. *CCpqCArpArq*

Exercises on page 282:

2. $P{\downarrow}P{:}{\downarrow}{:}P{\downarrow}P.{\downarrow}.P{\downarrow}P{:}.{\downarrow}{:}.P{\downarrow}P{:}{\downarrow}{:}P{\downarrow}P.{\downarrow}.P{\downarrow}P$
6. $P|Q.|.P|Q{:}|{:}P|P$
8. $P{\downarrow}P{:}{\downarrow}{:}P{\downarrow}P.{\downarrow}.P{\downarrow}P{:}.{\downarrow}{:}.P{\downarrow}P{:}{\downarrow}{:}P{\downarrow}P.{\downarrow}.P{\downarrow}P$

Exercises on page 300:

TH. 4. $\vdash (x)(P \supset Q) \supset [(\exists x)P \supset (\exists x)Q]$

Proof:	$(x)(P \supset Q) \vdash (x)(P \supset Q) \supset (P \supset Q)$	P 5
	$(x)(P \supset Q) \vdash (x)(P \supset Q)$	premiss
	$(x)(P \supset Q) \vdash P \supset Q$	R 1
	$(x)(P \supset Q) \vdash \sim Q \supset \sim P$	Ⓟ
	$(x)(P \supset Q) \vdash (x)(\sim Q \supset \sim P)$	R 2
	$(x)(P \supset Q) \vdash (x)\sim Q \supset (x)\sim P$	DR 5
	$(x)(P \supset Q) \vdash \sim(x)\sim P \supset \sim(x)\sim Q$	Ⓟ
	$(x)(P \supset Q) \vdash (\exists x)P \supset (\exists x)Q$	df.
	$\vdash (x)(P \supset Q) \supset [(\exists x)P \supset (\exists x)Q]$	D.T.

Exercises on pages 305–306:

TH. 13. $\vdash (\exists x)(P \cdot Q) \supset (\exists x)P \cdot (\exists x)Q$

Proof:	$\vdash [(x)\sim P \vee (x)\sim Q] \supset (x)(\sim P \vee \sim Q)$	Th. 10
	$\vdash \sim(x)(\sim P \vee \sim Q) \supset \sim[(x)\sim P \vee (x)\sim Q]$	Ⓟ
	$\vdash \sim(x)\sim(P \cdot Q) \supset [\sim(x)\sim P \cdot \sim(x)\sim Q]$	Duality Theorem, Ⓟ, and R.R.
	$\vdash (\exists x)(P \cdot Q) \supset (\exists x)P \cdot (\exists x)Q$	df.

TH. 21. $\vdash (x)(P \supset Q) \equiv (\exists x)P \supset Q$

Proof:	$\vdash (x)(\sim Q \supset \sim P) \equiv \sim Q \supset (x)\sim P$	Th. 20
	$\vdash (x)(P \supset Q) \equiv \sim(x)\sim P \supset \sim\sim Q$	Ⓟ and R.R.
	$\vdash (x)(P \supset Q) \equiv \sim(x)\sim P \supset Q$	Ⓟ and R.R.
	$\vdash (x)(P \supset Q) \equiv (\exists x)P \supset Q$	df.

DR 6. $P \supset Q \vdash (\exists x)P \supset Q$

Proof:	$P \supset Q$	premiss
	$(x)(P \supset Q)$	R 2
	$(\exists x)P \supset Q$	Th. 21 and R.R.

Exercises on page 313:

2. $(\exists x)(w)(z)(\exists u)(\exists v)(t)\{[F(u, v) \supset [G(t) \vee H(y)]] \cdot [[G(x) \vee H(y)] \supset F(w, z)]\}$
4. $(\exists y)(x)(\exists z)(w)[F(x, y) \supset F(z, w)]$

Exercises on page 315:

1. $(\exists t)(x)(y)(z)\{\{[H(x) \vee H(y)] \vee H(z)\} \cdot \sim[D(t) \cdot \sim D(t)]\}$
3. $(\exists t)(y)(z)(\exists x)\{[G(x, y) \vee F(z)] \cdot \sim[D(t) \cdot \sim D(t)]\}$

Exercises on page 318:

1. $(\exists x)(\exists z)(y)[F(x, y) \vee G(z)]$
3. $(\exists w)(\exists x)(\exists y)(t)\{\{[F(x, y) \cdot \sim[D(w) \cdot \sim D(w)]] \supset H(w, x)\} \supset H(w, t)\}$

Exercises on page 343:

1. $p \equiv \overline{p}$
   $(p \cdot \overline{p}) \vee (\overline{p} \cdot \overline{\overline{p}})$
   $p \cdot \overline{p}$

5. $p \equiv (q \cdot \bar{p})$
$(p \cdot q \cdot \bar{p}) \vee \bar{p} \cdot (\overline{q \cdot \bar{p}})$
$(p \cdot q \cdot \bar{p}) \vee [\bar{p} \cdot (\bar{q} \vee \bar{\bar{p}})]$
$(p \cdot q \cdot \bar{p}) \vee (\bar{p} \cdot \bar{q}) \vee (\bar{p} \cdot \bar{\bar{p}})$
$(p \cdot q \cdot \bar{p}) \vee (\bar{p} \cdot \bar{q})$
$(q \cdot p \cdot \bar{p}) \vee (p \cdot \bar{q})$
$(q \cdot p \cdot \bar{p}) \vee (q \cdot \bar{q}) \vee (\bar{p} \cdot \bar{q})$
$q \cdot [(p \cdot \bar{p}) \vee \bar{q}] \vee (\bar{p} \cdot \bar{q})$
$q \cdot [\bar{q} \vee (p \cdot \bar{p})] \vee (\bar{p} \cdot \bar{q})$
$(q \cdot \bar{q}) \vee (\bar{p} \cdot \bar{q})$
$(\bar{p} \cdot \bar{q}) \vee (q \cdot \bar{q})$
$\bar{p} \cdot \bar{q}$

10. $p \supset [(p \supset q) \supset q]$
$\bar{p} \vee (\overline{\bar{p} \vee q}) \vee q$
$\bar{p} \vee (\bar{\bar{p}} \cdot \bar{q}) \vee q$
$\bar{p} \vee (p \cdot \bar{q}) \vee q$
$[\bar{p} \cdot (q \vee \bar{q})] \vee (p \cdot \bar{q}) \vee [q \cdot (p \vee \bar{p})]$
$(\bar{p} \cdot q) \vee (\bar{p} \cdot \bar{q}) \vee (p \cdot \bar{q}) \vee (q \cdot p) \vee (q \cdot \bar{p})$
$(\bar{p} \cdot q) \vee (\bar{p} \cdot \bar{q}) \vee (p \cdot \bar{q}) \vee (p \cdot q) \vee (\bar{p} \cdot q)$
$(p \cdot q) \vee (p \cdot \bar{q}) \vee (\bar{p} \cdot q) \vee (\bar{p} \cdot q) \vee (\bar{p} \cdot \bar{q})$
$(p \cdot q) \vee (p \cdot \bar{q}) \vee (\bar{p} \cdot q) \vee (\bar{p} \cdot \bar{q})$
$[p \cdot (q \vee \bar{q})] \vee [\bar{p} \cdot (q \vee \bar{q})]$
$p \vee \bar{p}$

15. $(p \vee q) \supset [(p \supset q) \supset (\overline{q \vee \bar{q}})]$
$(\overline{p \vee q}) \vee (\overline{\bar{p} \vee q}) \vee (\bar{q} \cdot \bar{\bar{q}})$
$(\overline{p \vee q}) \vee (\overline{\bar{p} \vee q})$
$(\bar{p} \cdot \bar{q}) \vee (\bar{\bar{p}} \cdot \bar{q})$
$(\bar{q} \cdot \bar{p}) \vee (\bar{q} \cdot \bar{\bar{p}})$
$\bar{q} \cdot (\bar{p} \vee \bar{\bar{p}})$
$\bar{q}$

SPECIAL SYMBOLS

INDEX

B

C

F

J

K

L

O

P

Q

R

T

U

V

Rules of Inference

1. *Modus Ponens* (M.P.)
   $p \supset q$
   p
   $\therefore q$

2. *Modus Tollens* (M.T.)
   $p \supset q$
   $\sim q$
   $\therefore \sim p$

3. Hypothetical Syllogism (H.S.)
   $p \supset q$
   $q \supset r$
   $\therefore p \supset r$

4. Disjunctive Syllogism (D.S.)
   $p \vee q$
   $\sim p$
   $\therefore q$

5. Constructive Dilemma (C.D.)
   $(p \supset q) \cdot (r \supset s)$
   $p \vee r$
   $\therefore q \vee s$

6. Destructive Dilemma (D.D.)
   $(p \supset q) \cdot (r \supset s)$
   $\sim q \vee \sim s$
   $\therefore \sim p \vee \sim r$

7. Simplification (Simp.)
   $p \cdot q$
   $\therefore p$

8. Conjunction (Conj.)
   p
   q
   $\therefore p \cdot q$

9. Addition (Add.)
   p
   $\therefore p \vee q$

Rule of Replacement: Any of the following logically equivalent expressions can replace each other wherever they occur:

10. De Morgan's Theorems (De M.): $\sim(p \cdot q) \equiv (\sim p \vee \sim q)$
    $\sim(p \vee q) \equiv (\sim p \cdot \sim q)$
11. Commutation (Com.): $(p \vee q) \equiv (q \vee p)$
    $(p \cdot q) \equiv (q \cdot p)$
12. Association (Assoc.): $[p \vee (q \vee r)] \equiv [(p \vee q) \vee r]$
    $[p \cdot (q \cdot r)] \equiv [(p \cdot q) \cdot r]$
13. Distribution (Dist.): $[p \cdot (q \vee r)] \equiv [(p \cdot q) \vee (p \cdot r)]$
    $[p \vee (q \cdot r)] \equiv [(p \vee q) \cdot (p \vee r)]$
14. Double Negation (D.N.): $p \equiv \sim\sim p$
15. Transposition (Trans.): $(p \supset q) \equiv (\sim q \supset \sim p)$
16. Material Implication (Impl.): $(p \supset q) \equiv (\sim p \vee q)$
17. Material Equivalence (Equiv.): $(p \equiv q) \equiv [(p \supset q) \cdot (q \supset p)]$
    $(p \equiv q) \equiv [(p \cdot q) \vee (\sim p \cdot \sim q)]$
18. Exportation (Exp.): $[(p \cdot q) \supset r] \equiv [p \supset (q \supset r)]$
19. Tautology (Taut.): $p \equiv (p \vee p)$
    $p \equiv (p \cdot p)$

Quantification Rules

The expression '$\Phi\mu$' denotes any proposition or propositional function. The expression '$\Phi\nu$' denotes the result of replacing every free occurrence of μ in $\Phi\mu$ by ν, provided that if ν is a variable it must occur free in $\Phi\nu$ at all places that μ occurs free in $\Phi\mu$. (If $\Phi\mu$ contains no free occurrence of μ then $\Phi\nu$ and $\Phi\mu$ are identical, ν and μ may of course be the same variable: if they are, in this case too $\Phi\nu$ and $\Phi\mu$ are identical.)

1. UNIVERSAL INSTANTIATION

$$\textbf{UI: } \frac{(\mu)(\Phi\mu)}{\therefore \Phi\nu}$$

2. EXISTENTIAL GENERALIZATION

$$\textbf{EG: } \frac{\Phi\nu}{\therefore (\exists\mu)(\Phi\mu)}$$

3. EXISTENTIAL INSTANTIATION

$$\begin{array}{l} \textbf{EI: } (\exists\mu)(\Phi\mu) \\ \quad \rightarrow \Phi\nu \\ \quad\quad \vdots \\ \quad\quad p \\ \hline \therefore p \end{array}$$

provided that ν is a variable and does not occur free either in p or in any line preceding $\Phi\nu$.

4. UNIVERSAL GENERALIZATION

$$\textbf{UG: } \frac{\Phi\nu}{\therefore (\mu)(\Phi\mu)}$$

provided that ν is a variable and does not occur free either in $(\mu)(\Phi\mu)$ or in any assumption within whose scope $\Phi\nu$ lies.

QUANTIFIER NEGATION

$$\begin{aligned} \textbf{QN: } \quad (\nu)(\Phi\nu) &\equiv \sim(\exists\nu)\sim(\Phi\nu) \\ \sim(\nu)(\Phi\nu) &\equiv (\exists\nu)\sim(\Phi\nu) \\ (\nu)\sim(\Phi\nu) &\equiv \sim(\exists\nu)(\Phi\nu) \\ \sim(\nu)\sim(\Phi\nu) &\equiv (\exists\nu)(\Phi\nu) \end{aligned}$$

PRINCIPLE OF IDENTITY

$$\textbf{Id: } \frac{\begin{array}{c}\Phi\mu \\ \nu = \mu\end{array}}{\therefore \Phi\nu}, \qquad \frac{\begin{array}{c}\Phi\mu \\ \sim(\Phi\nu)\end{array}}{\therefore \sim(\nu = \mu)}, \qquad \frac{\nu = \mu}{\therefore \mu = \nu}, \qquad \text{and} \qquad \frac{p}{\therefore \mu = \mu}$$